A First Course in Predictive Control

Second Edition

A First Course in Predictive Control

Second Edition

J.A. Rossiter

CRC Press
Taylor & Francis Group
Boca Raton London New York

CRC Press is an imprint of the
Taylor & Francis Group, an **informa** business

CRC Press
Taylor & Francis Group
6000 Broken Sound Parkway NW, Suite 300
Boca Raton, FL 33487-2742

First issued in paperback 2022

© 2018 by Taylor & Francis Group, LLC
CRC Press is an imprint of Taylor & Francis Group, an Informa business

No claim to original U.S. Government works

ISBN-13: 978-1-138-09934-0 (hbk)
ISBN-13: 978-1-03-233916-0 (pbk)
DOI: 10.1201/9781315104126

Visit the Taylor & Francis Web site at
http://www.taylorandfrancis.com

and the CRC Press Web site at
http://www.crcpress.com

In the hope that students and lecturing staff will find this easy to learn from and in the spirit of sharing, creating a world that accepts and supports each other.

Contents

Overview and guidance for use

The main aim is to create a focussed and affordable textbook which is suitable for a single or beginners course on model predictive control (MPC). As a consequence:

1. In the early parts of each chapter, an attempt is made not to dwell too much on the mathematical details and proofs of interest to researchers but not to a typical student and instead focus as much as possible on the concepts and understanding needed to apply the method effectively.

2. Many books and resources on MPC cover the basic concepts far too briefly and focus on more advanced issues. This book does the opposite and explains the basics slowly and with numerous examples.

3. Numerous illustrative examples are included throughout and also problems which encourage readers to engage with and understand the concepts.

4. The topic coverage is deliberately limited to a range considered sufficient for taught modules.

5. Chapters are supported by MATLAB® files which are available on the open web (or with the book) so that readers can implement basic designs on-linear models. Numerous *student problems* encourage readers to learn by doing/designing.

6. In some cases short videos are available which support key parts of this book. [http://controleducation.group.shef.ac.uk/indexwebbook.html]

7. Chapters include *guidance for the lecturer* to help highlight which sections **can be used for a taught module** and which could be considered **broadening but non-essential**.

Insight is given in a non-theoretical way and there are a number of summary boxes to give a quick picture of the key results without the need to read through the detailed explanation. There is a strong focus on the philosophy of predictive control answering the questions, 'why?' and 'does it help me?' The basic concepts are introduced and then these are developed to fit different purposes: for instance, how to model, to predict, to tune, to handle constraints, to ensure feasibility, to guarantee stability and to consider what options there are with regard to models, algorithms, complexity versus performance, and so forth.

Research students who want to study predictive control in more depth are advised to make use of the research literature which is very extensive, but even for them I hope they find the focus on concepts in this book will be an invaluable foundation.

About the author

Dr. Rossiter has been researching predictive control since the late 1980s and he has published over 300 articles in journals and conferences on the topic. His particular contributions have focused on stability, feasibility and computational simplicity. He also has a parallel interest in developing good practice in university education. He has a Bachelor's degree and a Doctorate from the University of Oxford. He spent nine years as a lecturer at Loughborough University and is currently a reader at:

The University of Sheffield
Department of Automatic Control and Systems Engineering
Mappin Street
Sheffield, S1 3JD
UK
email: J.A.Rossiter@sheffield.ac.uk

Websites:
http://controleducation.group.shef.ac.uk/indexwebbook.html
https://www.sheffield.ac.uk/acse/staff/jar
https://www.youtube.com/channel/UCMBXZxd-j6VqrynykO1dURw

Assumptions about the reader

It is assumed that readers have taken at least a first course in control and thus are familiar with concepts such as system behaviours, Laplace transforms, z-transforms, state space models and simple closed-loop analysis. The motivation aspects of Chapter 1 in particular are premised on a reader's ability to reflect on the efficacy of basic control design techniques. Otherwise, most material is introduced as required.

MATLAB® is a registered trademark of The Mathworks Inc. For product information please contact; The Mathworks Inc., 3 Apple Hill Drive, Natick, MA, 01760 2098, USA. web: www.mathworks.com

Book organisation

Chapter 1: Gives basic motivation and background. This chapter explores, fairly concisely, control problems where classical approaches are difficult to apply or obviously suboptimal. It then gives some insight into how prediction forms a logical foundation with which to solve many such problems and the key components required to form a systematic control law.

Chapter 2: Considers how predictions are formed for a number of different model types. Predictions are a main building block of MPC and thus this is a core skill users will need. Also explains the concept of unbiased prediction and why this is important.

Chapter 3: Introduces a very simple to implement and widely applied MPC law, that is predictive functional control (PFC). Gives numerous illustrative examples to help readers understand tuning, where PFC is a plausible solution and where a more expensive MPC approach is more appropriate.

Chapter 4: Introduces the most common MPC performance index/optimisation used in industrial applications and shows how these are combined with a prediction model to form an effective control law.

Chapter 5: Gives insights and guidance on tuning finite horizon MPC laws. How do I ensure that I get sensible answers and well-posed decision making? What is the impact of uncertainty?

Chapter 6: Considers the stability of MPC and shows how a more systematic analysis of MPC suggests the use of so-called dual-mode predictions and infinite horizons as these give much stronger performance and stability assurances. Introduces the algebra behind these types of MPC approaches.

Chapter 7: Considers constraint handling and why this is an important part of practical MPC algorithms. Shows how constraints are embedded systematically into finite horizon MPC algorithms.

Chapter 8: Considers constraint handling with infinite horizon algorithms. It is noted that while such approaches have more rigour, they are also more complicated to implement and moreover are more likely to encounter feasibility challenges.

Chapter 9: Gives a concise conclusion and some indication of how a user might both choose from and design with the variety of MPC approaches available.

Appendix A: Provides extensive examples and guidance on tutorial questions, examination questions and assignments. Also includes some case studies that can be used as a basis for assignments. Outline solutions are available from the publisher to staff adopting the book.

Appendix B: Gives a brief taster for a number of issues deserving further study. Parametric methods are a tool for moving the online optimisation computations to offline calculations, thus improving transparency of the control law. It is demonstrated that feedforward information can easily be misused with a careless MPC design. There is also some discussion of common methods for handling uncertainty.

Appendix C: Summarises basic notation and common background used throughout the book.

MATLAB: Examples are supported throughout by MATLAB simulations; all the MATLAB files for this are available to the reader so they can reproduce and modify the scenarios for themselves. Sections at the end of each chapter summarise the MATLAB code available on an open website; also available from the publisher's website.

http://controleducation.group.shef.ac.uk/htmlformpc/introtoMPCbook.html

STUDENT PROBLEMS: Problems for readers to test their understanding and to direct useful private study/investigations are embedded throughout the book in the relevant sections rather than at chapter endings. In the main these are open-ended.

Not included: It is not the purpose of this book to write history but rather to state what is now understood and how to use this. Other books and many journal articles (e.g., [5, 16, 41, 94, 100, 177]) already give good historical accounts. Also, it is not the intention to be comprehensive but rather to cover the basics well, so many topics are excluded, a notable one being non-linear systems.

Acknowledgements

Enormous thanks to my wife Diane Rossiter who has performed an incredibly diligent and careful proof reading and editing as well as provided a number of useful critical comments.

Also thanks to all the people I have collaborated with over the years from whom I have learnt a lot which is now embedded within this work and with apologies to the many not listed here: Basil Kouvaritakis, Jesse Gossner, Scott Trimboli, Luigi Chisci, Liuping Wang, Sirish Shah, Mark Cannon, Colin Jones, Ravi Gondalaker, Jan Schurrmans, Lars Imsland, Jacques Richalet, Robert Haber, Bilal Khan, Guillermo Valencia Palomo, Il Seop Choi, Wai Hou Lio, Adham Al Sharkawi, Evans Ejegi, Shukri Dughman, Yahya Al-Naumani, Muhammad Abdullah, Bryn Jones and many more.

1

Introduction and the industrial need for predictive control

CONTENTS

1.1   Guidance for the lecturer/reader

This chapter is intended primarily as motivational and background material which need not be part of the assessment of a typical module on predictive control. As such, it is largely non-mathematical and readers should not struggle with reading any parts of it.

Nevertheless, this content could be assessed through an essay/bookwork type question within an exam, or if desired, aspects of this content could be included through case studies within an assignment.

1.2   Motivation and introduction

A useful starting point for any book is to motivate the reader: why do I need this book or topic? Consequently this chapter will begin by demonstrating the numerous scenarios where classical control approaches are inadequate. Not too much time is spent on this as this is motivational, but hopefully enough for the reader to be convinced that there are a significant number of real control problems where better approaches are required.

Having given some motivation, next this chapter gives some insight into how humans deal with such challenging control problems and demonstrates the underlying concepts, often prediction, which ultimately form the main building blocks of predictive control.

Predictive control is very widely implemented in industry and hence a key question to ask is why? In fact, more specifically, readers will want to know when predictive control is a logical choice from the many control approaches available to them. A key point to note is that predictive control is an approach, *not a specific algorithm*. The user needs to tailor the approach to their specific needs in order to produce an effective control law.

It is more important that readers understand key concepts than specific algorithms.

1. **Why is it done this way?**
2. **What is the impact of uncertainty?**
3. **How does this change with constraints?**
4. **What are the tuning parameters and how can I use them? Which choices are poor and why?**
5. **Etc.**

Consequently, although this book includes technical details, the underlying focus is on readers understanding the concepts and thus how to ensure the algorithm and tuning they choose is likely to lead to an effective control law for their specific context.

Section 1.3 will give a concise review of some classical control approaches before Section 1.4 demonstrates a number of scenarios where such approaches are difficult to tune effectively, or suboptimal. Section 1.5 then introduces arguments for a predictive control approach, followed by Section 1.6 which uses analysis of human behaviour to set out solid principles before we move into mathematical detail. The overall MPC philosophy is summarised in Section 1.7 followed by Section 1.9 which gives a concise summary of the structure of the book.

1.3 Classical control assumptions

It is not the purpose of this book to teach classical control or to give a comprehensive view of the control techniques used in industry. Nevertheless, it is useful to begin from a brief summary of the sorts of techniques which are commonly used (e.g., [32, 108]). For convenience and clarity, these will be presented in the simplest forms.

1.3.1 PID compensation

PID or proportional, integral and derivative control laws are the most commonly adopted structure in industry. Their ubiquity and success is linked to the three parameters being intuitive feedback parameters and simple to tune for many cases.

$$K(s) = K_p + \frac{K_I}{s} + K_d s \qquad (1.1)$$

1. Proportional or K_p: The magnitude of the control action is proportional to the size of the error. As the proportional is increased, the response to error becomes more aggressive leading to faster responses. If the proportional is too small, the response to error is very slow. Usually, at least for systems with simple dynamics, a proportional exists which gives the right balance between speed of response and input activity.

2. Integral or K_I: Proportional alone cannot enable output tracking as, in steady-state, the output of the proportional is only non-zero if the tracking error is non-zero. Integration of the error is an obvious (also human based strategy) and simple mechanism to enable a non-zero steady-state controller output, even when the steady-state error is zero. As with proportional, K_I must not be too large or it will induce oscillation and not too small or one will get very slow convergence.

3. Derivative or K_d: An obvious component to detect the rate at which the error is reducing and change the input accordingly - if the error is reducing

too fast it is highly likely that the control is too aggressive and a reduction in input is needed to avoid overshoot and oscillation. However, the derivative is often selected to be zero as it has a high gain at high frequency and thus can accentuate noise and lead to input chatter. Such discussions are beyond the remit of this book.

Several tuning rules for PID are given in the literature but actually, for a typical system, one could arrive at close to optimum values with very little trial and error using a simulation package.

1.3.2 Lead and Lag compensation

To some extent Lead and Lag (e.g., Equation (1.2)) discussions can be subsumed into PID compensation. A lead K_{lead} acts a little like a derivative, but with lower sensitivity to high frequency noise and thus can be used to reduce oscillation and overshoot. A Lag K_{lag} acts a little like an integral in that it increases low frequency gain, but this can be at the expense of bandwidth (speed of response). No more will be said here.

$$K_{lag} = K\frac{s+w}{s+w/\alpha}; \quad K_{lead} = K\frac{s+w/\alpha}{s+w}; \quad \alpha > 1 \qquad (1.2)$$

1.3.3 Using PID and lead/lag for SISO control

Many real systems have relatively simple dynamics and there is one very important observation that is used by control practitioners and theoreticians alike.

For single input single output (SISO) systems, if the system dynamic can be well approximated by a second-order system which is not noticeably under-damped, then the control you can achieve from a PID compensator is usually quite close to the best you can get from optimal control or any other advanced strategy. In other words, you need a good reason not to use a PID type of approach!

Of course this begs the question, when would you deviate from advocating PID and hence the next few sections give some examples.

1.3.4 Classical control analysis

Some parts of this chapter assume that readers are familiar with analysis techniques such as root-loci, Nyquist diagrams and the like. We will not use these to any great depth but readers will find it useful to undertake a quick revision of those topics if necessary. Suitable materials are well covered in all mainstream classical control textbooks and also there are some videos and summary resources on the author's website [http://controleducation.group.shef.ac.uk/indexwebbook.html].

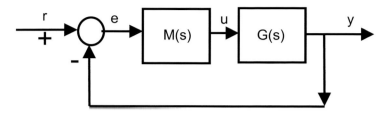

FIGURE 1.1
Feedback loop structure.

Summary: Classical control is not part of this book, but some understanding of the basics will help readers to understand the motivation for adopting predictive control.

1.4 Examples of systems hard to control effectively with classical methods

This section will assume a simple feedback structure as given in Figure 1.1 where M represents a compensator and G the process, r is the target, u the system input, e the tracking error and y the system output.

1.4.1 Controlling systems with non-minimum phase zeros

Non-minimum phase systems are quite common.

- One of the most well-known non-minimum phase responses is boiler drum level. Steam drums in large boilers are a saturated mixture of liquid and vapour. When steam out of the drum increases due to demand, level control must increase feedwater flow, which, initially, cools and collapses bubbles in the saturated mixture in the drum causing level to temporarily drop (shrink) before rising, with also an inverse response going the other direction (decreasing feedwater causes temporary level increase, or swell). The phenomenon is sometimes referred to as "shrink/swell".

- A boat turning right or left: a boat steered to the left (port) will initially move starboard (right) and then swing the boat to the left (port) because the initial forces on the rudder are opposite to the direction of the forces on the prow of the boat. This is why you MUST have a tugboat to move large ships away from docks, since the ship must initially swing into the dock to turn away from dock after an initial transient. The same phenomenon happens with rear-steered passenger vehicles (or front-steered vehicles that are backing up).

• Balancing a pole on your hand is non-minimum phase - in order to tilt the pole to the right, you initially have to swing your hand to the left and then to the right. This is a great example of using of non-minimum-phase zeros in the control loop to stabilize an otherwise unstable system. This effect also comes into play with rocket launches, since a rocket engine must balance the rocket above it during take-off.

• Another example is the feeding of bacteria (their number is evaluated through a mean value accounting for births and deaths within the overall population). When you feed bacteria they start to eat and then *forget* to reproduce themselves. The mean number of bacteria first decreases (they are still dying with the same rate) and then increases (as they are stronger) until a new equilibrium.

ILLUSTRATION: Impact of right half plane zeros

Consider a SISO system:

$$G(s) = \frac{s-2}{(s+1)(s+4)} \qquad (1.3)$$

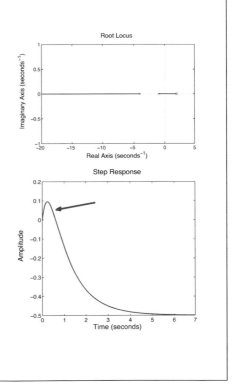

The key point is that the zero is in the right half plane (RHP). From the root-loci plot it is clear that for large values of gain, the closed-loop will have a pole in the RHP and this is totally unavoidable!

Next look at the open-loop step response. The system has a non-minimum phase characteristic (see arrow in the figure). This means that a high gain control law can easily lead to instability because the initial response is opposite to what might be expected and thus the normal control logic can get confused; it is important to be patient (cautious or low gain) before responding to observations to ensure one does not respond to the non-minimum phase characteristic.

Summary: Although you can find a PI or PID compensator to give reasonable closed-loop behaviour for non-minimum phase systems, it is less straightforward to tune and critically, much harder to get high bandwidth responses.

STUDENT PROBLEM

Produce some examples using classical control design methods which demonstrate how systems with non-minimum phase charactersitics (RHP poles) achieve much lower closed-loop bandwidths in general than systems with equivalent poles but only left half plane (LHP) zeros.

ILLUSTRATION: Car example and impact of delay

A reader can immediately see the dangers of delay using the simple analogy of driving a car. How would your driving be affected if you had to wait 2 seconds between observing something (in effect a 2 second measurement delay) and making a change to the controls (accelerator, brake, steering). Clearly, this delay would cause numerous accidents, dead pedestrians, going through red lights and so forth if driving at normal speeds.

One could only avoid such incidents by driving slow enough so that you were guaranteed not to hit anything within the next 3-4 seconds, thus leaving 1-2 seconds for any required action. The key point here is that you have to DRIVE VERY SLOWLY and thus sacrifice closed-loop bandwidth/performance. The larger the delay, the more performance is sacrificed.

1.4.2 Controlling systems with significant delays

This section demonstrates why delays within a process can have catastrophic effects on closed-loop behaviour. Delays can be caused by problems with measurement as some things cannot be measured instantaneously (such as blood tests in hospitals). They may also be caused by transport delays, for example a delay between a demanded actuation and the impact affecting the process; for example this would be typical where a conveyor belt is being used.

It is always worth asking whether one can change the process in order to reduce or remove the delay. Sometimes actions such as moving a sensor are possible and can give substantial benefits. It is always better to have as little delay as possible.

Here we are less interested in the cause of the delay than its existence and will use a simple delay model based on Laplace transforms as follows:

- Undelayed process $G(s)$ and delayed process $e^{-sT}G(s)$ where T is the delay time.

- The implied phase shift (that is $-wT$) caused by the delay in the Nyquist diagram is derived from: $e^{-sT} \rightarrow e^{-jwT}$; $\angle e^{-jwT} = -wT$.

Nyquist diagrams are useful for illustrating the impact of delays on the gain and phase margins and thus indirectly on closed-loop behaviour. A delay acts like a phase rotation within the Nyquist diagram and specifically, rotates the diagram clockwise, thus reducing margins. The larger the delay, the more the rotation.

1.4.3   ILLUSTRATION: Impact of delay on margins and closed-loop behaviour

Consider the following two processes (one with and one without delay):

$$G(s) = \frac{3}{s(s+1)(s+4)}; \quad H(s) = \frac{3e^{-sT}}{s(s+1)(s+4)} \tag{1.4}$$

The Bode diagrams (a) for systems $G(s), H(s)$ are given below with $T = 1$. The phase margin has dropped from 50° to around 15° as a consequence of the delay. The gain plot is unaffected by the delay. The corresponding closed-loop step responses (b) and Nyquist diagrams (c) emphasise that adding delay causes a significant degradation in performance. With a delay only a little bigger than one, the Nyquist diagram encircles the -1 point and the system is closed-loop unstable.

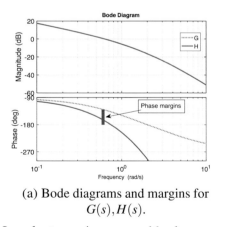

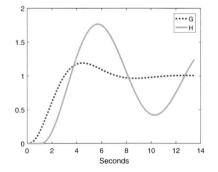

(a) Bode diagrams and margins for $G(s), H(s)$.

(b) Closed-loop step responses with unity negative feedback for $G(s), H(s)$ with $T = 1$.

   In order to regain a reasonable phase margin, a significant reduction in gain is required (d) and thus a significant delay requires a significant loss in bandwidth in order to retain reasonable behaviour, limited overshoot/oscillations and good margins.

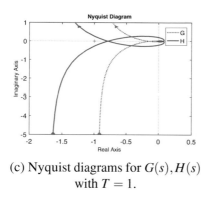

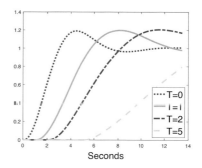

(c) Nyquist diagrams for $G(s), H(s)$ with $T = 1$.

(d) Closed-loop step responses with proportional gain selected to ensure a 50° phase margin with varying T.

Smith predictor

A common solution to significant delays is the so-called Smith predictor which has the structure below.

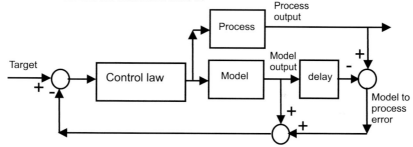

It is not the purpose of this book to explore Smith predictor techniques, but two observations are in order:

- The technique makes use of prediction and thus can be considered in the class of predictive control laws, although the design methodology is not close to modern MPC methods.

- In simple terms, the compensator is designed on a *delay-free* model with a correction for the difference between the model and process outputs. It is assumed that, if the model behaves well, so will the process but the overall design can be sensitive to errors in the assumed delay.

Summary: Classical design techniques which rely on simple gain changes are often not appropriate for systems with delays.

STUDENT PROBLEM

Produce some examples which demonstrate how delays impact upon achievable performance using classical design methods. Standard MATLAB® tools such as *feedback.m, nyquist.m* and *step.m* can be used. Delay may be entered through the *tf.m* block, for example with:

$$G = tf(1, [1, 2, 5],'IODelay', 3) \quad \Rightarrow \quad G(s) = \frac{e^{-3s}}{s^2 + 2s + 5}$$

ILLUSTRATION: Constraints have only a small impact on behaviour

In some cases, the presence of a constraint may have a relatively small impact. One such case would be upper input constraints in conjunction with a process having benign dynamics. In this case, the best control one can achieve is often to saturate the input. Consider the system and lead compensator pairing, using the loop structure of Figure 1.1, and upper input limit in Eqn.(1.5).

$$G = \frac{1}{s(s+1)^2}; \quad M(s) = 4\frac{s+0.2}{s+2}; \quad u(t) \le 0.5 \qquad (1.5)$$

The figure compares the behaviour with and without constraints from which it is clear that the simple use of saturation has led to rather sluggish performance, especially when compared to the achievable constrained behaviour with an alternative approach (shown in dashed lines). Nevertheless, the constrained behaviour is still acceptable.

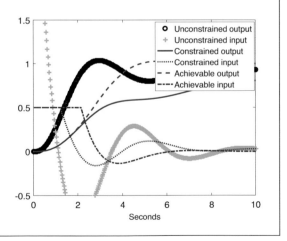

1.4.4 Controlling systems with constraints

Classical control techniques assume linear analysis is valid. This includes any discussion of robustness to uncertainty and the presence of constraints.

ILLUSTRATION: Constraints have a major impact on behaviour

In some cases, the presence of a constraint may have a huge impact on behaviour if not properly accounted for. Here an illustration of a process with almost 1st order dynamics is used to demonstrate the point. This example (Eqn.(1.6)), again using the loop structure of Figure 1.1, includes a rate constraint on the input.

$$G = \frac{2s+4}{(s+1)(s+4)}; \quad M(s) = \frac{s+1}{s}; \quad -0.5 \le u(t) \le 4; \quad \left\| \frac{du}{dt} \right\| \le 0.2 \quad (1.6)$$

The corresponding closed-loop responses are shown here. It is immediately clear that although the unconstrained responses are excellent, the constrained responses are unacceptable and show no sign of converging to the desired target of one in a reasonable manner. Indeed, the responses seem to enter some form of oscillation which is quite unexpected from the underlying linear dynamics.

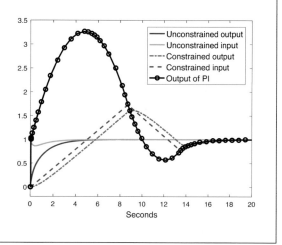

However, all real processes include constraints such as limits in absolute values and rates for actuators (inputs), desired safety limits on outputs and states, desired quality limits on outputs (linked to profit) and so forth. Whenever a system comes up against a constraint, then the overall system behaviour is highly likely to become non-linear and therefore any linear analysis is no longer valid. Indeed, one can easily come up with examples where a linear feedback system is supposedly very robust to parameter uncertainty, but the inclusion of a constraint causes instability. This section gives some illustrations of the dangers of constraints but does not discuss classical control solutions for such problems.

The problem in the second illustration above is caused by so-called integral saturation or windup, that is where the integral term keeps increasing even though the input has saturated. The PI is proposing to use the signal in dotted line, but the implemented input is the signal in dashed line. Hence, even though the output reaches and passes the target around $t=6$, the integral term is up at around 3.2 and therefore a significant period of negative error is required before the integration brings this input value back down to the sorts of steady-state input values required. Consequently, the saturated input continues to be implemented even though one really wants this to start reducing immediately the output passes the target at around $t = 6$. The output only begins reducing again once the output from the PI drops sufficiently to bring this well below the saturation level, and this is not until around $t = 9$. In this case it is the rate limit which has dominated the behaviour but one can easily create examples where the absolute limit is equally problematic.

A common solution to input constraints is to put some form of anti-windup, which detects and uses the observation that the actual input is not the same as the controller output. Such approaches may not be easy to tune in general. Moreover, in the modern era with ready access to online computing and indeed the expectation that even routine PI compensators are implemented via a computer, it is possible to be a little more systematic to reset the integral as required. This discussion is not part of the current book, suffice to say that doing this systematically is still a major challenge.

STUDENT PROBLEMS

1. Produce some further examples which demonstrate how including constraints, even for SISO systems, can lead to a challenging control design.
A m-file and Simulink pair (**openloopunstable.m, openloopunstablesim.slx** *) are provided as a template to enable you to do this more quickly. Enter your parameters into the m-file which calls the Simulink file to provide a simulation. It is assumed that you can create a suitable classical design for the unconstrained case.*
2. This section has dealt solely with undelayed SISO systems for simplicity. Readers who want a bigger challenge might like to produce some examples which demonstrate how combining delays and constraints leads to an even more challenging control design.
3. Lecturers who wish to focus on classical methods could ask students to code and demonstrate the efficacy of a variety of anti-windup techniques, but these topics are beyond the remit of this book.

Summary: There is a need for control approaches which enable systematic and straightforward constraint handling.

1.4.5   Controlling multivariable systems

Classical control techniques such as root-loci, Bode and Nyquist are targeted at SISO systems. While some extensions to the multi-input-multi-output (MIMO) have appeared in the literature, in the main these are clumsy and awkward to use and often do not lead to systematic or satisfactory control designs. For example, one popular technique [93, 95] is multivariable Nyquist whereby pre- and post-compensators (K_{pre}, K_{post}) are used to diagonalise the process $G(s)$. A diagonal process can be treated as a set of SISO systems with no interaction between loops and thus normal SISO techniques can be used on each loop. [Connect input 1 with output 1 and so forth, noting that inputs and outputs are based on the pre- and post-compensated system rather than actual inputs and outputs.]

$$G = \begin{bmatrix} g_{11} & g_{12} & \cdots & g_{1n} \\ g_{21} & g_{22} & \cdots & g_{2n} \\ \vdots & \vdots & \ddots & \vdots \\ g_{n1} & g_{n2} & \cdots & g_{nn} \end{bmatrix}; \quad K_{post} G K_{pre} \approx \begin{bmatrix} h_{11} & 0 & \cdots & 0 \\ 0 & h_{22} & \cdots & 0 \\ \vdots & \vdots & \ddots & \vdots \\ 0 & 0 & \cdots & h_{nn} \end{bmatrix} \quad (1.7)$$

The following illustrations demonstrate the consequences of using conventional PI type approaches on a MIMO process which are not nearly diagonal.

It is clear that while a simple approach can work at times, at other times the closed-loop behaviour is difficult to predict and significant detuning may be required to ensure stability and smooth behaviour. In practice, even with pre- and post-compensation, it is not possible to diagonalise a process completely, but the hope is that the off diagonal elements will be small compared to the diagonal elements, that is:

$$H = K_{post} G K_{pre} = \begin{bmatrix} h_{11} & h_{12} & \cdots & h_{1n} \\ h_{21} & h_{22} & \cdots & h_{2n} \\ \vdots & \vdots & \ddots & \vdots \\ h_{n1} & h_{n2} & \cdots & h_{nn} \end{bmatrix}; \quad \begin{matrix} |h_{11}| \gg |h_{12}| + \ldots + |h_{1n}| \\ \vdots \\ |h_{nn}| \gg |h_{1n}| + \ldots + |h_{n,n-1}| \end{matrix}$$

$$(1.8)$$

In this case, one can do a design based just on the diagonal elements and this is likely to give a reasonable result, but the bigger the off diagonal elements the less applicable this approach will be. We will not discuss these approaches further as:

- They have limited applicability.

- Identifying suitable pre- and post-compensation is non-trivial in general.

- Pre- and post-compensation make it harder to manage constraints on the actual inputs.

In practice, a more common practice in industry is experience of a given process whereby over time operators have learnt a control structure which works well enough for the MIMO process in question. It is of course quite possible that the corresponding structure results in cautious control.

ILLUSTRATION: Example of MIMO system with mild interaction

Consider the system/compensator pair (1.9) where the system is close to diagonally dominant. Defining $G(s), M(s)$ as in Figure 1.1:

$$G = \begin{bmatrix} \frac{1}{(s+1)^2} & \frac{0.2}{s+3} \\ \frac{0.4}{s+0.4} & \frac{s+5}{(s+2)(s+3)} \end{bmatrix} ; \quad M = \begin{bmatrix} \frac{s+1}{s} & 0 \\ 0 & \frac{s+2}{s} \end{bmatrix} \quad (1.9)$$

The PI design is done ignoring interaction and would give the closed-loop step responses appearing marked SISO (column 1 for changes in target 1 and column 2 for changes in target 2). It can be seen that when applied to the full MIMO system the diagonal outputs of y_{11}, y_{22} are almost the same whereas the off diagonal elements y_{12}, y_{21} are almost zero and hence a SISO approach to the design has been reasonably effective. Readers can reproduce this with file **mimodiagonal.m**.

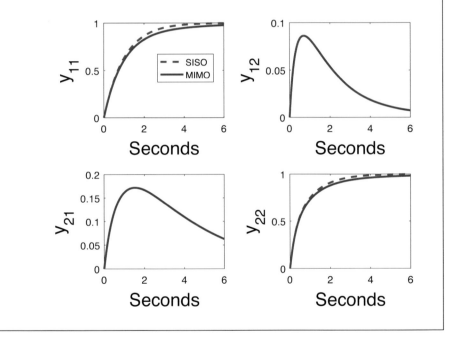

ILLUSTRATION: Example of MIMO system with significant interaction

Next consider the system/compensator pair of (1.10) where the system is not close to diagonally dominant.

$$G = \begin{bmatrix} \frac{1}{(s+1)^2} & \frac{0.5}{(s+0.2)(s+3)} \\ \frac{0.3}{(s+1)(s+0.4)} & \frac{s+5}{(s+2)(s+3)} \end{bmatrix} ; \quad M = \begin{bmatrix} \frac{s+1}{s} & 0 \\ 0 & \frac{s+2}{s} \end{bmatrix} \quad (1.10)$$

The PI design is done ignoring interaction and would give the closed-loop step responses appearing marked SISO (column 1 for changes in target 1 and column

2 for changes in target 2). It can be seen that when applied to the full MIMO system the diagonal outputs begin almost the same as the SISO case, but rapidly the interaction from the non-diagonal elements begins to have an effect and here the system is actually closed-loop unstable (note that the off diagonal y_{12} is clearly not converging). Readers can reproduce this with file **mimodiagonal2.m**.

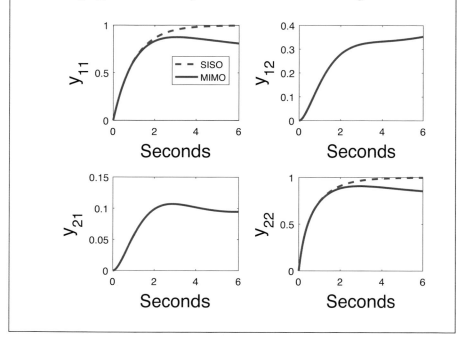

STUDENT PROBLEMS

1. Consider the example in (1.11) which is a simplified state space model of a power generation facility (see Section A.3.2). The inputs **u** are governor valve position and fuel flow and the outputs **y** are steam pressure (bar) and power (MW). Investigate the efficacy of a classical control approach on this system.

$$\dot{\mathbf{x}} = A\mathbf{x} + B\mathbf{u}; \quad \mathbf{y} = C\mathbf{x} + D\mathbf{u}; \tag{1.11}$$

$$A = \begin{bmatrix} 0.0043 & 0 & 0.005 \\ 0.02 & -0.1 & 0 \\ 0 & 0 & 0.1 \end{bmatrix};$$

$$B = \begin{bmatrix} 0 & -0.0041 \\ 0 & 0.019 \\ 0.1 & 0 \end{bmatrix}; \quad C = \begin{bmatrix} 1.75 & 13.2 & 0 \\ 0.87 & 0 & 0 \end{bmatrix}; \quad D = \begin{bmatrix} 0 & 1.66 \\ 0 & -0.16 \end{bmatrix}$$

2. This chapter has dealt with constraints and interactive multivariable systems in separate sections. Produce some examples which demonstrate how including constraints in a multivariable problem increases the difficulty of finding an effective control strategy.

Summary: Stability and performance analysis for a MIMO process is nontrivial using classical techniques although for stable processes which are near diagonal, PID compensation is often still used and is reasonably effective. Hence, there is a need for control approaches which enable systematic and straightforward handling of interactive multivariable systems.

1.4.6   Controlling open-loop unstable systems

Open-loop unstable systems are not particularly common, although systems which include integral action are more so. Both cases are more challenging for traditional control design due to an open-loop pole/poles being in the RHP or on the imaginary axis. As a consequence, low gain control is usually unacceptable as the poles would then remain close to the imaginary axis or in the RHP. This observation is in conflict with, for example, lag compensation, where the initial design step is to reduce the gain to obtain good margins. Moreover, lag compensators move the asymptotes of the root-loci to the right, thus giving little, if any, space for a system to achieve good closed-loop poles. In consequence, unstable open-loop systems almost invariably require some form of lead compensation, which moves asymptotes to the left and provides phase uplift; unsurprisingly lead compensation is high gain at high frequency and this is consistent with low gain control being unacceptable.

A common problem with high gain compensation is that it is far more sensitive to uncertainty of all forms. High gain at high frequency means that any noise components are amplified. Also, modelling uncertainty tends to be larger at larger frequencies and these frequencies now correspond to regions near the critical point of the Nyquist plot. Moreover, interaction with constraints is also more critical in that, failure to anticipate limits on the inputs can easily put the system into an unstabilisable state; controllability and stable closed-loop pole analysis are only valid if the desired inputs can be delivered!

ILLUSTRATION: Unstable example

Consider the example of Equation (1.12) and the scenarios with and without input constraints.

$$G = \frac{s+2}{s^2+3s-4}; \quad M(s) = \frac{6s+3}{s}; \quad -2.5 \le u \le 1 \qquad (1.12)$$

In the former case, the input can be delivered and the closed-loop behaviour appears excellent. However, with input constraints activated, the closed-loop behaviour is divergent as the initial inputs drive the system into an unstabiliseable state as large enough inputs to stabilise the system are not possible.

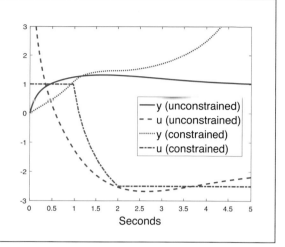

Summary: Classical design techniques which rely on simple gain changes are often not appropriate for open-loop unstable systems as high gain compensation including lead characteristics are often required, but these in turn may not be robust to uncertainty. Moreover, stability analysis is predicated on the availability of *large* inputs which may not be possible.

STUDENT PROBLEM

Investigate the impact of input constraints (rate and absolute) on some open-loop unstable examples of your choice.

*A m-file and Simulink pair (***openloopunstable.m, openloopunstablesim.slx***) are provided as a template to enable you to do this more quickly. Enter your parameters into the m-file which calls the Simulink file to provide a simulation. It is assumed that you can create a suitable classical design for the unconstrained case.*

1.5   The potential value of prediction

The previous section has highlighted a number of systems where classical control can struggle to give good solutions, that is it will not achieve closed-loop behaviour close to performance that is possible. This is unsurprising as classical control has a limited structure and limited parameters and thus may not have the flexibility to cater for challenging dynamics where a more involved decision making process is required. Nevertheless, for many of these scenarios, a human operator (or controller) is able to maintain high quality performance. Consequently it is interesting to ask what is different about the control decision making humans deploy?

1.5.1   Why is predictive control logical?

It is recognised that as humans we are very good at controlling the world around us and thus it is logical to analyse human behaviours and ask how we manage to be so effective – what underpins our decision making and makes this decision making more or less effective?

In each example next, humans use anticipation, that is prediction, to help determine effective control strategies.

EXAMPLE 1: Driving a car

1.   Drivers look ahead and anticipate future targets or demands.

2.   Change in the road, pedestrians, other vehicles, change in speed limit, etc.

EXAMPLE 2: Filling a vessel

1.   We observe the change in depth and anticipate the future changes.

2.   We modify the input flow to ensure the future depth does not exceed the target.

EXAMPLE 3: Racquet sports

1.   Players plan several shots ahead in order to move their opponent into a weaker position, or to prevent themselves being put in such a position.

2.   They predict the impact of different shot choices, and select the ones which lead to the most desirable outcome.

STUDENT PROBLEM

Come up with your own examples of where humans use prediction to support effective decision making and control.

1.5.2   Potential advantages of prediction

When humans use predictions (equivalently anticipation) to plan the best decision, it is implicit that many aspects are entered into the decision making process.

1. Constraints (or expected avoidance of constraints) are ensured for all valid decisions.

2. When there is interaction between different inputs and outputs, this is modelled into the predictions and thus accounted for systematically rather than in an ad hoc manner.

3. When we have advance information about future targets and disturbances, due allowance is made.

4. Difficult dynamics such as non-minimum phase behaviours and delays are automatically catered for.

Consequently, predictive control approaches have become the de facto norm in industry in scenarios where aspects such as constraints are particularly important and/or where small improvements in performance can mean big increases in profit, thus justifying the extra expense of a systematic design approach.

> **Summary:** On a superficial level at least, it is clear that using prediction enables humans to be highly effective controllers of the world around them and thus this indicates that it is a sensible direction to follow for automatic control.

1.6   The main components of MPC

Having established that the use of prediction (anticipation) can be an aid to effective decision making, this section reviews the key components in a predictive control law, that is what underpins the construction of the predictions and the ultimate decision making. These concepts form a useful foundation for the mathematical developments later in the book and also gives insight into the action of different algorithms and hence how to modify a given algorithm to achieve specific aims. It is noted at the outset that, by definition, it is assumed that all the following discussions apply equally to MIMO and SISO systems.

In order to structure the understanding of the core concepts, these can be usefully summarised in the following table, although, to some extent, there is overlap in these which will be apparent in the following discussion.

Summary of main components and some issues to consider					
Degrees of freedom			Receding horizon		
	Multivariable				Prediction
Constraint handling			Performance index		
	Modelling				Tuning
Future demands			optimisation		

1.6.1   Prediction and prediction horizon

Some key questions are as follows:

- **Why is prediction important?**

- **How far ahead should we predict?**

- **How do we predict?**

- **What are the consequences of not predicting?**

- **How accurate do predictions need to be?**

A number of simple analogies will help to explore these question and give the viewer an understanding of good answers.

1.6.2   Why is prediction important?

Any parent will get *fed up* telling their children to think of possible consequences before they act. For example:

Jumped off shed roof → broken ankle.	Used knife incorrectly → badly cut finger.
Spoke back at a teacher → lots of detentions.	Did not consult bus timetable before going out → had a long wait at 11pm at night!

The prediction horizon, that is, how far ahead does one predict/anticipate is an often misunderstood concept in MPC and treated as a tuning parameter by many misguided users. However, in terms of normal human behaviour, we all know how far ahead we should predict and the consequences of getting this wrong.

ILLUSTRATIONS: How far ahead should I predict?	
When driving, what prediction horizon is used and why? What happens when it is foggy?	*One always predicts beyond the safe braking distance, otherwise one cannot give a reasonable guarantee of avoiding a crash*
When heating a house, what prediction horizon is used and why?	*One always considers the time at which the house needs to be warm enough, and we turn the heating on sufficiently far in advance, that is beyond the settling time.*
When moving a very heavy item, what prediction horizon is used?	*One considers the whole trajectory, lifting, carrying and putting down again. Otherwise one could drop the item at a place which causes serious damage.*

A common observation in all the above is that the prediction horizon should be greater than the settling time (or key system dynamics), otherwise one cannot allow effectively for significant possible outcomes.

Most control laws, say PID, do not explicitly consider the future implication of current control actions and instead future behaviour is implied by the expected closed-loop dynamics, gain and phase margins and so forth. A consequence of this is an inability to anticipate *all possible consequences* and take preventative action in good time. Think about crossing the road. It is not sufficient that the road has no cars on it between you and the other side; you also check whether there are cars still some distance away that will cross in front of you soon. That is, you predict whether at any time during your crossing, you are likely to be hit by a car. If the answer is yes, you wait at the kerb. Moreover, all the time you are crossing, you keep looking, that is updating your predictions, so that if necessary your trajectory across the road can be altered.

STUDENT PROBLEM

Choose a number of scenarios and explain how, as a human, you would select the appropriate *prediction horizon* before deciding how to act.

Summary: Prediction is invaluable for avoiding otherwise unforeseen disaster. For this to be effective, the prediction must consider all significant dynamics and future behaviours.

ILLUSTRATION: Receding horizon concepts in a maze

Consider the route planning through a maze problem shown here.

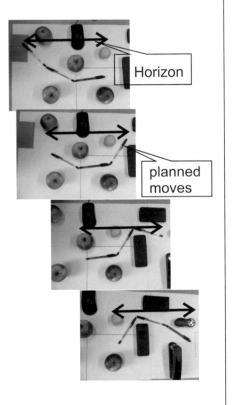

1. The user can plan 3 steps at a time (indicated by the black pens and/or horizon double arrow).

2. As one step has occurred, they are able to re-plan and thus the 3 steps take them further through the maze.

Receding horizon means that the furthest part ahead of the maze we look at is always the same distance away from our current position; for example, you could say we always look 3 moves ahead. Thus, this *far horizon* changes or moves with us as we move forward.

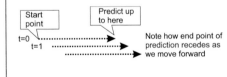

1.6.3 Receding horizon

Consider again the process of driving. It is usual to consider the road several hundred yards ahead and to anticipate any potential dangers. As we move along the road, the part of the road within our vision moves with us so that we can always see the next few hundred yards. This in essence is the receding horizon, that is the limit of our vision (say 200m away) is always moving away at the same speed we are, that is, it is receding. This means that we are continually picking up new information from the far horizon and this information is used to update our control actions (decisions).

Predictive control works just like this: it considers the predicted behaviour over some horizon into the future and therefore at each successive sampling instant, it predicts one further sample into the future. As new information comes available the input trajectory is automatically modified to take account of it.

A receding horizon means that we continually update our predictions and decision making to take account of the most recent target and measurement data. The effect is that the prediction horizon is always relative to the current position, and thus recedes away from the viewer as the viewer moves forward.

Readers may hear the argument that the receding horizon introduces feedback. In fact, one could be more precise and say that the continual update of predictions and decision making to take account of the most recent target and measurement data introduces feedback.

Summary:

1. Measurement is a core part of a feedback loop.

2. Decisions based on measurement and upto date target information are the 2nd core part.

Predictive control incorporates both elements of measurement and a related decision.

1.6.4   Predictions are based on a model

A core part of predictive control is the prediction, but we need to ask: how are these predictions determined?

1. While humans are very good at predicting outcomes, these predictions are based on a lot of experience which could be difficult to untangle and automate.

2. In order to automate predictions, implicitly we are modelling system behaviour and thus a model is required. However, it is not immediately obvious how we go about defining an appropriate prediction model.

Ideally we want a model from which it is easy to form predictions (ideally linear), easy to identify model parameters and gives accurate predictions. Of course these are all somewhat vague. What do we mean by accurate: steady-state, fast transients, mid-response, or ... ? What parameter identification algorithm is being proposed and does this give a model which is most accurate? Most simple black-box methods are based on one-step ahead prediction errors, thus fast transients only! In summary:

1. The simplest model that gives accurate enough predictions is usually best.

2. Accurate enough is ill-defined, but in practice predictions can often be 10-20% out in the steady-state and still be highly effective as long as they also capture the key dynamic changes during transients.

3. It is rarely beneficial to spend excessive effort improving accuracy as this is expensive, can give high order models but may have little impact on behaviour. Feedback will correct for small modelling errors.

4. The model must show the dependence of the output on the current measured variable and the current/future inputs.

5. Make sure you check the ability of your model to give good long range prediction.

Fit-for-purpose model

In predictive control, the model is used solely to compute system output predictions, so the model is fit-for-purpose *if it gives accurate enough predictions. The effort and detail put into the modelling stage should reflect this. There may be no need to model all the physics, chemistry and internal behaviour of the process in order to get a model that gives reliable prediction, and in fact one should not model all this detail if it is not required. A basic rule base states that one should use the simplest model one can get away with as this also eases maintenance and ongoing tuning by local staff.*

It is noted that the model does not have to be linear (e.g., transfer function, state space) and in fact can be just about anything. Indeed, as humans, we often use fuzzy models and yet achieve very accurate control; for instance, if I am in third gear, doing 40 mph and not going up a steep incline, then depressing the accelerator should give good acceleration. The key point to note here is that a precise model is not always required to get tight control because the decisions are updated regularly and this will deal with some model uncertainty in a fairly fast time scale. In the driving example above, failure to achieve good acceleration would be noted very quickly by comparison of actual behaviour with predicted model behaviour.

STUDENT PROBLEM

How accurate a model is needed for use, as humans, to control the world around us? What types of inaccuracies are acceptable and what types are not? How do these *specifications* vary with the scenario?

In practice most model predictive control (MPC) algorithms use linear models because the dependence of the predictions on future control choices is then linear and this facilitates easier optimisation as well as off-line analysis of expected closed-loop behaviour. However, non-linear models can be used where the implied computational burden is not a problem and linear approximations are not accurate enough; non-linear MPC is not covered in this book as that is beyond the remit of typical first courses.

> **Summary:** A model is used to generate system predictions including the dependence on future control decisions/inputs. One should use the simplest model possible which is *fit for purpose*, that is, gives accurate enough predictions.

1.6.5   Performance indices

We need an objective measure of performance in order to select between alternative control strategies. This measure should be quantitative so that it can be used by a computer based control law. However, for convenience, the choice of performance index will always contain some arbitrariness because the user would like measures that are easy to optimise, insensitive to small changes in parameters (robust), easy to design and so forth.

1. How should the performance index (also called cost function, performance measure, ...) be designed?

2. How can we handle trade-offs between optimal and safe/robust performance?

3. How do performance indices change to reflect the scenario?

General guidance suggests that simpler definitions are better as they lead to better conditioned and simpler optimisations. As the optimisation is carried out online, with corresponding risks, this should only have increased complexity where the benefits are clear. Typically quadratic performance indices are used as these give well-conditioned optimisations with a unique minimum and generally smooth behaviours but it should be emphasised that, in principle, users can choose whichever performance index they like. Nevertheless, the users' options are also severely restricted by the accuracy with which they can predict.

Because MPC is usually implemented by computer, this requires a *numerical* definition of performance index or *cost* so that a precise calculation can be made, that is, which predicted input trajectory gives the lowest numerical value to the *cost*. The main requirement is that the *cost* depends on the future choices for the system input and that low values of *cost* imply good closed-loop performance - good being defined for the process in question. Of course the choice of the *cost* affects the complexity of the implied optimisation and this is also a consideration. Selection of the *cost* is an area of both engineering and theoretical judgement. However, as long as some basic guidelines are followed, the actual choice of *cost* often has little effect on closed-loop performance, in that what might be considered *large* changes in performance index parameters may have relatively little impact on the resulting control. For this reason 2-norm measures are popular, as the optimisation is then straightforward.

Racquet sports analogy

In order to control a process very accurately, we need a very accurate model. Some insight into this can be given by racquet sports. When you are learning the sport, your mind has only an imprecise model of how arm movements, etc. affect the trajectory of the ball. As a result, control, direction, etc. are poor. As you practice, your internal model becomes more accurate and hence the quality of your play improves. A novice uses simple strategies, as they have a simple model. Their goal is to keep the ball in play; they think only about the current move. An expert on the other hand may think several moves ahead and aim the ball very precisely in order to construct an opening with which to win the point.

Application to MPC

1. **Low quality prediction models:** If one's prediction model is imprecise, then it is not possible to specify that the future predictions are precise. As a consequence, the performance index (or desired performance) must be suitably cautious to reflect this uncertainty. Simple objectives tend to lead to robust feedback control laws and thus will often be favoured in practice.

2. **High quality prediction models:** Where one is able to predict future behaviour very precisely, then it is possible to set target performances/output trajectories which are very precise, and thus demand a highly tuned loop. Indeed it may be sacrificing potential not to do so. However, the downside of demanding very precise trajectories is a greater risk of failure and thus implicitly, a greater sensitivity to uncertainty.

STUDENT PROBLEM

Come up with your own examples of the sorts of performance indices humans use to support effective decision making and control. How easily can these performance indices be used in an automated system?

Summary: You can only set performance objectives as precisely as you can model! If you want a highly tuned controller with correspondingly demanding performance objective, then you need a very accurate model.

1.6.6   Degrees of freedom in the predictions or prediction class

The user has to decide which class of predictions they will use to optimise performance. This class of predictions could be very simple such as a couple of changes in the input signal, or highly involved with many variables to be selected. A key point here is that the user must match their definition of degrees of freedom within the prediction to their desired output prediction, otherwise the computer is left with a nonsense optimisation to do.

ILLUSTRATION: Nonsense optimisation 1

Find the best straight line, or quadratic, which follows a sinusoid over several periods. Clearly no sensible solution exists to this problem and therefore it is an example of an ill-posed optimisation. In order to match a sinusoid closely, a different class of functions (and different degrees of freedom) are needed.

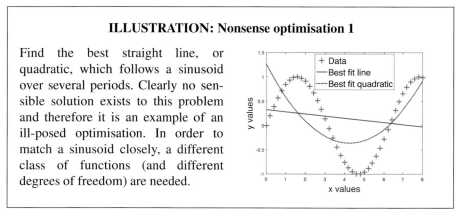

ILLUSTRATION: Nonsense optimisation 2

A driver is asked to select a steering wheel input which must be activated now, in order to steer around a corner which is in 50m time. Again this is clearly madness as the degree of freedom, that is the move of the steering wheel, needs to be coincident with the corner rather than at another point in time. Again, any solution to such an optimisation will not be useful and likely drive the car off the road.

STUDENT PROBLEM

Come up with your own examples of how humans quantify the degrees of freedom available to support effective decision making and control.

Summary: As with the performance index, the number of degrees of freedom and the complexity of the functions (or future input trajectories for predictive control) used for fitting, must be sensibly matched to the accuracy of the prediction model otherwise the implied predictions will not be reliable and thus any associated optimisation will not be representative of reality.

1.6.7 Tuning

There has been historical interest in the topic of tuning predictive control laws to ensure stability, much as there is much knowledge on how to tune PID. However, with MPC it is now understood that tuning is better allied to an appropriate specification of desired performance as discussed in the previous subsection. If you get the *cost* function right, stability and tuning will look after themselves, as by definition you are optimising a *cost* which can only be small for good performance. When sensible guidelines are used, MPC will always give stable control (at least for the nominal) so the important judgements are how to get a balance between the performance in different loops, good sensitivity and also a balance between input activity and speed of response.

Classically these balances are achieved by weighting matrices, that is, putting different emphasis on performance in different loops according to their importance. It is doubtful, however, that one could construct such weights systematically from a financial/operational viewpoint and in general on-line tuning is required until the balance looks right. A typical guideline to give a sensible initial value for the weights would be to normalise all the signals (input and output) so that the range 0 to 1 is equally important for each signal; then use unity weights on all the loops. It is difficult to generalise beyond this because each process has different priorities which may be too subtle to include in general guidelines.

Some authors [120] take the view that there should be no weighting of the inputs at all, as the goal is to drive the outputs to the desired point with whatever inputs are required. In this case it is paramount to set up the cost function carefully to avoid overtuned control laws and inversion of non-minimum phase characteristics. For instance, do not request unrealistically fast changes in the output and restrict the space of allowable input trajectories to those which move smoothly.

Summary: Tuning is often straightforward if one can define the relative importance of performance of the different outputs and inputs.

1.6.8 Constraint handling

Real systems contain numerous constraints such as actuator limits (input constraints), pressure/temperature limits (state constraints) and many others the reader can think of. While some constraints can be violated, the consequences are often undesirable such as loss of profit, excess wear and tear and even equipment damage. Other constraints cannot be violated as they represent hard limits on equipment movement.

In view of this it is obvious that a sensible control strategy should take account of constraints. A feedback law with excellent gain/phase margins and good robustness in the absence of constraints could be disastrous when constraints are active. For example, a control law could take you to a point of no return (nuclear reactors, inverted pendulum/space rockets, overfilling tanks causing spillage, etc). Anyone who drives a car will recognise the necessity of planning strategies which avoid all future constraints as well as current constraints; the consequence of not doing so is likely to be a crash. In order to handle constraints effectively and systematically, they need to be embedded throughout and not added as an afterthought.

- Predictive control will not propose input flows that allow the tank to overflow. This may result in slightly slower rise times during transients, but it will be safe.

- Predictive control will be aware that the input is limited to 100% and thus will not allow earlier input choices which make the system not stabilisable thereafter.

In general, embedding constraints (awareness of kerbs while driving, flow or power limits in process control, speed limits for a robot, etc.) will ensure that the proposed input strategies are optimised and the tuning changes, if need be, for different operating points.

Racing driver analogy

A simple well-tuned but unconstrained law may be very effective in a small range, but disastrous if used over a wider operating range. For example, a racing driver optimises speed, subject to the constraint that the car remains on the track. If they optimised speed without taking explicit account of the track, their lap times would actually be far slower because they would end up coming off the track and into the gravel, or worse still a barrier.

The details of how constraints are incorporated are very much dependent on the algorithm deployed and are discussed later. The algorithm selected will depend upon money available (i.e., related to potential increase in profit) and sampling time. The more you can spend, the closer to the true constrained optimum your algorithm is likely to get.

STUDENT PROBLEM

Come up with your own examples which demonstrate that it is often, but not always, essential to embed constraints in any planning and control activities.

Summary: A well designed predictive control law should take systematic account of constraints, both current ones and those in the future. This facilitates better overall performance.

1.6.9   Multivariable and interactive systems

There are many examples of processes which have numerous inputs and outputs.

- A car has 2 main inputs (throttle and steering) and 2 main outputs (speed and direction).

- An aeroplane has numerous control surfaces and moves in 3D space (or 4D space if time is also considered as a core dimension).

- A chemical reaction could be affected by flow rates of reactants, heat supply, pressure and the outputs could be speed of reaction (production rate), quality, purity, etc.

A key observation is that for multivariable processes, often changing one input changes all the outputs and therefore an effective control law has to consider all inputs and outputs simultaneously.

Simple aeroplane analogy

When a pilot pulls back on the control stick and thus changes the angle of attack of the aircraft, the aircraft will begin to climb. However, this has the inevitable consequence of a drop in speed as some of the thrust is now being used to facilitate climb (and also drag may increase with the increased angle of attack).

STUDENT PROBLEM

Come up with your own examples which demonstrate the difficulties humans face in controlling interactive multivariable systems. To what extent are the solutions we adopt useful for automation?

Summary: One advantage of predictive control methods is that they provide a framework (performance index, degrees of freedom and optimisation) that automatically takes account of interaction, although of course we need to express the algorithm in mathematical language and get a computer to do the number crunching.

1.6.10   Systematic use of future demands

When driving, if you see a corner ahead, you adjust the state (speed, gear, etc.) of the car so that you can take the corner safely. Typically a PID type of control law deals with the corner only once you reach it with the possible danger of going off the road (overshoot) due to excess entry speed. Feedforward can be added but the design could be somewhat ad hoc.

Conversely MPC automatically incorporates feedforward and moreover does this in a systematic way. The optimisation of the *performance index* can take account of future changes in the desired trajectory and future measurable disturbances and so include them as an intrinsic part of the overall control design/decision making. Be careful however; the default feedforward from MPC is not always a good one [134].

STUDENT PROBLEM

Come up with your own examples to demonstrate how humans use advance information about future targets and how is our efficacy affected when this information is not available?

Summary: Predictive control methods automatically enable systematic feedforward design which is also integrated with the constraint handling.

1.7   MPC philosophy in summary

This book takes the view that we should concentrate our efforts on understanding the principles which unite most model-based predictive control (MPC) algorithms and not get distracted by the fine details. Usually the details are self-evident given a clear understanding of the philosophy. We will use the acronym MPC to denote all types of predictive control laws, for which many other abbreviations exist, e.g., IDCOM

[120], DMC[28], GPC[25], QDMC[40], IMC[39], MUSMAR[49], to name just a few.

Philosophically MPC reflects human behaviour whereby we select control actions which we think will lead to the <u>best</u> (*where best needs to be defined*) predicted outcome (or output) over some limited horizon. Hence a predictive control law has the following core components:

1. The control law depends on predicted behaviour and moreover, one should consider the entire prediction trajectory capturing all key dynamics.

2. The output predictions are computed using a process model; the simpler the model the easier the design will be although it should be complicated enough to capture the core dynamic behaviour. Modelling efforts should be focussed on efficacy for prediction, including dependence on any degrees of freedom (d.o.f.).

3. The current input is determined by *optimising* some measure of predicted performance; the resulting behaviour is closely linked to the *measure* adopted. Performance indices must be realistic and matched to model accuracy.

4. Decisions, or input choices, are restricted to those whose corresponding predictions satisfy system constraints (both now and in the future); that is, constraints are embedded throughout.

5. Ideally the class of predictions, which is linked to the choice of input predictions, should include the desired closed-loop behaviour. If not, the optimisation of the performance index could be ill-posed in that it will not be able to deliver the desired behaviour.

6. The receding horizon: the control input, or decision making, is updated at every sampling instant and makes use of any new measurements and feed forward information.

7. Efficient computation requires linear models and simple parameterisations of the d.o.f..

Summary: Predictive control encapsulates common human behaviour and thus is expected to deliver effective control strategies.

1.8 MATLAB files from this chapter

openloopunstable.m, openloopunstablesim.slx	Templates to study the impact of constraints on classical control loops.
mimodiagonal.m	Closed-loop simulation of a MIMO system with low interaction
mimodiagonal2.m	Closed-loop simulation of a MIMO system with high interaction

1.9 Reminder of book organisation

The reader can revisit the preface where a more detailed overview of the book organisation is provided. It should be apparent how the different chapters embody the core concepts in Section 1.7.

Chapter 2 covers prediction, Chapters 3 and 4 introduce basic performance indices for translating predictions into control decisions, Chapters 5, 6 consider tuning, performance and stability and their links to the performance indices, Chapters 7, 8 discuss how to include constraints into the decision making, and finally chapter 9 attempts to bring some order and insight into the many alternative choices users need to make. Appendix A gives a number of tutorial questions and suggestions on assignments.

Appendix B gives food for thought on what the book does not cover and common notation and generic background are summarised in appendix C.

2

Prediction in model predictive control

CONTENTS

2.1 Introduction

MPC algorithms make use of predictions of the system behaviour. In this chapter it will be shown how to compute those predictions for various types of linear models. For the reader who is not interested in the detailed derivations, simply read the results which will be highlighted at the end of each section.

> **Main observation:** For linear models, the predictions of future outputs are affine in the current state and the future input increments.

2.2 Guidance for the lecturer/reader

This is a long chapter because the ability to predict is the foundation on which predictive control algorithms are formulated and hence it is essential that students have a good understanding of both how to predict for different model formulations and also, the properties of those predictions. Depending on the scenario, different types of prediction might be appropriate and potential users should understand the choices available to them.

For a basic course in predictive control, I would suggest a focus on Section 2.3 for broad concepts and then depending on which type of model format is preferred one or two of the following options:

1. Sections 2.4.1 and 2.4.2 for state space models.

2. Section 2.5 and in particular Section 2.5.2 for transfer function models (the Subsection 2.5.4 on T-filtering is probably best done by numerical illustration and discussion rather than including the algebra).

3. Section 2.8.1 for FIR models.

The later sections covering topics such as recursive methods, independent model prediction and closed-loop prediction give more insight into practical requirements/details and the associated algebra which will be useful for a larger or second course and research students who want to design controllers for new scenarios, but are likely to be unnecessary distractions for beginners without reinforcing core concepts.

As throughout this book, the author would favour supplying students with basic code to create and simulate designs, as writing and validating code for a whole MPC algorithm will take too long to incorporate into most modules at the expense of other learning outcomes. The relevant MATLAB® code is summarised in Section 2.11 and related tutorial exercises encouraging students to learn by doing are spread throughout the chapter. More comprehensive tutorial, examination and assignment questions are in Appendix A.

2.2.1 Typical learning outcomes for an examination assessment

Some typical exam questions are placed throughout the chapter and in Appendix A, but in the author's view the main assessment focus should be on algebra rather than tedious number crunching given that prediction will **always** be carried out by computer, and not by hand. Thus, students can be asked to demonstrate how to form predictions for different scenarios: (i) different model types; (ii) in terms of inputs or input increments; (iii) to be unbiased in the steady-state; (iv) for different horizons. However, in general, number crunching should not be asked for in an exam.

2.2.2 Typical learning outcomes for an assignment/coursework

For assignments, the author would be less likely to assess prediction in itself and the focus is more likely to be on the application of MPC to a specific process. Consequently, effective and correct prediction would be assumed rather than tested. Indeed, the author would normally provide students with suitable MATLAB code for use in assignments, so they can focus on the higher level concepts rather than spending excessive time debugging.

2.3 General format of prediction modelling

Many papers concentrate on the details to such an extent that the *simplicity* of the main result is lost. This section is intended to draw attention to it.

2.3.1 Notation for vectors of past and future values

It is convenient to have a notation for a vector of future/past values of given variables. The notation of arrows pointing right is used for strictly future (not including current value) and arrows pointing left for past (including current value). The subscript denotes the sample taken as the base point, e.g.,

$$
\underset{\rightarrow k+1}{\mathbf{x}} = \begin{bmatrix} \mathbf{x}_{k+1} \\ \mathbf{x}_{k+2} \\ \vdots \\ \mathbf{x}_{k+n} \end{bmatrix} ; \quad \underset{\leftarrow k}{\mathbf{x}} = \begin{bmatrix} \mathbf{x}_k \\ \mathbf{x}_{k-1} \\ \vdots \\ x_{k-m} \end{bmatrix} \tag{2.1}
$$

The length of these vectors (that is, the time into the future n or past m) is not defined here and is usually implicit in the context.

2.3.2 Format of general prediction equation

Define the system's current state as $\mathbf{x}_k$ (this could consist of past input/output data), output $\mathbf{y}_k$ and input $\mathbf{u}_k$, then a general form of future predictions is

$$
\underset{\rightarrow k+1}{\mathbf{y}} = H\Delta\underset{\rightarrow k}{\mathbf{u}} + P\underset{\leftarrow k}{\mathbf{x}} \tag{2.2}
$$

- H is the Toeplitz matrix of the system step response.

- P is a matrix whose coefficients depend on the model parameters in a straightforward but nonlinear manner.

- $\Delta\underset{\rightarrow k}{\mathbf{u}}$ is a vector of future values of input increments ($\Delta\mathbf{u}_k = \mathbf{u}_k - \mathbf{u}_{k-1}$).

- $\underset{\rightarrow k+1}{\mathbf{y}}$ is a vector of future output predictions.

For many users the details of how to compute H, P are not important and it is sufficient to accept that this is relatively straightforward with linear models.

> **Summary:** Once you are happy that Equation (2.2) is true, you can move to the next chapter. Continue with this chapter to find out more details. A key point is the linear dependence of output predictions $\underset{\rightarrow k+1}{\mathbf{y}}$ on future choices of the input $\Delta\underset{\rightarrow k}{\mathbf{u}}$ and the current system state $\mathbf{x}_k$.

2.3.3 Double subscript notation for predictions

Sometimes the meaning of the notation is obvious from the context, for example a sample k is typically taken to be the present, $k+1$ to be one sample into the future and $k-1$ one sample into the past. However, at times, it is useful to be precise. This section introduces a common notation used in the literature to clearly distinguish between past and future and also what sample predictions are being made.

A popular notation uses a double subscript with two samples marked. The second value is the sample at which a prediction was made and the first subscript is the sample the prediction applies too.

$\mathbf{x}_{k+4|k}$ Prediction of state $\mathbf{x}_{k+4}$ made at the kth sample.

$\mathbf{y}_{k+5|k-2}$ Prediction of output $\mathbf{y}_{k+5}$ made at the (k 2)th sample.

This notation is used where needed to add clarity, but is avoided as cumbersome when implicit. As an illustration of applying this notation, one would alter the prediction Equation (2.10) as follows:

$$
\begin{aligned}
\mathbf{x}_{k+n|k} &= A^n\mathbf{x}_k + A^{n-1}B\mathbf{u}_{k|k} + A^{n-2}B\mathbf{u}_{k+1|k} + \cdots + B\mathbf{u}_{k+n-1|k} \\
\mathbf{y}_{k+n|k} &= C[A^n\mathbf{x}_k + A^{n-1}B\mathbf{u}_{k|k} + A^{n-2}B\mathbf{u}_{k+1|k} + \cdots + B\mathbf{u}_{k+n-1|k}] + \mathbf{d}_{k+n|k}
\end{aligned}
\tag{2.3}
$$

Readers will note that $\mathbf{u}_{k|k}$ is treated as future even though it is **now** because the current input can be chosen. Conversely, the current state $\mathbf{x}_k$ is considered as past because it cannot be altered.

2.4   Prediction with state space models

This section looks at basic open-loop prediction using a state space model. The system is assumed to be multivariable. Dimensions are not stated as taken to be implicit in the state $\mathbf{x}_k$, input $\mathbf{u}_k$ and output $\mathbf{y}_k$ and of course corresponding matrices A, B, C, D (this book takes $D = 0$ throughout as that is normal for strictly proper systems). Also, examples are restricted to square systems as non-square systems introduce a number of additional issues which would distract from the core presentation of MPC concepts.

Hence define the discrete state space model:

$$
\mathbf{x}_{k+1} = A\mathbf{x}_k + B\mathbf{u}_k; \quad \mathbf{y}_k = C\mathbf{x}_k + \mathbf{d}_k
\tag{2.4}
$$

Readers will note an additional term $\mathbf{d}_k$ used to denote any output disturbances.

2.4.1   Prediction by iterating the system model

One-step ahead prediction: A discrete state space model is, in effect, a one-step ahead prediction model as one obtains a prediction for $\mathbf{x}_{k+1}$ from the current state and input $(\mathbf{x}_k, \mathbf{u}_k)$. This in turn can be used to estimate a one-step ahead prediction for the output as follows.

$$
\left.\begin{aligned}
\mathbf{x}_{k+1} &= A\mathbf{x}_k + B\mathbf{u}_k; \\
\mathbf{y}_{k+1} &= C\mathbf{x}_{k+1} + \mathbf{d}_{k+1}
\end{aligned}\right\} \quad \Rightarrow \quad \mathbf{y}_{k+1} = CA\mathbf{x}_k + CB\mathbf{u}_k + \mathbf{d}_{k+1}
\tag{2.5}
$$

Two-step ahead prediction: To get the two-step ahead prediction one simple uses the same prediction model as above, but at the next sampling instant and substituting in the predicted value for $\mathbf{x}_{k+1}$ from (2.5).

$$\left.\begin{array}{l} \mathbf{x}_{k+2} = A\mathbf{x}_{k+1} + B\mathbf{u}_{k+1}; \\ \mathbf{y}_{k+2} = C\mathbf{x}_{k+2} + \mathbf{d}_{k+2} \end{array}\right\} \quad \Rightarrow \quad \left\{\begin{array}{l} \mathbf{x}_{k+2} = A\mathbf{x}_{k+1} + B\mathbf{u}_{k+1} \\ \mathbf{x}_{k+2} = A[A\mathbf{x}_k + B\mathbf{u}_k] + B\mathbf{u}_{k+1} \\ \mathbf{x}_{k+2} = A^2\mathbf{x}_k + AB\mathbf{u}_k + B\mathbf{u}_{k+1} \end{array}\right. \quad (2.6)$$

$$\mathbf{y}_{k+2} = C[A^2\mathbf{x}_k + AB\mathbf{u}_k CB\mathbf{u}_{k+1}] + \mathbf{d}_{k+2} \quad (2.7)$$

Three-step ahead prediction: In a similar way one can find the three-step ahead prediction.

$$\left.\begin{array}{l} \mathbf{x}_{k+3} = A\mathbf{x}_{k+2} + B\mathbf{u}_{k+2}; \\ \mathbf{y}_{k+3} = C\mathbf{x}_{k+3} + \mathbf{d}_{k+3} \end{array}\right\} \quad \Rightarrow \quad \left\{\begin{array}{l} \mathbf{x}_{k+3} = A\mathbf{x}_{k+2} + B\mathbf{u}_{k+2} \\ \mathbf{x}_{k+3} = A[A^2\mathbf{x}_k + AB\mathbf{u}_k + B\mathbf{u}_{k+1}] + B\mathbf{u}_{k+2} \\ \mathbf{x}_{k+3} = A^3\mathbf{x}_k + A^2B\mathbf{u}_k + AB\mathbf{u}_{k+1} + B\mathbf{u}_{k+2} \end{array}\right.$$
$$(2.8)$$
$$\mathbf{y}_{k+3} = C[A^3\mathbf{x}_k + A^2B\mathbf{u}_k + AB\mathbf{u}_{k+1} + B\mathbf{u}_{k+2}] + \mathbf{d}_{k+3} \quad (2.9)$$

Many-step ahead prediction: The general pattern should now be clear. One can set up a recursion to find four-step ahead, five-step ahead predictions and so on. Hence, by inspection, one can write:

$$\begin{array}{rl} \mathbf{x}_{k+n} =& A^n\mathbf{x}_k + A^{n-1}B\mathbf{u}_k + A^{n-2}B\mathbf{u}_{k+1} + \cdots + B\mathbf{u}_{k+n-1} \\ \mathbf{y}_{k+n} =& C[A^n\mathbf{x}_k + A^{n-1}B\mathbf{u}_k + A^{n-2}B\mathbf{u}_{k+1} + \cdots + B\mathbf{u}_{k+n-1}] + \mathbf{d}_{k+n} \end{array} \quad (2.10)$$

2.4.2   Predictions in matrix notation

It is convenient to stack all the state/output predictions into a single vector as in (2.1). Hence one can form the whole vector of future predictions up to a horizon n_y as follows:

$$\underbrace{\begin{bmatrix} \mathbf{x}_{k+1} \\ \mathbf{x}_{k+2} \\ \mathbf{x}_{k+3} \\ \vdots \\ \mathbf{x}_{k+n_y} \end{bmatrix}}_{\underrightarrow{\mathbf{x}}_{k+1}} = \underbrace{\begin{bmatrix} A \\ A^2 \\ A^3 \\ \vdots \\ A^{n_y} \end{bmatrix}}_{P_x} \mathbf{x}_k + \underbrace{\begin{bmatrix} B & 0 & 0 & \cdots \\ AB & B & 0 & \cdots \\ A^2B & AB & B & \cdots \\ \vdots & \vdots & \vdots & \vdots \\ A^{n_y-1}B & A^{n_y-2}B & A^{n_y-3}B & \cdots \end{bmatrix}}_{H_x} \underbrace{\begin{bmatrix} \mathbf{u}_k \\ \mathbf{u}_{k+1} \\ \mathbf{u}_{k+2} \\ \vdots \\ \mathbf{u}_{k+n_y-1} \end{bmatrix}}_{\underrightarrow{\mathbf{u}}_k}$$
$$(2.11)$$

and

$$
\underbrace{\begin{bmatrix} \mathbf{y}_{k+1} \\ \mathbf{y}_{k+2} \\ \mathbf{y}_{k+3} \\ \vdots \\ \mathbf{y}_{k+n_y} \end{bmatrix}}_{\substack{\mathbf{y} \\ \rightarrow k+1}} = \underbrace{\begin{bmatrix} CA \\ CA^2 \\ CA^3 \\ \vdots \\ CA^{n_y} \end{bmatrix}}_{P} \mathbf{x}_k + \underbrace{\begin{bmatrix} CB & 0 & 0 & \cdots \\ CAB & CB & 0 & \cdots \\ CA^2B & CAB & CB & \cdots \\ \vdots & \vdots & \vdots & \vdots \\ CA^{n_y-1}B & CA^{n_y-2}B & CA^{n_y-3}B & \cdots \end{bmatrix}}_{H} \underset{\rightarrow k}{\mathbf{u}} + \begin{bmatrix} \mathbf{d}_{k+1} \\ \mathbf{d}_{k+2} \\ \mathbf{d}_{k+3} \\ \vdots \\ \mathbf{d}_{k+n_y} \end{bmatrix}
$$

$$(2.12)$$

ILLUSTRATION: Prediction example

For the following A, B, C matrices

$$
A = \begin{bmatrix} 0.9 & -0.5 \\ 0 & 0.8 \end{bmatrix}; \quad B = \begin{bmatrix} 1 & 1 \\ -2 & 0 \end{bmatrix}; \quad C = \begin{bmatrix} 2 & 0.5 \\ -1 & 1 \end{bmatrix}
$$

one can define the prediction matrices for $n_y = 4$ and defining H, H_x giving the dependence on the first two future controls $\mathbf{u}_k, \mathbf{u}_{k+1}$ only, as follows:

$$
P_x = \begin{bmatrix} 0.9 & -0.5 \\ 0 & 0.8 \\ 0.81 & -0.85 \\ 0 & 0.64 \\ 0.729 & -1.085 \\ 0 & 0.512 \\ 0.6561 & -1.2325 \\ 0 & 0.4096 \end{bmatrix}; \quad H_x = \begin{bmatrix} 1 & 1 & 0 & 0 \\ -2 & 0 & 0 & 0 \\ 1.9 & 0.9 & 1 & 1 \\ -1.6 & 0 & -2 & 0 \\ 2.51 & 0.81 & 1.9 & 0.9 \\ -1.28 & 0 & -1.6 & 0 \\ 2.899 & 0.729 & 2.51 & 0.81 \\ -1.024 & 0 & -1.28 & 0 \end{bmatrix}
$$

$$
P = \begin{bmatrix} 1.8 & -0.6 \\ -0.9 & 1.3 \\ 1.62 & -1.38 \\ -0.81 & 1.49 \\ 1.458 & -1.914 \\ -0.729 & 1.597 \\ 1.3122 & -2.2602 \\ -0.6561 & 1.6421 \end{bmatrix}; \quad H = \begin{bmatrix} 1 & 2 & 0 & 0 \\ -3 & -1 & 0 & 0 \\ 3 & 1.8 & 1 & 2 \\ -3.5 & -0.9 & -3 & -1 \\ 4.38 & 1.62 & 3 & 1.8 \\ -3.79 & -0.81 & -3.5 & -0.9 \\ 5.286 & 1.458 & 4.38 & 1.62 \\ -3.923 & -0.729 & -3.79 & -0.81 \end{bmatrix}
$$

Remark 2.1 *It is commonplace to assume that the* unknown *disturbance term is constant in the future, that is* $\mathbf{d}_{k+i} = \mathbf{d}_k, \forall i > 0$. *A matrix L is defined to simplify the inclusion of this assumption. The dimensions of L are varied as required as* n_y

changes.

$$
\begin{bmatrix} \mathbf{d}_{k+1} \\ \mathbf{d}_{k+2} \\ \mathbf{d}_{k+3} \\ \vdots \\ \mathbf{d}_{k+n_y} \end{bmatrix} = \underbrace{\begin{bmatrix} I \\ I \\ I \\ \vdots \\ I \end{bmatrix}}_{L} \mathbf{d}_k \tag{2.13}
$$

Summary: The predictions for discrete state space model (2.4) are

$$
\begin{aligned}
\underset{\rightarrow k+1}{\mathbf{x}} &= P_x \mathbf{x}_k + H_x \underset{\rightarrow k}{\mathbf{u}} \\
\underset{\rightarrow k+1}{\mathbf{y}} &= P \mathbf{x}_k + H \underset{\rightarrow k}{\mathbf{u}} + L \mathbf{d}_k
\end{aligned} \tag{2.14}
$$

STUDENT PROBLEMS

*1. MATLAB code (named **sspredmat.m**) is available to form the matrices of Eqn.(2.14) for any system. Readers are encouraged to try this code and verify it gives the correct answers. An example of the usage is shown here with a user choice of n_y.*

[H,P,L,Hx,Px]=sspredmat(A,B,C,ny)

*2. Verify numerically using random examples of different dimensions that the prediction Equations (2.14) give the same result as simulating the system dynamics forward one sample at a time. The file **sssim.m** is available to save time in simulating the system for any given sequence of future inputs.*
You are advised to use examples with stable poles (eigenvalues of A inside the unit circle) to avoid ill-conditioning with large n_y.
3. In practice, for computational reasons, prediction often makes the following assumption:

$$
\mathbf{u}_{k+i|k} = \mathbf{u}_{k+n_u-1|k}, \quad \forall i \geq n_u
$$

What impact does this assumption have on your prediction equations (2.14)? Formulate a set of prediction equations in terms of the d.o.f. only, that is the future inputs $\mathbf{u}_{k|k}, \cdots, \mathbf{u}_{k+n_u-1|k}$.

Remark 2.2 *There was an implicit assumption in this section that the state $\mathbf{x}_k$ is known/measureable, whereas in practice one may need an observer for this. It is also assumed that a disturbance estimate $\mathbf{d}_k$ is available. Observer design is not covered in this book.*

2.4.3 Unbiased prediction with state space models

In this case, unbiased is taken to mean that the prediction model gives correct predictions in steady-state, notwithstanding the presence of parameter uncertainty and a non-zero, but constant and unknown disturbance. Using $(.)_p$ to denote the process and $(.)_m$ to denote model, let the true process be:

$$\mathbf{x}_p(k+1) = A_p\mathbf{x}_p(k) + B_p\mathbf{u}_k; \quad \mathbf{y}_p(k) = C_p\mathbf{x}_p(k) + \mathbf{d}_p(k) \tag{2.15}$$

Simulate a model with state $\mathbf{x}_m$ in parallel with the process, using the same input signal, as shown in Figure 2.1. The prediction model given in (2.11) uses model state $\mathbf{x}_m$. This is used in conjunction with a definition for the estimate for $\mathbf{d}_k$:

$$\mathbf{d}(k) = \mathbf{y}_p(k) - C_m\mathbf{x}_m(k) \tag{2.16}$$

to substitute into (2.14). Therefore the estimated process system predictions (that is $E[\underset{\rightarrow p}{\mathbf{y}}(k+1)]$) are given from:

$$
\begin{aligned}
\underset{\rightarrow m}{\mathbf{x}}(k+1) &= P_x\mathbf{x}_m(k) + H_x\underset{\rightarrow}{\mathbf{u}}(k) \\
\underset{\rightarrow m}{\mathbf{y}}(k+1) &= P\mathbf{x}_m(k) + H\underset{\rightarrow}{\mathbf{u}}(k) \\
E[\underset{\rightarrow p}{\mathbf{y}}(k+1)] &= P\mathbf{x}_m(k) + H\underset{\rightarrow}{\mathbf{u}}(k) + L\mathbf{d}(k)
\end{aligned}
\tag{2.17}
$$

Notably, these predictions use the *known* model state alongside a measured disturbance estimate.

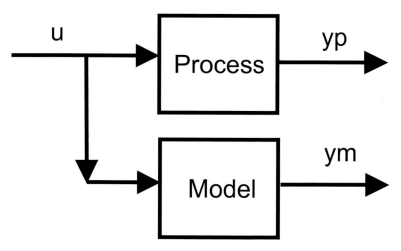

FIGURE 2.1
Example structure of simulation of model in parallel with process to obtain a disturbance estimate from (2.16).

STUDENT PROBLEM

Generate a plant and process which are slightly different $A_p \neq A_m, B_p \neq B_m, C_P \neq C_M$ and simulate both with the same input signal until steady-state has been achieved (restrict yourself to stable open-loop processes for now). Prove, numerically, that using the disturbance estimate of (2.16) ensures that the predictions from the model (2.17) and true process (2.15) match, irrespective of the steady-state value of $\mathbf{u}_k$ and real plant disturbance $\mathbf{d}_p(k)$.

HINT: *You can use the corresponding prediction equations from (2.14) or use the provided MATLAB file* **sspredmat_unbiased.m.**

ILLUSTRATION: Steady-state predictions plant and model

Assume that the process is at steady-state and the input and disturbance are constant. Therefore, the actual system behaviour will evolve as follows (as steady-state is assumed, the output evolution will be constant):

$$\left. \begin{array}{l} \mathbf{x}_p(k+1) = A_p\mathbf{x}_p(k) + B_p\mathbf{u}(k) \\ \mathbf{x}_p(k+1) = \mathbf{x}_p(k) \end{array} \right\} \Rightarrow \left\{ \begin{array}{l} \mathbf{x}_p(k) = [I - A_p]^{-1} B_p\mathbf{u}(k) \\ \mathbf{y}_p(k+1) = \mathbf{y}_p(k) = C_p\mathbf{x}_p(k) + \mathbf{d}_p(k) \end{array} \right.$$
(2.18)

Using $(.)_m$ to denote model, the model predictions will take the following form. It can also be assumed that these will also be constant because, by definition, the input is constant and any dynamics have settled.

$$\left. \begin{array}{l} \mathbf{x}_m(k+1) = A_m\mathbf{x}_m(k) + B_m\mathbf{u}(k) \\ \mathbf{x}_m(k+1) = \mathbf{x}_m(k) \end{array} \right\} \Rightarrow \left\{ \begin{array}{l} \mathbf{x}_m(k) = [I - A_m]^{-1} B_m\mathbf{u}_k \\ E[\mathbf{y}_p(k+1)] = C_m\mathbf{x}_m(k+1) + \mathbf{d}(k) \end{array} \right.$$
(2.19)

where here $\mathbf{d}(k)$ is the disturbance estimated added to ensure $\mathbf{y}_m = \mathbf{y}_p$. Clearly, the states $\mathbf{x}_p(k), \mathbf{x}_m(k)$ of the plant and model may be different. However, the output predictions will still be the same because, using (2.16), $\mathbf{x}_m(k+1) = \mathbf{x}_m(k)$ and assumptions $\mathbf{d}_p(k+1) = \mathbf{d}_p(k), \mathbf{d}(k+1) = \mathbf{d}(k)$:

$$\left. \begin{array}{c} \mathbf{y}_p(k) = \mathbf{y}_p(k+1) \\ E[\mathbf{y}_p(k+1)] = C_m\mathbf{x}_m(k+1) + \mathbf{d}(k+1) \\ \mathbf{d}(k) = \mathbf{y}_p(k) - C_m\mathbf{x}_m(k) \end{array} \right\} \Rightarrow E[\mathbf{y}_p(k+1)] = \mathbf{y}_p(k) = \mathbf{y}_p(k+1)$$
(2.20)

STUDENT PROBLEM

Using numerical examples of your own, verify that predictions using disturbance estimates are indeed unbiased in the steady-state, irrespective of the parameter/disturbance uncertainty of the real process.

Summary: Predictions from model (2.14) will be unbiased in the steady-state if the disturbance estimate is computed using

$$\mathbf{d}(k) = \mathbf{y}_p(k) - C_m \mathbf{x}_m(k). \tag{2.21}$$

It is implicit therefore that the model is simulated in parallel (see Figure 2.1) with the process to make the upto date model state $\mathbf{x}_m(k)$ available.

2.4.4 The importance of unbiased prediction and deviation variables

Readers will have noted the arguments in the introductory chapter which indicated how humans control the world around them quite effectively with relatively crude or inaccurate predictions. However, there is one notable exception and that is steady-state. Humans tend to automatically default to using deviation variables and base predictions as being relative to the current state rather than computed in absolute terms.

1. *A car driver controlling speed thinks in terms of increases and decreases to the current accelerator position and takes little account of its absolute position.*

2. *A person desiring a comfortable shower adjusts the hot and cold flows relative to the current tap positions and not relative to absolute positions.*

The advantage of this technique is that, when at the correct steady-state, the relative movement of the input to remain at the correct steady-state is zero and thus its numerical estimation does not depend in any way on having an accurate model. Hence, in the steady-state our predictions are unbiased, that is, they are correct.

If, alternatively, we were to attempt to choose the absolute value of the input to retain the correct steady-state and this computation was based on model parameters which are invariably approximate; then our choice of input would be wrong and the corresponding output would then deviate from the correct steady-state. In other words, our predictions would be biased (incorrect) in the steady-state.

Summary: In general unbiased predictions are essential to support effective control. This is often achieved by using deviation variables rather than absolute values.

STUDENT PROBLEM

Using numerical examples of your own, verify that predictions using deviation variables are indeed unbiased in the steady-state, irrespective of the parameter/disturbance uncertainty of the real process.

2.4.5   State space predictions with deviation variables

It is often convenient to use deviation variables, that is to measure states, inputs and outputs relative to some steady-state. An obvious trivial example of this is the scale degrees centigrade which is a deviation from 273 Kelvin or pressures being measured relative to atmospheric pressure. Such a practice makes sense in that, the states and outputs in deviation form represent *distances* from their desirable values and thus values of zero are the target! [1]

Definition of deviation variables relative to desired output

Let the desired steady-state output be given by:

$$\mathbf{y}_k = \mathbf{r} \tag{2.22}$$

For a given output disturbance $\mathbf{d}_k$, the corresponding steady-state input $\mathbf{u}_{ss}$ and states $\mathbf{x}_{ss}$ must satisfy:

$$\left\{ \begin{array}{l} \mathbf{r} = C\mathbf{x}_{ss} + \mathbf{d} \\ \mathbf{x}_{ss} = A\mathbf{x}_{ss} + B\mathbf{u}_{ss} \end{array} \right\} \tag{2.23}$$

These are simple simultaneous equations and give a solution of the form

$$\left[\begin{array}{c} \mathbf{x}_{ss} \\ \mathbf{u}_{ss} \end{array} \right] = \left[\begin{array}{cc} K_{xr} & K_{xd} \\ K_{ur} & K_{ud} \end{array} \right] \left[\begin{array}{c} \mathbf{r} \\ \mathbf{d} \end{array} \right] \tag{2.24}$$

where matrices $K_{xr}, K_{xd}, K_{ur}, K_{ud}$ depend solely on matrices A, B, C

---

[1]Using zero as the target can make the inclusion of integral action simpler, but this discussion is omitted.

representing the model ($A = A_m$, $B = B_m$, $C = C_m$). Deviation variables $\tilde{\mathbf{x}}_k, \tilde{\mathbf{u}}_k$ are then defined as:

$$\tilde{\mathbf{x}}_k = \mathbf{x}_k - \mathbf{x}_{ss}; \quad \tilde{\mathbf{u}}_k = \mathbf{u}_k - \mathbf{u}_{ss} \tag{2.25}$$

NOTE: In this case, that is model (2.23), $K_{xr} = -K_{xd}, K_{ur} = -K_{ud}$.

Remark 2.3 *Different disturbance models such as input disturbances will lead to slightly different equations [107] but the same principles apply so those computations are left for the reader.*

Theorem 2.1 *Prediction equations in terms of deviation variables take the same format as those in terms of the full states and inputs.*

Proof: A simplistic proof could simply state that this follows automatically from linearity of the model. However, to clarify some further detail is given here where predictions with the full states and with deviation variables are set side by side.

$$\left\{ \begin{array}{rcl} \underset{\rightarrow k+1}{\mathbf{x}} & = & P_x \mathbf{x}_k + H_x \underset{\rightarrow k}{\mathbf{u}} \\ \underset{\rightarrow k+1}{\mathbf{y}} & = & P\mathbf{x}_k + H\underset{\rightarrow k}{\mathbf{u}} + L\mathbf{d}_k \end{array} \right\} ; \quad \left\{ \begin{array}{rcl} \underset{\rightarrow k+1}{\tilde{\mathbf{x}}} & = & P_x \tilde{\mathbf{x}}_k + H_x \underset{\rightarrow k}{\tilde{\mathbf{u}}} \\ \underset{\rightarrow k+1}{\tilde{\mathbf{y}}} & = & P\tilde{\mathbf{x}}_k + H\underset{\rightarrow k}{\tilde{\mathbf{u}}} \end{array} \right\} \tag{2.26}$$

Next, substituting in from (2.25):

$$\begin{array}{rcll} \underset{\rightarrow k+1}{\tilde{\mathbf{x}}} = \underset{\rightarrow k}{\mathbf{x}} - L\mathbf{x}_{ss} & = & P_x[\mathbf{x}_k - \mathbf{x}_{ss}] + H_x[\underset{\rightarrow k}{\mathbf{u}} - L\mathbf{u}_{ss}] \\ \underset{\rightarrow k+1}{\tilde{\mathbf{y}}} = \underset{\rightarrow k+1}{\mathbf{y}} - L\mathbf{y}_{ss} & = & P[\mathbf{x}_k - \mathbf{x}_{ss}] + H[\underset{\rightarrow k}{\mathbf{u}} - L\mathbf{u}_{ss}] \end{array} ; \tag{2.27}$$

It is clear that the same prediction equations are implicit as, by definition, $L\mathbf{x}_{ss} - P_{xx}\mathbf{x}_{ss} - H_x L\mathbf{u}_{ss} = 0, L\mathbf{y}_{ss} - P\mathbf{x}_{ss} - HL\mathbf{u}_{ss} = L\mathbf{d}_k$. □

ILLUSTRATION: Integral action with state feedback

Stabilising state feedback based on deviation variables ensures offset-free tracking, and thus implicitly includes integral action.

$$\left. \begin{array}{rcl} \mathbf{u}_k - \mathbf{u}_{ss} & = & -K(\mathbf{x}_k - \mathbf{x}_{ss}) \\ \mathbf{x}_{k+1} & = & A\mathbf{x}_k + B\mathbf{u}_k \end{array} \right\}$$

$$\Rightarrow \left\{ \begin{array}{l} \mathbf{x}_{k+1} = A\mathbf{x}_k + B[-K(\mathbf{x}_k - \mathbf{x}_{ss}) + \mathbf{u}_{ss}] \\ \mathbf{x}_{k+1} - \mathbf{x}_{ss} = \underbrace{[A - BK]}_{\Phi}[\mathbf{x}_k - \mathbf{x}_{ss}] - \mathbf{x}_{ss} + A\mathbf{x}_{ss} + B\mathbf{u}_{ss} \end{array} \right. \tag{2.28}$$

Now, noting that by definition $-\mathbf{x}_{ss} + A\mathbf{x}_{ss} + B\mathbf{u}_{ss} = 0$, Φ has stable eigenvalues and then computing the asymptotic limit as times goes to infinity gives:

$$\lim_{k \to \infty}[\mathbf{x}_k - \mathbf{x}_{ss}] = \lim_{k \to \infty} \Phi^k[\mathbf{x}_0 - \mathbf{x}_{ss}] = 0 \tag{2.29}$$

offset-free tracking follows as long as $\mathbf{x}_k = \mathbf{x}_{ss} \Rightarrow \mathbf{y}_k = \mathbf{r}_k$ (implicit from (2.23)).

Summary: Deviation variables about a specified steady-state $\mathbf{x}_{ss}, \mathbf{u}_{ss}$ are convenient and easy to include in prediction algebra so that, in practice, we will often use the predictions:

$$\begin{aligned} \underset{\rightarrow k+1}{\tilde{\mathbf{x}}} &= P_x \tilde{\mathbf{x}}_k + H_x \underset{\rightarrow k}{\tilde{\mathbf{u}}} \\ \underset{\rightarrow k+1}{\tilde{\mathbf{y}}} &= P \tilde{\mathbf{x}}_k + H \underset{\rightarrow k}{\tilde{\mathbf{u}}} \end{aligned} \quad ; \quad \begin{aligned} \tilde{\mathbf{x}}_k &= \mathbf{x}_k - \mathbf{x}_{ss} \\ \tilde{\mathbf{u}}_k &= \mathbf{u}_k - \mathbf{u}_{ss} \end{aligned} \qquad (2.30)$$

It is implicit therefore that there is a requirement that one is able to obtain good estimates of $\mathbf{x}_{ss}, \mathbf{u}_{ss}$ (which notably depend on the disturbance estimate $\mathbf{d}_k$).

2.4.6   Predictions with state space models and input increments

The standard state space model is defined in terms of absolute inputs. However, in predictive control it is often convenient to express predictions in terms of a dependence on changes to the input rather than absolute values. Consequently, there is a need for a minor modification of the state space model to support this algebra. One simple method is to define the input as an additional state with update equation as follows:

$$\mathbf{u}_k = \mathbf{u}_{k-1} + \Delta\mathbf{u}_k; \quad \Delta(z) = 1 - z^{-1} \qquad (2.31)$$

Clearly $\Delta\mathbf{u}_k = \mathbf{u}_k - \mathbf{u}_{k-1}$ is defined as the change in the input and z^{-1} is the unit delay operator. The corresponding augmented state space model, beginning from (2.4) takes the following form:

$$\underbrace{\begin{bmatrix} \mathbf{x}_{k+1} \\ \mathbf{u}_k \end{bmatrix}}_{\mathbf{w}_{k+1}} = \underbrace{\begin{bmatrix} A & B \\ 0 & I \end{bmatrix}}_{A_a} \begin{bmatrix} \mathbf{x}_k \\ \mathbf{u}_{k-1} \end{bmatrix} + \underbrace{\begin{bmatrix} B \\ I \end{bmatrix}}_{B_a} \Delta\mathbf{u}_k \qquad (2.32)$$

$$\mathbf{y}_k = \underbrace{[C \quad 0]}_{C_a} \begin{bmatrix} \mathbf{x}_k \\ \mathbf{u}_{k-1} \end{bmatrix} + \mathbf{d}_k$$

Hence, in terms of the augmented matrices and states:

$$\mathbf{w}_{k+1} = A_a \mathbf{w}_k + B_a \Delta\mathbf{u}_k; \quad \mathbf{y}_k = C_a \mathbf{w}_k + \mathbf{d}_k \qquad (2.33)$$

Augmented models and deviation variables

In practice, due to the desire for offset-free tracking and unbiased prediction, it is often appropriate to combine the augmented model of (2.33) with the deviation variables of (2.30). The prediction equations will take the following form:

$$\begin{aligned} \underset{\rightarrow k+1}{\tilde{\mathbf{w}}} &= P_w \tilde{\mathbf{w}}_k + H_w \underset{\rightarrow k}{\Delta\tilde{\mathbf{u}}} \\ \underset{\rightarrow k+1}{\tilde{\mathbf{y}}} &= P \tilde{\mathbf{w}}_k + H \underset{\rightarrow k}{\Delta\tilde{\mathbf{u}}} \end{aligned} \qquad (2.34)$$

where prediction matrices P_w, H_w, P, H are defined using model parameters A_a, B_a, C_a. However, one may further note that, as by definition $\Delta \mathbf{u}_{ss} = 0$, one could equally write:

$$
\begin{aligned}
\underset{\rightarrow k+1}{\tilde{\mathbf{w}}} &= P_w \tilde{\mathbf{w}}_k + H_w \underset{\rightarrow k}{\Delta \mathbf{u}} \\
\underset{\rightarrow k+1}{\tilde{\mathbf{y}}} &= P \tilde{\mathbf{w}}_k + H \underset{\rightarrow k}{\Delta \mathbf{u}}
\end{aligned} \tag{2.35}
$$

STUDENT PROBLEM

Using numerical examples of your own, verify that predictions using input increments alongside deviation variables are unbiased in the steady-state, irrespective of the parameter/disturbance uncertainty of the real process.

Summary: The use of input increments with state space models is not particularly common in the literature, but there are times when formulating the d.o.f. this way rather than with absolute inputs is advantageous.

2.5 Prediction with transfer function models – matrix methods

This section will demonstrate matrix methods for finding systems predictions for transfer function/CARIMA (C.5) and matrix fraction description (C.8) models. The cases of $T(z) = 1$ and $T(z) \neq 1$ are tackled in turn.

Readers should note that there are several different ways of deriving the prediction equations for transfer function models (e.g., [127]) but some are more transparent than others. This book will not contrast the different methods in detail because they give the same prediction equations, so which to use reduces to personal preference. Often papers in the academic journals make use of Diophantine identities to form the prediction equations. However, this procedure tends to obscure what is actually going on, that is prediction, and hence can be confusing for the newcomer. Moreover the historical reason for such a preference is more likely following previous authors rather than any clear rational reason.

2.5.1   Ensuring unbiased prediction with transfer function models

Readers are reminded of an underlying objective discussed in the state space section, which is to achieve unbiased prediction in the steady-state. This book will use the CARIMA model representation (C.5) of a transfer function. Such a model has the sensible attribute of a built-in disturbance effect and this is exploited to give unbiased prediction in the steady-state. The original model is given as:

$$a(z)y_k = b(z)u_k + \frac{T(z)}{\Delta(z)}\zeta_k \qquad (2.36)$$

where ζ is a zero-mean random variable. The parameters $a(z), b(z), T(z)$ may be determined by a standard identification algorithm ([109]) although that discussion is not part of this book. However model (2.36) is represented in terms of absolute variables and hence, due to parameter uncertainty, would lead to biased prediction in general, even if the disturbance term $\frac{T(z)}{\Delta(z)}\zeta_k$ were known exactly. One can modify this model to one based on *pseudo-deviation* variables by multiplying through by the Δ term and hence:

$$a(z)\Delta(z)y_k = b(z)\Delta(z)u_k + T(z)\zeta_k \qquad (2.37)$$

The implied states of $\Delta y, \Delta u$ are changes in the output and input (not strictly deviation variables as not defined to be relative to a given steady-state) and thus this model will give unbiased prediction in the steady-state, as, the best future assumption for the other term is that $E[\zeta_k] = 0$.

Summary: Unbiased prediction can be ensured by using the following equation for prediction.

$$a(z)\Delta(z)y_k = b(z)\Delta(z)u_k \qquad (2.38)$$

In practice, **for convenience**, it is common-place to group the $\Delta(z)$ with the $a(z)$ and the input term, and hence:

$$\underbrace{[a(z)\Delta(z)]}_{A(z)}y_k = b(z)[\Delta u_k] \qquad (2.39)$$

$$\begin{aligned} A(z) &= 1 + A_1 z^{-1} + A_2 z^{-2} + \cdots + A_{n+1} z^{-n-1} \\ b(z) &= b_0 + b_1 z^{-1} + b_2 z^{-2} + \cdots + b_n z^{-n} \end{aligned}$$

It is assumed hereafter that for a strictly proper system $b_0 = 0$.

Observation: A matrix fraction description for MIMO processes is equivalent to a transfer function and thus a similar statement can be made where now the original model of (C.8) becomes:

$$\underbrace{[D(z)\Delta(z)]}_{A(z)}\mathbf{y}_k = N(z)[\Delta \mathbf{u}_k] \qquad (2.40)$$

ILLUSTRATION: Adding $\Delta(z)$ to polynomials in MATLAB

- **SISO case:** Adding $\Delta(z)$ to a SISO polynomial is straightforward using the MATLAB function **conv.m** for polynomial multiplication. Hence, for example:

$$a(z)\Delta(z) = [1 - 1.7z^{-1} + 0.8z^{-2}][1 - z^{-1}] = 1 - 2.7z^{-1} + 2.5z^{-2} - 0.8z^{-3}$$

In MATLAB: $a=[1,-1.7,0.8], A=conv(a,[1,-1]);$

- **MIMO case:** This is similar to the SISO case with one minor exception. There is a need to form a new matrix polynomial $\Delta(z)D(z)$; this requires multiplication of two matrix polynomials, albeit one is simple, $\Delta(z) = I - z^{-1}I$. The author has provided code for this in **convmat.m** (assumes underlying system is square and the authors' notation for storing the matrix parameters). For example consider how to find $D(z)N(z)$ when:

$$D(z) = \begin{bmatrix} 1 & 0 \\ 0 & 1 \end{bmatrix} + \begin{bmatrix} -1.4 & 0 \\ 0.2 & 1.2 \end{bmatrix} z^{-1} + \begin{bmatrix} 0.52 & 0 \\ -0.1 & 0.6 \end{bmatrix} z^{-2}$$

$$N(z) = \begin{bmatrix} 1 & 0.3 \\ -2.1 & 0 \end{bmatrix} z^{-1} + \begin{bmatrix} -0.2 & 1.1 \\ 1 & 0.4 \end{bmatrix} z^{-2}$$

In MATLAB (see appendix C for author's notation for MFD models):
$D = [1, 0, -1.4, 0, 0.52, 0; 0, 1, 0.2, 1.2, -0.1, 0.6];$
$N = [1, 0.3, -0.2, 1.1; -2.1, 0, 1, 0.4];$
$DN = convmat(D,N);$ % *Does matrix polynomial multiplication for $D(z)N(z)$*
$Delta = [eye(2),-eye(2)];$ % *MIMO version of $\Delta(z)$*
$A = convmat(D,Delta);$ % *Forms $A(z) = D(z)\Delta(z)$*

STUDENT PROBLEM

Use MATLAB to form the polynomials $A(z) = a(z)\Delta(z)$ and $D(z)\Delta(z)$
for the following examples and validate with pen and paper working.

$$a(z) = 1 + 2z^{-1} + 2.4z^{-2} + 1.5z^{-3} + 0.5z^{-4}; \quad a(z) = 1 - 0.3z^{-1} + 1.4z^{-2} - 0.54z^{-3};$$

$$D(z) = \begin{bmatrix} 1 & 0 \\ 0 & 1 \end{bmatrix} + \begin{bmatrix} -1.2 & 0.3 \\ 0.42 & 1.25 \end{bmatrix} z^{-1} + \begin{bmatrix} 0.82 & 0.34 \\ -0.4 & 0.93 \end{bmatrix} z^{-2} + \begin{bmatrix} 0.12 & -0.3 \\ 0.1 & 0.23 \end{bmatrix} z^{-3}$$

2.5.2 Prediction for a CARIMA model with $T(z) = 1$: the SISO case

A one-step ahead prediction model based on (2.38) is given as:

$$y_{k+1} = -A_1 y_k - \cdots - A_{n+1} y_{k-n} + b_1 \Delta u_k + \cdots + b_{n-1} \Delta u_{k-n+1} \qquad (2.41)$$

In Section 2.4.1 on prediction for state space models, the basic technique was to iterate the one-step ahead prediction equations for n_y samples; exactly the same technique can be used here.

1. Thus, write Equation (2.41) for the next n_y samples:

$$
\begin{aligned}
y_{k+1} + A_1 y_k + \cdots + A_{n+1} y_{k-n} &= b_1 \Delta u_k + b_2 \Delta u_{k-1} + \cdots + b_n \Delta u_{k-n+1} \\
y_{k+2} + A_1 y_{k+1} + \cdots + A_{n+1} y_{k-n+1} &= b_1 \Delta u_{k+1} + b_2 \Delta u_k + \cdots + b_n \Delta u_{k-n+2} \\
&\vdots \\
y_{k+n_y} + \cdots + A_{n+1} y_{k+n_y+1-n} &= b_1 \Delta u_{k+n_y-1} + \cdots + b_n \Delta u_{k+n_y-n}
\end{aligned}
$$

$$(2.42)$$

2. These can be placed in the following compact matrix/vector form where, for convenience, the data has been separated into past and future.

$$
\underbrace{\begin{bmatrix} 1 & 0 & \cdots & 0 \\ A_1 & 1 & \cdots & 0 \\ A_2 & A_1 & \cdots & 0 \\ \vdots & \vdots & \vdots & \vdots \end{bmatrix}}_{C_A}
\underbrace{\begin{bmatrix} y_{k+1} \\ y_{k+2} \\ \vdots \\ y_{k+n_y} \end{bmatrix}}_{\underset{\rightarrow k+1}{y}}
+
\underbrace{\begin{bmatrix} A_1 & A_2 & \cdots & A_{n+1} \\ A_2 & A_3 & \cdots & 0 \\ A_3 & A_4 & \cdots & 0 \\ \vdots & \vdots & \vdots & \vdots \end{bmatrix}}_{H_A}
\underbrace{\begin{bmatrix} y_k \\ y_{k-1} \\ \vdots \\ y_{k-n} \end{bmatrix}}_{\underset{\leftarrow k}{y}}
$$

$$
= \underbrace{\begin{bmatrix} b_1 & 0 & \cdots & 0 \\ b_2 & b_1 & \cdots & 0 \\ b_3 & b_2 & \cdots & 0 \\ \vdots & \vdots & \vdots & \vdots \end{bmatrix}}_{C_{zb}}
\underbrace{\begin{bmatrix} \Delta u_k \\ \Delta u_{k+1} \\ \vdots \\ \Delta u_{k+n_y-1} \end{bmatrix}}_{\underset{\rightarrow k}{\Delta u}}
+
\underbrace{\begin{bmatrix} b_2 & b_3 & \cdots & b_n \\ b_3 & b_4 & \cdots & 0 \\ b_4 & b_5 & \cdots & 0 \\ \vdots & \vdots & \vdots & \vdots \end{bmatrix}}_{H_{zb}}
\underbrace{\begin{bmatrix} \Delta u_{k-1} \\ \Delta u_{k-2} \\ \vdots \\ \Delta u_{k-n+1} \end{bmatrix}}_{\underset{\leftarrow k-1}{\Delta u}}
$$

$$(2.43)$$

Matrices denoted C_A, C_{zb} are lower triangular and striped and have the coefficients of $A(z), b(z)$ in each diagonal. Matrices denoted H_A, H_{zb} have an upper triangular structure, again based on the coefficients of $A(z), b(z)$. Consequently, these matrices are easy to define.

3. Using the Toeplitz/Hankel notation one can simplify (2.43) to

$$C_A \underset{\rightarrow k+1}{y} + H_A \underset{\leftarrow k}{y} = C_{zb} \underset{\rightarrow k}{\Delta u} + H_{zb} \underset{\leftarrow k-1}{\Delta u} \qquad (2.44)$$

4. Hence, from (2.44) the output predictions are

$$\underset{\rightarrow k+1}{y} = C_A^{-1} [C_{zb} \underset{\rightarrow k}{\Delta u} + H_{zb} \underset{\leftarrow k-1}{\Delta u} - H_A \underset{\leftarrow k}{y}] \qquad (2.45)$$

5. For convenience one may wish to represent (2.45) as

$$\underset{\rightarrow k+1}{y} = H\Delta\underset{\rightarrow k}{u} + P\Delta\underset{\leftarrow k-1}{u} + Q\underset{\leftarrow k}{y} \tag{2.46}$$

where $H = C_A^{-1}C_{zb}, P = C_A^{-1}H_{zb}, Q = -C_A^{-1}H_A$.

Remark 2.4 C_A^{-1} *can be computed very efficiently if required as* $C_A^{-1} = C_{1/A}$. *So one can simply use the coefficients of the expansion* $1/A(z)$.

ILLUSTRATION: Prediction with a CARIMA model

Consider the difference equation model (ignoring ζ_k):

$$y_{k+1} = 1.3y_k - 0.4y_{k-1} + u_k - 2u_{k-1} \tag{2.47}$$

First write this in incremental form of Eqn.(2.37) as

$$y_{k+1} = 2.3y_k - 1.7y_{k-1} + 0.4y_{k-2} + \Delta u_k - 2\Delta u_{k-1} \tag{2.48}$$

Then substitute into expressions (2.49) using:

$$C_A = \begin{bmatrix} 1 & 0 & 0 & 0 \\ -2.3 & 1 & 0 & 0 \\ 1.7 & -2.3 & 1 & 0 \\ -0.4 & 1.7 & -2.3 & 1 \end{bmatrix}; \quad H_A = \begin{bmatrix} -2.3 & 1.7 & -0.4 \\ 1.7 & -0.4 & 0 \\ -0.4 & 0 & 0 \\ 0 & 0 & 0 \end{bmatrix}$$

$$C_{zb} = \begin{bmatrix} 1 & 0 & 0 & 0 \\ -2 & 1 & 0 & 0 \\ 0 & -2 & 1 & 0 \\ 0 & 0 & -2 & 1 \end{bmatrix}; \quad H_{zb} = \begin{bmatrix} -2 \\ 0 \\ 0 \\ 0 \end{bmatrix}$$

Hence (given $n_u = 2$):

$$H = \begin{bmatrix} 1 & 0 \\ 0.3 & 1 \\ -1.01 & 0.3 \\ -2.433 & -1.01 \end{bmatrix}; \quad P = \begin{bmatrix} -2 \\ -4.6 \\ -7.18 \\ -9.494 \end{bmatrix}; \quad Q = \begin{bmatrix} 2.3 & -1.7 & 0.4 \\ 3.59 & -3.51 & 0.92 \\ 4.747 & -5.183 & 1.436 \\ 5.7351 & -6.6339 & 1.8988 \end{bmatrix}$$

Summary: Output predictions, with $T = 1$, are given by

$$\underset{\rightarrow k+1}{y} = H\Delta\underset{\rightarrow k}{u} + P\Delta\underset{\leftarrow k-1}{u} + Q\underset{\leftarrow k}{y}$$
$$H = C_A^{-1}C_{zb}, \ P = C_A^{-1}H_{zb}, \ Q = -C_A^{-1}H_A \tag{2.49}$$

Hence, the predictions can be written down explicitly in terms of the model coefficients and this is very simple to code in the SISO case.

ILLUSTRATION: Prediction for CARIMA models with MATLAB

It is notable that the definition of matrices C_A, H_A, C_{zb}, H_{zb} requires no multiplication or addition and thus is trivial to code. In consequence, the MATLAB coding is very short and is illustrated with the following snippet in **tfpredexample.m** (file **caha.m** is available online).

```
a=[1 -1.4 0.52 0.24];
b=[1,0.3,-0.2,1.1]; % Assumes first parameter is zero so excluded
[CA,HA]=caha(conv(a,[1 -1]),1,5);   % forms A = aΔ and uses ny = 5
[Cb,Hb]=caha(b,1,5);   % The argument '1' is for system dimension
CAi=inv(CA);
H=CAi*Cb
P=CAi*Hb
Q=-CAi*HA
```

STUDENT PROBLEMS

Equation 2.43 defined the notation for so-called Toeplitz and Hankel matrices. Find the corresponding matrices for the following examples and then verify your answers by comparing what you have written with the output of **caha.m**.

1. *Find* C_g, H_g *with:*

$$g(z) = 1 + 2z^{-1} + 2.4z^{-2} + 1.5z^{-3} + 0.5z^{-4}; \quad n_y = 6$$

   HINT: MATLAB command will be **[Cg,Hg]=caha ([1,2,2.4,1.5,0.5],1,6)**.

2. *Find* $C_{m\Delta}, H_{m\Delta}$ *with:*

$$m(z) = 1 - 0.4z^{-1} + 0.25z^{-2}; \quad n_y = 8$$

   HINT: **[Cmd,Hmd]=caha(conv([1,-0.4,0.25],[1 -1]),1,8)**.

3. *Find* C_{zb}, H_{zb} *with:*

$$b(z) = 0.4z^{-1} + 0.25z^{-2}; \quad n_y = 7$$

   HINT: **[Cb,Hb]=caha([0.4,0.25],1,7)** *and* $zb(z) = 0.4 + 0.25z^{-1}$.

4. *Find the prediction matrices in the form of (2.49) for the following system with* $n_y = 10$.

$$G(z) = \frac{0.4z^{-1} + 0.25z^{-2}}{1 + 2z^{-1} + 2.4z^{-2} + 1.5z^{-3} + 0.5z^{-4}}$$

Check your answers by using the following command.
[H,P,Q] = mpc_predmat([1,2,2.4,1.5,0.5],[0.4,0.25],10)

5. *Find the prediction matrices in the form of (2.49) for the following system with $n_y = 6$.*

$$G(z) = \frac{0.4z^{-4} + 0.25z^{-5}}{1 + 1.4z^{-1} + 0.7z^{-2}}$$

Check your answers by using the following command.
[H,P,Q] = mpc_predmat([1,1.4,0.7],[0,0,0,0.4,0.25],6)

2.5.3 Prediction with a CARIMA model and $T = 1$: the MIMO case

Assume here that a matrix fraction description (MFD) model $D(z)\mathbf{y}_k = N(z)\mathbf{u}_k$ is going to be deployed. One can reuse the development of the previous subsection without the need for any further complication apart from the extra dimensionality. That is, the algebra has the same complexity as given in (2.42). In place of (2.41) use the difference equation (for simplicity of notation assume $D(z)\Delta(z) = I + A_1 z^{-2} + \cdots + A_n z^{-n} = A(z)$):

$$\mathbf{y}_{k+1} + A_1\mathbf{y}_k + \cdots + A_{n+1}\mathbf{y}_{k-n} = N_1\Delta\mathbf{u}_k + N_2\Delta\mathbf{u}_{k-1} + \cdots + N_n\Delta\mathbf{u}_{k-n+1} \quad (2.50)$$

Then replicating all the steps of the previous subsection with the exception that the terms in the corresponding Toeplitz/Hankel matrices are matrix coefficients rather than scalar coefficients, one can jump straight to an equivalent matrix/vector form:

$$C_A \underset{\rightarrow k+1}{\mathbf{y}} + H_A \underset{\leftarrow k}{\mathbf{y}} = C_{zN}\underset{\rightarrow k}{\Delta\mathbf{u}} + H_{zN}\underset{\leftarrow k-1}{\Delta\mathbf{u}} \quad (2.51)$$

Hence, by inspection

$$\underset{\rightarrow k+1}{\mathbf{y}} = H\underset{\rightarrow k}{\Delta\mathbf{u}} + P\underset{\leftarrow k-1}{\Delta\mathbf{u}} + Q\underset{\leftarrow k}{\mathbf{y}} \quad (2.52)$$

where $H = C_A^{-1}C_{zN}$, $P = C_A^{-1}H_{zN}$, $Q = -C_A^{-1}H_A$.

ILLUSTRATION: Prediction for MFD models with MATLAB

This is almost identical to the SISO case although **convmat.m** can be used to form $D(z)\Delta(z)$ (see Appendix C.2.2). The provided code **caha.m** works with matrix polynomials and an example is given in **tfpredexample2.m**.

$$D(z) = \begin{bmatrix} 1 & 0 \\ 0 & 1 \end{bmatrix} + \begin{bmatrix} -1.4 & 0 \\ 0.2 & 1.2 \end{bmatrix} z^{-1} + \begin{bmatrix} 0.52 & 0 \\ -0.1 & 0.6 \end{bmatrix} z^{-2}$$

$$N(z) = \begin{bmatrix} 1 & 0.3 \\ -2.1 & 0 \end{bmatrix} z^{-1} + \begin{bmatrix} -0.2 & 1.1 \\ 1 & 0.4 \end{bmatrix} z^{-2}$$

D=[1 0 -1.4 0 0.52 0;0 1 0.2 1.2 -0.1 0.6];
N=[1,0.3,-0.2,1.1;-2.1 0 1 0.4];
A = convmat(D,[eye(2) -eye(2)]); % Does matrix polynomial multiplication
[CA,HA]=caha(A,2,5); % the second argument '2' gives the system dimension
[CN,HN]=caha(N,2,5);   % the 3rd argument is the horizon n_y
CAi=inv(CA);
H=CAi*CN
P=CAi*HN
Q=-CAi*HA

ALTERNATIVE CODE: **[H,P,Q] = mpc_predmat(D,N,5)**

Summary: The algebraic complexity for the MIMO case is identical to the SISO case. Predictions are given as

$$\underset{\rightarrow k+1}{\mathbf{y}} = H\underset{\rightarrow k}{\Delta\mathbf{u}} + P\underset{\leftarrow k-1}{\Delta\mathbf{u}} + Q\underset{\leftarrow k}{\mathbf{y}} \qquad (2.53)$$

Remark 2.5 *Using matrix inversion as in (2.49) is neat and simple for paper and pen exercises and also effective for low values of n_y. However, for very large prediction horizons readers should use more computationally efficient and numerically robust methods [127] or Section 2.6 in any computer coding.*

STUDENT PROBLEMS

1. *Find C_A, H_A and C_{zN}, H_{zN} for $n_y = 5$ with $A = D\Delta$ and:*

$$D(z) = \begin{bmatrix} 1 & -0.5 \\ 0.2 & 1 \end{bmatrix} + \begin{bmatrix} -1.23 & 0.6 \\ 0.25 & -1.2 \end{bmatrix} z^{-1} + \begin{bmatrix} 0.2 & 0.4 \\ -0.5 & 0.67 \end{bmatrix} z^{-2}$$

$$N(z) = \begin{bmatrix} 1 & 0.35 \\ -1.1 & 0.2 \end{bmatrix} z^{-1} + \begin{bmatrix} -0.3 & 1.14 \\ 0.7 & 0.41 \end{bmatrix} z^{-2}$$

   Use MATLAB to construct the same matrices and validate the answers are the same.

2. *Find the prediction matrices in the form of (2.53) for a system with the $D(z), N(z)$ above, $n_y = 10$ and:*

$$G(z) = D(z)^{-1}N(z); \quad \mathbf{y}_k = G(z)\mathbf{u}_k \qquad (2.54)$$

Check your answers using **mpc_predmat.m**

3. *Construct a 3 by 3 MIMO MFD system of your choice and find the prediction matrices.*

2.5.4   Prediction equations with $T(z) \neq 1$: the SISO case

It has been noted by many authors that the $T(z)$ part of the CARIMA model [25] is often essential for practical applications of MPC based on transfer function models. Hence, it is necessary to consider how this changes the prediction equations.

> *To skip straight to the solution, go to the summary at the end of Section 2.5.4.2.*

Here we treat only the SISO case as the algebra for the MIMO and SISO case is identical, as long as matrices replace scalar coefficients as appropriate.

2.5.4.1   Summary of the key steps in computing prediction equations with a T-filter

Including a T-filter the CARIMA model is given as:

$$A(z)y_k = b(z)\Delta u_k + T(z)\zeta_k \qquad (2.55)$$

The assumption is made that $E[\zeta_k] = 0$ and thus, ideally, the prediction equation would be set up so that ζ_k was isolated and thus can reasonably be ignored. Hence, rearrange (2.55) to an alternative in which the unknown term is zero mean and unknown simply by filtering left- and right-hand sides by $T(z)$, i.e.,

$$\frac{A(z)}{T(z)}y_k = \frac{b(z)}{T(z)}\Delta u_k + \zeta_k \qquad (2.56)$$

If one uses (2.56) for prediction, then the bias due to past values of ζ_k is removed hence improving prediction accuracy. Hereafter the ζ_k term is ignored.

Next, pair the filter with the signals rather than the model parameters, that is, rearrange (2.56) as follows:

$$A(z)[\frac{y_k}{T(z)}] = b(z)[\frac{\Delta u_k}{T(z)}] \quad \Rightarrow \quad A(z)\tilde{y}_k = b(z)\Delta\tilde{u}_k \qquad (2.57)$$

where filtered variables are defined as:

$$\tilde{y}_k = \frac{y_k}{T(z)}; \quad \Delta\tilde{u}_k = \frac{\Delta u_k}{T(z)} \qquad (2.58)$$

Intuitive interpretation of the T-filter

The T-filter in effect filters the input/output (I/O) data in the one-step ahead prediction model of (2.57). Typically $1/T(z)$ is a low-pass filter so the filtering reduces the transference of high frequency noise into the predictions. It is unsurprising therefore that general observations are that a T-filter is essential in environments where output measurements are noisy as in such a case some form of low-pass filter would be required.

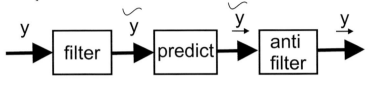

The benefits of the filtering are obtained by using (2.57) for prediction in place of (2.41), the only difference being the substitution of filtered signals for unfiltered signals. Hence, by inspection (from Eqn.(2.49)) one can write the prediction equation for filtered variables as

$$\underset{\rightarrow k+1}{\tilde{y}} = H\Delta\underset{\rightarrow k}{\tilde{u}} + P\Delta\underset{\leftarrow k-1}{\tilde{u}} + Q\underset{\leftarrow k}{\tilde{y}} \tag{2.59}$$

where H, P, Q are defined identically as in (2.49). Unfortunately this is no use because we want predictions for $\underset{\rightarrow k+1}{y}$ in terms of $\Delta\underset{\rightarrow k}{u}$; that is, future signals must be unfiltered.

Summary:

1. The T-filter is usually a low-pass filter and hence improves prediction accuracy in the high frequency range by reducing the transference of high frequency noise.

2. However, use of (2.57, 2.59) gives predictions in terms of future filtered variables and thus is not yet useful.

2.5.4.2 Forming the prediction equations with a T-filter beginning from predictions (2.59)

The weakness of prediction Equation (2.59) is that the future predictions are in the filtered space whereas we want to know the values of the future outputs and the corresponding future inputs. Nevertheless, given relationship (2.58) it is straightforward to find the future output predictions from the future filter output predictions.

Lemma 2.1 *The relationship between predicted filtered variables and predicted unfiltered variables is given as follows:*

$$C_T \underset{\rightarrow k+1}{\tilde{y}} + H_T \underset{\leftarrow k}{\tilde{y}} = \underset{\rightarrow k+1}{y}; \quad C_T \underset{\rightarrow k}{\Delta \tilde{u}} + H_T \underset{\leftarrow k-1}{\Delta \tilde{u}} = \underset{\rightarrow k}{\Delta u} \quad (2.60)$$

or equivalently

$$\underset{\rightarrow k+1}{y} = C_T^{-1} \lfloor \underset{\rightarrow k+1}{y} - H_T \underset{\leftarrow k}{y} \rfloor; \quad \underset{\rightarrow k}{\Delta \tilde{u}} = C_T^{-1} \lfloor \underset{\rightarrow k}{\Delta u} - H_T \underset{\leftarrow k-1}{\Delta \tilde{u}} \rfloor \quad (2.61)$$

Proof: This follows from an analogous development to Section 2.5.2 but using the update equations $y_k = T(z)\tilde{y}_k, \Delta u_k = T(z)\Delta \tilde{u}_k$. $\qquad \square$

Theorem 2.2 *Prediction equations for $\underset{\rightarrow k}{y}$ in terms of $\underset{\rightarrow k}{\Delta u}$ and beginning from (2.59) take the following form.*

$$\underset{\rightarrow k+1}{y} = H \underset{\rightarrow k}{\Delta u} + \tilde{P} \underset{\leftarrow k-1}{\Delta \tilde{u}} + \tilde{Q} \underset{\leftarrow k}{\tilde{y}}; \quad \begin{cases} \tilde{P} = [C_T P - HH_T] \\ \tilde{Q} = [H_T + C_T Q] \end{cases} \quad (2.62)$$

Proof: Substituting (2.61) into (2.59) gives:

$$\underbrace{C_T^{-1} [\underset{\rightarrow k+1}{y} - H_T \underset{\leftarrow k}{\tilde{y}}]}_{\underset{\rightarrow k+1}{\tilde{y}}} = H \underbrace{C_T^{-1} [\underset{\rightarrow k}{\Delta u} - H_T \underset{\leftarrow k-1}{\Delta \tilde{u}}]}_{\underset{\rightarrow k}{\Delta \tilde{u}}} + P \underset{\leftarrow k}{\Delta \tilde{u}} + Q \underset{\leftarrow k}{\tilde{y}} \quad (2.63)$$

Multiplying left and right by C_T and then grouping common terms gives the result. $\square$

Summary: Derivation of prediction equations is very similar with and without $T(z)$. The differences when $T \neq 1$ are:

1. Filtering on past data is used which can have the benefit of reducing the impact of noise on the predictions.
2. Matrices are changed as: $P \Rightarrow \tilde{P}, \quad Q \Rightarrow \tilde{Q}$.
3. If $T = 1$, then $H_T = 0$, $C_T = I$ so $P = \tilde{P}, \quad Q = \tilde{Q}$.

Without a T-filter, predictions are:	**With a T-filter the predictions are:**
$\underset{\rightarrow k+1}{y} = H \underset{\rightarrow k}{\Delta u} + P \underset{\leftarrow k-1}{\Delta u} + Q \underset{\leftarrow k}{y}$	$\underset{\rightarrow k+1}{y} = H \underset{\rightarrow k}{\Delta u} + \tilde{P} \underset{\leftarrow k-1}{\Delta \tilde{u}} + \tilde{Q} \underset{\leftarrow k}{\tilde{y}}$
$H = C_A^{-1} C_{zb},$	$\tilde{y}_k = y_k / T(z), \quad \Delta \tilde{u}_k = \Delta u_k / T(z)$
$P = C_A^{-1} H_{zb},$	$\tilde{P} = [C_T P - HH_T];$
$Q = -C_A^{-1} H_A$	$\tilde{Q} = [H_T + C_T Q]$

STUDENT PROBLEM

Prove that the prediction equations of (2.62) can equivalently be written down directly in terms of model parameters as follows.

$$\underset{\rightarrow k+1}{y} = H\Delta\underset{\rightarrow k}{u} + \tilde{P}\Delta\underset{\leftarrow k-1}{\tilde{u}} + \tilde{Q}\underset{\leftarrow k}{\tilde{y}}; \qquad \left\{ \begin{array}{l} H = C_A^{-1}C_{zb} \\ \tilde{P} = C_A^{-1}[C_T H_{zb} - C_{zb}H_T] \\ \tilde{Q} = C_A^{-1}[C_A H_T - C_T H_A] \end{array} \right. \qquad (2.64)$$

STUDENT PROBLEMS

Find predictions in the form of (2.53) for the following systems for various n_y.

$$G(z) = \frac{0.4z^{-1} + 0.25z^{-2}}{1 + 2z^{-1} + 2.4z^{-2} + 1.5z^{-3} + 0.5z^{-4}}; \quad T(z) = 1 - 0.8z^{-1}$$

$$G(z) = \frac{0.4z^{-4} + 0.25z^{-5}}{1 + 1.4z^{-1} + 0.7z^{-2}}; \quad T(z) = (1 - 0.8z^{-1})^2$$

HARDER: *Find the prediction equations for the MFD system of Eqn.(2.54) with $T = (1 - 0.85z^{-1})I$.*
HINT: You can validate your answers using the MATLAB file **mpc_predtfilt.m** *in conjunction with* **mpc_predmat.m**.

2.6 Using recursion to find prediction matrices for CARIMA models

This section is solely here for completeness because it gives an efficient means of computing the predictions for coding purposes. You are advised to *skip it* unless computational efficiency is more important than having compact equations. The solutions given are identical to those that arise from Diophantine methods (e.g., [25]). Although it complicates notation a little, for generality the MIMO case will be given here with an MFD model.

1. Assume an underlying difference equation for the one-step ahead prediction:

$$\mathbf{y}_{k+1} + A_1\mathbf{y}_k + \cdots + A_{n+1}\mathbf{y}_{k-n} = N_1\Delta\mathbf{u}_k + N_2\Delta\mathbf{u}_{k-1} + \cdots + N_n\Delta\mathbf{u}_{k-n+1}$$
$$(2.65)$$

2. Introduce notation for prediction horizon $(.)^{[i]}$ to denote i-step ahead prediction such that in general.

$$\mathbf{y}_{k+i} = H^{[i]}\Delta\underset{\rightarrow k}{\mathbf{u}} + P^{[i]}\Delta\underset{\leftarrow k-1}{\mathbf{u}} + Q^{[i]}\underset{\leftarrow k}{\mathbf{y}} \qquad (2.66)$$

3. Initialise (2.66) for $i = 1$ from (2.65):

$$H^{[1]} = [N_1, 0, 0, 0, \ldots]; \quad P^{[1]} = [N_2, N_3, \ldots, N_n]; \quad Q^{[1]} = [-A_1, \ldots, -A_{n+1}]$$
$$(2.67)$$

4. Use recursive substitution to find $\mathbf{y}_{k+i+1}$ in terms of $\mathbf{y}_{k+i}$ and $\mathbf{y}_{k+1}$.

   (a) Given the coefficients of the prediction equation for $\mathbf{y}_{k+i}$, compute $\mathbf{y}_{k+i+1}$ as follows:

$$\begin{aligned}
\mathbf{y}_{k+i} &= H^{[i]}\Delta\underset{\rightarrow k}{\mathbf{u}} + P^{[i]}\Delta\underset{\leftarrow k-1}{\mathbf{u}} + Q^{[i]}\underset{\leftarrow k}{\mathbf{y}} \\
&\Downarrow \\
\mathbf{y}_{k+i+1} &= H^{[i]}\Delta\underset{\rightarrow k+1}{\mathbf{u}} + P^{[i]}\Delta\underset{\leftarrow k}{\mathbf{u}} + Q^{[i]}\underset{\leftarrow k+1}{\mathbf{y}}
\end{aligned} \qquad (2.68)$$

   (b) Note that

$$\begin{aligned}
Q^{[i]}\underset{\leftarrow k+1}{\mathbf{y}} &= [Q^{[i]}, 0]\begin{bmatrix} \mathbf{y}_{k+1} \\ \underset{\leftarrow k}{\mathbf{y}} \end{bmatrix} &= Q_1^{[i]}\mathbf{y}_{k+1} + [Q_2^{[i]}, \ldots, Q_n^{[i]}, 0]\underset{\leftarrow k}{\mathbf{y}} \\
P^{[i]}\Delta\underset{\leftarrow k}{\mathbf{u}} &= [P^{[i]}, 0]\begin{bmatrix} \Delta\mathbf{u}_k \\ \Delta\underset{\leftarrow k-1}{\mathbf{u}} \end{bmatrix} &= P_1^{[i]}\Delta\mathbf{u}_k + [P_2^{[i]}, \ldots, P_{n-1}^{[i]}, 0]\Delta\underset{\leftarrow k-1}{\mathbf{u}}
\end{aligned} \qquad (2.69)$$

   (c) Substitute observations (2.69) into prediction (2.68b)

$$\begin{aligned}
\mathbf{y}_{k+i+1} = [P_1^{[i]}, H^{[i]}]\Delta\underset{\rightarrow k}{\mathbf{u}} + [P_2^{[i]}, \ldots, P_{n-1}^{[i]}, 0]\Delta\underset{\leftarrow k-1}{\mathbf{u}} \\
+ [Q_2^{[i]}, \ldots, Q_n^{[i]}, 0]\underset{\leftarrow k}{\mathbf{y}} + Q_1^{[i]}\mathbf{y}_{k+1}
\end{aligned} \qquad (2.70)$$

   (d) Substitute $\mathbf{y}_{k+1}$ from (2.66) into (2.70), to give

$$\begin{aligned}
\mathbf{y}_{k+i+1} = [P_1^{[i]}, H^{[i]}]\Delta\underset{\rightarrow k}{\mathbf{u}} + [P_2^{[i]}, \ldots, P_{n-1}^{[i]}, 0]\Delta\underset{\leftarrow k-1}{\mathbf{u}} + [Q_2^{[i]}, \ldots, Q_n^{[i]}, 0]\underset{\leftarrow k}{\mathbf{y}} \\
+ Q_1^{[i]}[H^{[1]}\Delta\underset{\rightarrow k}{\mathbf{u}} + P^{[1]}\Delta\underset{\leftarrow k-1}{\mathbf{u}} + Q^{[1]}\underset{\leftarrow k}{\mathbf{y}}]
\end{aligned} \qquad (2.71)$$

(e) Finally, group common terms to give the form of (2.66) again

$$\mathbf{y}_{k+i+1} = \underbrace{[P_1^{[i]} + Q_1^{[i]}H_1^{[1]}, H^{[i]}]}_{H^{[i+1]}} \Delta\underset{\rightarrow k}{\mathbf{u}}$$
$$+ \underbrace{\{[P_2^{[i]}, ..., P_{n-1}^{[i]}, 0] + Q_1^{[i]}P^{[1]}\}}_{P^{[i+1]}} \Delta\underset{\leftarrow k-1}{\mathbf{u}} \qquad (2.72)$$
$$+ \underbrace{\{[Q_2^{[i]}, ..., Q_n^{[i]}, 0] + Q_1^{[i]}Q^{[1]}\}}_{Q^{[i+1]}} \underset{\leftarrow k}{\mathbf{y}}$$

Remark 2.6 *Recursion (2.72) has the same computational complexity of the matrix approach given earlier. However it may be preferable in packages that do not support matrix algebra.*

Remark 2.7 *The overall prediction equation analogous to (2.49) is determined using*

$$H = \begin{bmatrix} H^{[1]} & 0 & 0 & 0 & \cdots \\ & H^{[2]} & 0 & 0 & \cdots \\ & & H^{[3]} & 0 & \cdots \\ \vdots & \vdots & \vdots & \vdots \end{bmatrix}; \quad P = \begin{bmatrix} P^{[1]} \\ P^{[2]} \\ P^{[3]} \\ \vdots \end{bmatrix}; \quad Q = \begin{bmatrix} Q^{[1]} \\ Q^{[2]} \\ Q^{[3]} \\ \vdots \end{bmatrix} \qquad (2.73)$$

Note that $H^{[i+1]}$ has one more non-zero column than $H^{[i]}$. This was clear from the earlier observation that H is a Toeplitz matrix $C_{G/\Delta}$ of the open-loop step response.

Summary: Given a generic prediction equation

$$\mathbf{y}_{k+i} = H^{[i]}\Delta\underset{\rightarrow k}{\mathbf{u}} + P^{[i]}\Delta\underset{\leftarrow k-1}{\mathbf{u}} + Q^{[i]}\underset{\leftarrow k}{\mathbf{y}} \qquad (2.74)$$

One can initialise with $i = 1$ from the model difference equation and then find the remaining prediction coefficients from the recursion:

$$\begin{aligned} H^{[i+1]} &= [P_1^{[i]} + Q_1^{[i]}H_1^{[1]}, H^{[i]}] \\ P^{[i+1]} &= [P_2^{[i]}, ..., P_{n-1}^{[i]}, 0] + Q_1^{[i]}P^{[1]} \\ Q^{[i+1]} &= [Q_2^{[i]}, ..., Q_n^{[i]}, 0] + Q_1^{[i]}Q^{[1]} \end{aligned} \qquad (2.75)$$

MATLAB code with this procedure is available in **mpc_predmat.m**.

ILLUSTRATION: Numerical example of using MATLAB to find prediction matrices for an MFD model

This is the same example as in Subsection 2.5.3. Code is stored in **tfpredexample3.m** but results are not displayed here due to the size of the matrices.

```
D = [1,0,−1.4,0,0.52,0;0,1,0.2,1.2,−0.1,0.6];
N = [1,0.3,−0.2,1.1;−2.1,0,1,0.4];
[H,P,Q] = mpc_predmat(D,N,5); % Prediction horizon of 5
```

STUDENT PROBLEM

Verify with numerical examples of your choice on MATLAB that the prediction matrices derived with the techniques in Subsection 2.5.3 and Section 2.6 give the same numerical answers.

2.7 Prediction with independent models

Readers should note that this section will not introduce any more prediction algebra and indeed will utilise the results already derived in this chapter. The fundamental difference with an *independent model* prediction is the handling of bias (see discussion in Section 2.4.3).

This section will first give two brief subsections, the first introducing what we mean by independent model prediction and the second showing how bias, or the need for offset-free prediction, is handled.

Computationally efficient prediction with higher-order models

The third subsection will illustrate a technique commonly used in predictive functional control (PFC) [123] which simplifies prediction algebra considerably in some cases, but can only be used with an independent transfer function format. This is thus useful for coding on processors with limited capabilities, as well as facilitating transparent coding of prediction.

2.7.1   Structure of an independent model and predictions

The structure of an independent model (IM) is illustrated in Figure 2.2. In simple terms, the model is simulated in parallel with the process, but with the same system input, so that two separate outputs are obtained.

1.   The real process output (sometimes denoted $\mathbf{y}_p$).

2.   The model output (sometimes denoted $\mathbf{y}_m$).

The reader will not be surprised to learn that, in general, these outputs will differ due to uncertainty in the model parameters and system disturbances, and indeed, in general a real process cannot be represented exactly by a low order linear (or even nonlinear) model.

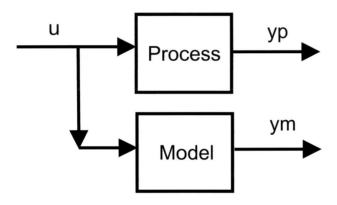

FIGURE 2.2
Simulation of independent model in parallel with the process with same input **u**.

Nevertheless, the independent model is useful because it can be used to provide estimates of the real process behaviour, especially where measurement of $\mathbf{y}_p$ is slow or severely corrupted by noise.

The independent model can be selected to take whatever format is convenient, such as state space, transfer function, FIR and so forth; predictions of future behaviour are obtained using the corresponding prediction equations, e.g., (2.14,2.49). However, a key point of note is that in the first instance the prediction is based *solely* on the model states and output $\mathbf{y}_m$ and thus, will suffer from bias. So for example, the predictions for a CARIMA model would be:

$$\underset{\rightarrow}{\mathbf{y}_m}(k+1) = H\underset{\rightarrow k}{\Delta \mathbf{u}} + P\underset{\leftarrow k-1}{\Delta \mathbf{u}} + Q\underset{\leftarrow}{\mathbf{y}_m}(k) \tag{2.76}$$

where the key point of note is the use of model output $\mathbf{y}_m$ to find predictions for future model outputs.

Summary: A simple independent model output prediction based on Figure 2.2 is simple to compute but clearly will suffer from bias compared to the actual process output.

2.7.2 Removing prediction bias with an independent model

The independent model prediction needs some correction to remove bias, that is the error between y_m and y_p. The method used here is analogous to that deployed with state space prediction in Section 2.4.3. In simple terms, assume that the error in the predictions is constant, that is:

$$\underset{\rightarrow p}{\mathbf{y}}(k+1) - \underset{\rightarrow m}{\mathbf{y}}(k+1) = L\mathbf{d}_k; \quad \mathbf{d}_k = \mathbf{y}_p(k) - \mathbf{y}_m(k) \tag{2.77}$$

where L has the usual meaning. The offset or correction term $\mathbf{d}$ is easily computed as seen in Figure 2.3.

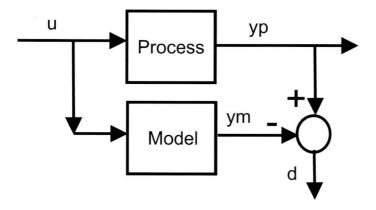

FIGURE 2.3
Offset or bias computation with an independent model.

STUDENT PROBLEM

Verify with numerical examples of your choice on MATLAB that the predictions using an independent model are indeed unbiased in the steady-state.
Compare the predictions to those which are derived with a CARIMA model. What is different?

Summary: Predictions with an independent model are computed by finding the *exact* predictions for the model (which is possible as the model is known exactly) and then adding a bias correction term to each prediction. For example:

1. If the independent model is a state space model, the process predictions are given using (2.14) and the disturbance estimate:

$$\underset{\rightarrow}{\mathbf{Y}}_p(k+1) = \underbrace{P\mathbf{x}_m(k) + H\underset{\rightarrow k}{\Delta\mathbf{u}}}_{\underset{\rightarrow}{\mathbf{Y}}_m(k+1)} + L[\mathbf{y}_p(k) - \mathbf{y}_m(k)] \qquad (2.78)$$

2. If the independent model is a MFD form, the process predictions are given from prediction (2.49) and the disturbance estimate:

$$\underset{\rightarrow}{\mathbf{Y}}_p(k+1) = \underbrace{H\underset{\rightarrow k}{\Delta\mathbf{u}} + P\underset{\leftarrow k-1}{\Delta\mathbf{u}} + Q\underset{\leftarrow}{\mathbf{y}}_m(k)}_{\underset{\rightarrow}{\mathbf{Y}}_m(k+1)} + L[\mathbf{y}_p(k) - \mathbf{y}_m(k)] \quad (2.79)$$

It is noted with an IM there is no obvious benefit from using a T-filter, as the predictions are based primarily on model outputs which consequently are not subject to measurement noise.

Where possible hereafter, *that is if the meaning is obvious, we will avoid the subscripts* $(.)_m, (.)_p$ *to simplify notation.*

Remark 2.8 *A key insight that is clear from Figure 2.2 is that the output predictions of the independent model (e.g., (2.76))* **depend solely** *on the past inputs and the model parameters. Given the model is known precisely, any model output information can be inferred exactly from the past inputs.*

2.7.3 Independent model prediction via partial fractions for SISO systems

The reader will have noted by now that in the main prediction involves significant algebra and some matrix manipulation. There is however a way to avoid this, in some cases (systems with only real roots), which has been used extensively by PFC practitioners [123]. The technique is based on the observation that for a first-order model, predictions can be written down by inspection as long as the future input is assumed constant.

2.7.3.1 Prediction for SISO systems with one pole

A first-order model G_m is presented in discrete time as:

$$y_m(k) = \frac{bz^{-1}}{1 + az^{-1}} u(k) = G_m(z)u(k) \qquad (2.80)$$

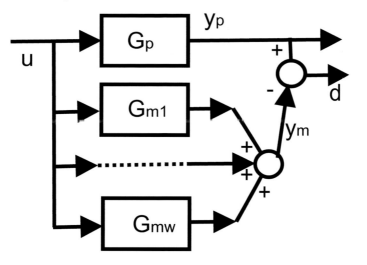

FIGURE 2.4
Independent model structure with partial fraction expansion.

The predicted model output for n_y steps ahead assuming a constant future $u(k)$ is:

$$y_m(k+n_y|k) = (-a)^{n_y} y_m(k) + K_m[1 - (-a)^{n_y}]u(k); \quad K_m = \frac{b}{1+a} \qquad (2.81)$$

2.7.3.2 PFC for higher-order models having real roots

In order to maintain simple coding, PFC overcomes the complexity of prediction algebra by using partial fractions to express a model with real and distinct roots as a sum of first-order models [122, 71, 57] and hence:

$$G_m(z) = \underbrace{\frac{b_1 z^{-1}}{1 + a_1 z^{-1}}}_{G_{m1}} + \cdots + \underbrace{\frac{b_2 z^{-1}}{1 + a_w z^{-1}}}_{G_{mw}} + \cdots \qquad (2.82)$$

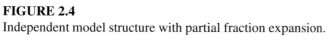

Consequently the model output can be constructed as:

$$y_m(k) = G_{m1}(z)u(k) + \cdots + G_{mw}(z)u(k) = y_m^{(1)}(k) + \cdots + y_m^{(w)}(k) \qquad (2.83)$$

An equivalent block diagram is indicated in Figure 2.4.

Assuming the future input is constant, the predictions for models expressed in the form of (2.82) are trivial and can be written down explicitly as follows:

1. The n_y step ahead predictions for a single model $y_m^{(i)}(k+1) + a_i y_m^{(i)}(k) =$

$b_i u(k)$ are:

$$y_m^{(i)}(k+n_y|k) = (-a_i)^{n_y} y_m^{(i)}(k) + K_i[1-(-a_i)^{n_y}]u(k); \quad K_i = \frac{b_i}{1+a_i}$$
$$(2.84)$$

2. The total n_y step ahead model prediction is thus:

$$y_m(k+n_y|k) = y_m^{(1)}(k+n_y|k) + y_m^{(2)}(k+n_y|k) + \cdots \qquad (2.85)$$

3. The process prediction is formed in the normal way by adding the bias correction term d_k, so:

$$y_p(k+n_y|k) = y_m(k+n_y|k) + d_k \qquad (2.86)$$

Remark 2.9 *Although modifications to deal with complex roots exist (e.g., [197]), this is not so straightforward and thus not included in this book.*

STUDENT PROBLEM

Verify with numerical examples of your choice on MATLAB that the predictions using an independent model prediction such as (2.85, 2.86) gives the same answer as for example using (2.79).

Compare the complexity of the code required to form the predictions for each case.

Summary: Prediction with independent models is straightforward using the same prediction tools as for any model form but is based on simulated data rather than measured data. The prediction is then corrected for the offset between the model and process output to remove bias.

2.8   Prediction with FIR models

Finite impulse response (FIR) models are common models utilised in commercial MPC packages where the restriction to stable processes is rarely an obstacle. This section considers how to obtain predictions with an FIR based model. An FIR model is defined as follows:

$$\mathbf{y}(z) = G(z)\mathbf{u}(z); \quad G(z) = \sum_{i=0}^{\infty} G_i z^{-1} \qquad (2.87)$$

Typically $G_0 = 0$ for strictly proper systems.

FIR models are equivalent to independent models

When using an FIR model, the predictions will be based solely on past input information along with a correction for offset. As noted in Remark 2.8, this is exactly analogous to the scenario when using IM models. In consequence, the use of an FIR model for prediction is exactly equivalent to the use of an IM model and, assuming the FIR model is an equivalent representation of a state space or transfer function model, the use of either will give equivalent sensitivity in the predictions (and indeed the resulting feedback loop based on these predictions).

2.8.1 Impulse response models and predictions

An impulse response model is an FIR model in the form of (2.87). For stable processes without any integrators, the following property holds:

$$\lim_{i \to \infty} G_i = 0 \tag{2.88}$$

Consequently, it would be normal to approximate the FIR model along the lines of

$$\mathbf{y}(z) = G(z)\mathbf{u}(z); \quad G(z) \approx \sum_{i=0}^{n} G_i z^{-1} \tag{2.89}$$

where n is finite and chosen such that

$$\sum_{i=n+1}^{\infty} |G_i| < \varepsilon$$

for ε a suitable small number.

The n-step ahead process prediction with an impulse response model can be stated by inspection using the difference equation equivalent of (2.89) and adding the offset correction term $\mathbf{d}_k = \mathbf{d}_{k+i}, i \geq 0$. As with the prediction equations for CARIMA models, it is convenient to separate the inputs into past inputs and future inputs.

$$
\begin{aligned}
\mathbf{y}_{k+j} &= \sum_{i=1}^{n} G_i \mathbf{u}_{k+j-i} + \mathbf{d}_k \\
\mathbf{y}_{k+j} &= \underbrace{\sum_{i=1}^{j} G_i \mathbf{u}_{k+j-i}}_{future} + \underbrace{\sum_{i=j+1}^{n} G_i \mathbf{u}_{k+j-i} + \mathbf{d}_k}_{past}
\end{aligned}
\tag{2.90}
$$

Summary: The matrix/vector equation representing a vector of output predictions based on an impulse response model takes the following form:

$$\underset{\rightarrow k+1}{\mathbf{y}} = C_G \underset{\rightarrow k}{\mathbf{u}} + H_G \underset{\leftarrow k-1}{\mathbf{u}} + L\mathbf{d}_k \tag{2.91}$$

$$C_G = \begin{bmatrix} G_1 & 0 & \cdots & 0 \\ G_2 & G_1 & \cdots & 0 \\ G_3 & G_2 & \cdots & 0 \\ \vdots & \vdots & \vdots & \vdots \end{bmatrix}; \quad H_G = \begin{bmatrix} G_2 & G_3 & G_4 & \cdots & G_n \\ G_3 & G_4 & G_5 & \cdots & 0 \\ \vdots & \vdots & \vdots & \vdots & \vdots \end{bmatrix}; \quad L = \begin{bmatrix} I \\ I \\ \vdots \end{bmatrix}$$

The offset term $\mathbf{d}_k$ is computed as follows:

$$\mathbf{d}_k = \mathbf{y}_p(k) - \underbrace{\sum_{i=1}^{n} G_i \mathbf{u}_{k-i}}_{\mathbf{y}_m(k)}$$

STUDENT PROBLEMS

1. Find an FIR model based on the following $G(z)$ and then find the corresponding C_G, H_G from (2.91). Comment on the dimensions of C_G, H_G in comparison to the equivalent H, P, Q determined from (2.49).

$$G(z) = \frac{0.2z^{-1} + 0.5z^{-2}}{1 - 0.92z^{-1}}$$

Repeat this for a number of other open-loop stable examples of your choice.

HINT: *The file* **caha.m** *will work for any length of polynomial in principle and, once truncated $G(z)$ is in effect an nth order polynomial.*

2. In practice, for computational reasons, prediction often makes the following assumption:

$$\mathbf{u}_{k+i|k} = \mathbf{u}_{k+n_u-1|k}, \quad \forall i \geq n_u$$

What impact does this assumption have on your prediction equations (2.91). Formulate a set of prediction equations in terms of the d.o.f. only, that is the future inputs $\mathbf{u}_{k|k}, \cdots, \mathbf{u}_{k+n_u-1|k}$.

2.8.2 Prediction with step response models

It is often easier in a real industrial scenario to determine the process step response as opposed to the impulse response, because a step in the input is relatively straightforward to implement whereas an impulse is more of a mathematical concept than something which can be realised. The impulse response $G(z)$ and step response $H(z)$ are related as follows.

$$H(z) = \frac{G(z)}{\Delta(z)}; \quad H(z) = \sum_{i=1}^{\infty} H_i z^{-i}$$

A typical problem with a step response model is that the values do not converge to zero, but rather to the system steady-state gain, that is:

$$\lim_{i \to \infty} H_i = G(0) = \sum_{i=0}^{\infty} G_i \tag{2.92}$$

This means we cannot use the same prediction approach as in (2.91) as we cannot truncate the sequence after n terms.

Basic prediction equation for a step response model: Consequently, prediction with a step response model is more easily based on input increments (as for the CARIMA model). Hence, construct a prediction model as follows:

$$\mathbf{y}_k = \frac{G(z)}{\Delta(z)} \Delta(z) \mathbf{u}_k = H(z) \Delta \mathbf{u}_k \tag{2.93}$$

Difference equation format for prediction equation for a step response model: Analogously to (2.90), and for now ignoring the fact that the coefficients H_i do not converge to zero, the model predictions are given as (assume $H_0 = 0$):

$$\mathbf{y}_m(k+i) = \sum_{j=1}^{\infty} H_j \Delta \mathbf{u}_{k+i-j|k} \tag{2.94}$$

Estimate of bias or offset term with a step response model: The best estimate of $\mathbf{d}_k$ uses prediction model (2.94) to determine $\mathbf{y}_m(k)$ and hence:

$$\left\{ \mathbf{y}_m(k) \approx \sum_{j=1}^{\infty} H_j \Delta \mathbf{u}_{k-j} \right\} \quad \Rightarrow \mathbf{d}_k = \mathbf{y}_p(k) - \sum_{j=1}^{\infty} H_j \Delta \mathbf{u}_{k-j} \tag{2.95}$$

Combining offset term with the model prediction: The process predictions are formed by adding $\mathbf{d}_k$ to $\mathbf{y}_m(k+i)$ and hence

$$\mathbf{y}_p(k+i) = \sum_{j=1}^{\infty} H_j \Delta \mathbf{u}_{k+i-j|k} + \mathbf{y}_p(k) - \sum_{j=1}^{\infty} H_j \Delta \mathbf{u}_{k-j} \tag{2.96}$$

Separating into terms containing future and past inputs: This is a simple rearrangement.

$$\mathbf{y}_p(k+i) = \mathbf{y}_p(k) + \underbrace{\sum_{j=1}^{i} H_j \Delta u_{k+i-j}}_{future} + \underbrace{\sum_{j=1}^{\infty} [H_{j+i} - H_j] \Delta u_{k-j}}_{past} \tag{2.97}$$

Reducing summation to a finite number of terms: It is known from (2.92) that asymptotically

$$\lim_{i \to \infty} H_i - H_{i-1} = 0 \tag{2.98}$$

Hence, there exists an n large enough that the terms $[H_{n+i+1} - H_{n+i}]$ are negligible and hence the prediction equation is simplified too:

$$\mathbf{y}_p(k+i) = \mathbf{y}_p(k) + \underbrace{\sum_{j=1}^{i} H_j \Delta u_{k+i-j}}_{future} + \underbrace{\sum_{j=1}^{n} [H_{j+i} - H_j] \Delta u_{k-j}}_{past} \tag{2.99}$$

This is similar in structure to that of (2.90) except that here the neglected terms are $[H_{n+i+1} - H_{n+i}], i \geq 0$.

Summary: For a step response model, the system predictions take the form

$$\underset{\rightarrow k+1}{\mathbf{y}_p} = C_H \underset{\rightarrow k}{\Delta \mathbf{u}} + L \mathbf{y}_p(k) + M \underset{\leftarrow k-1}{\Delta \mathbf{u}} \tag{2.100}$$

$$C_H = \begin{bmatrix} H_1 & 0 & \cdots & 0 \\ H_2 & H_1 & \cdots & 0 \\ H_3 & H_2 & \cdots & 0 \\ \vdots & \vdots & \vdots & \vdots \end{bmatrix}; \quad L = \begin{bmatrix} I \\ I \\ \vdots \end{bmatrix};$$

$$M = \begin{bmatrix} H_2 - H_1 & H_3 - H_2 & H_4 - H_3 & \cdots & H_{n+1} - H_n \\ H_3 - H_1 & H_4 - H_2 & H_5 - H_2 & \cdots & 0 \\ \vdots & \vdots & \vdots & \vdots & \vdots \end{bmatrix}$$

STUDENT PROBLEMS

1. Find a step response model based on the following $G(z)$ and then find the corresponding C_H, M from (2.100). Comment on the dimensions of C_H, M in comparison to the equivalent H, P, Q determined *from (2.49).*

$$G(z) = \frac{0.2z^{-1} + 0.5z^{-2}}{1 - 0.92z^{-1}}$$

Repeat this for a number of other open-loop stable examples of your choice.

2. In practice, for computational reasons, prediction often makes the following assumption:

$$\Delta \mathbf{u}_{k+i|k} = 0, \quad \forall i \geq n_u$$

What impact does this assumption have on your prediction Equations (2.100)?

2.9   Closed-loop prediction

Early MPC algorithms used open-loop prediction as described in this chapter so far. However, more recent MPC approaches tend to use what is in effect a closed-loop prediction. This section gives some brief descriptions of how these predictions are formed. However, first some explanation is given for scenarios when closed-loop prediction might be recommended, irrespective of the desired algorithm the reader wishes to use.

2.9.1   The need for numerically robust prediction with open-loop unstable plant

This section takes its content from two main publications [129, 138]. The key warnings follow.

- Open-loop prediction of unstable processes often leads to poor numerical conditioning because even with relatively small horizons the prediction matrices contain a wide range of numerical values (very small to very large). This issue is worse where the processor has a limited accuracy. *Normal best practice in MPC dictates the use of large horizons, so ill-conditioning is almost inevitable unless this best practice guidance is ignored.*

- Poor conditioning of the prediction matrices can result in significant errors for any computations depending upon those matrices and thus the computations may not be trustworthy.

- The general advice in the community is to avoid open-loop prediction when a process has unstable poles. Even if your controller seems to be working, you may have embedded *suboptimality* or faulty decision making due to the poor conditioning.

- Consider the following system (exaggerated for effect so that the numbers can be displayed concisely).

$$G(z) = \frac{z^{-1}}{1 - 5z^{-1}} \tag{2.101}$$

The first column of the associated matrix H (see (2.49)) takes the following form.

$$H(:,1) = [1, 6, 31, 156, 781, 3906, 19531, 97656, 488821, 2441406, \cdots]^T \tag{2.102}$$

It is apparent that even with a relatively small horizon, the range of numbers is becoming unmanageable unless the processor can handle a large number of significant figures. A typical operation for predictive control is the following:

$$(H^T H + I)^{-1} H^T \tag{2.103}$$

So if H has very large range of coefficient magnitudes and poor conditioning, the matrix $(H^T H + I)$ will be twice as bad due to the product terms and cannot be inverted reliably.

If the system is open-loop unstable, then you must use *pseudo-closed-loop predictions*; that is, prestabilise the plant before predicting. The choice of prestabilisation is not critical though different choices have different advantages. The mechanics/algebra for this is also fairly straightforward and will be explained next.

Key message: The proposal given in this section is to avoid prediction based on unstable open-loop processes. There are two simple alternatives in the literature.

1. Prestabilise the system *by hook or by crook* using some inner control loop and thus, in effect, deploy a cascade system.

2. Prestabilise the predictions using algebra to select only the future input sequences which *cancel out* the unstable modes. The algebra is embedded before the computation of the prediction matrices [52, 116, 132, 161].

The former of these proposals is simpler to implement and the generally accepted solution in the literature and so the latter will not be discussed further.

2.9.2 Pseudo-closed-loop prediction

The word *pseudo* has been added here because, in practice, the inner controller used to stabilise the system predictions is conceptual, that is embedded in the algebra, but never hard-wired onto the system. The basic structure is as follows:

1. First prestabilise the process using an inner control law.

2. Use the inputs to the prestabilised loop as the d.o.f. in the predictions.

A conceptual illustration of the implied structure is given in Figure 2.5 where the d.o.f. are the loop input signal c_k although it will be noted in the detailed algebra that this structure will be varied slightly in practice in order to accommodate both online prediction and non-zero targets.

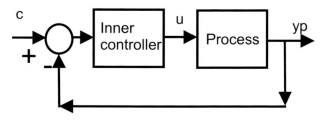

FIGURE 2.5
Example structure of prestabilised loop used for predictions with d.o.f. c_k.

Remark 2.10 *Readers may notice a link with reference governor strategies [43, 46, 47]. This is unsurprising, as reference governor strategies are often MPC algorithms using an alternative parameterisation of the d.o.f. to simplify computation [144].*

Summary: The basic principle is to use the new loop input c_k as the d.o.f. in the predictions. The inner controller is designed to ensure the loop has stable poles and, *ideally*, good behaviour.

Exploiting the inner closed-loop in prediction

Two different paradigms are common in the literature.

1. One assumes that the predictions are open-loop for n_y samples and then embeds a *pseudo-closed-loop* thereafter (e.g., [171]). This choice is made so that the d.o.f. in the predictions are in effect the first n_u system inputs. Denote these as open-loop paradigm predictions (OLP).

2. The second (e.g., [138]) embeds the closed-loop or inner controller throughout the predictions. In the writer's view, this latter choice is more sensible, even though the d.o.f. are no longer the actual system inputs. Dcnote these as closed-loop paradigm predictions (CLP).

2.9.3   Illustration of prediction structures with the OLP and CLP

It will be easier to understand the previous two sections after seeing the prediction structure so this is illustrated next. The open-loop paradigm predictions (OLP) and closed-loop paradigm predictions (CLP) prediction structures, for the state space case, are given together to aid comparison.

Assume a nominal state-space model and inner loop state feedback as follows:

$$\mathbf{x}_{k+1} = A\mathbf{x}_k + B\mathbf{u}_k; \quad \mathbf{u}_k = -K\mathbf{x}_k; \quad \Phi = A - BK \qquad (2.104)$$

Open loop input predictions   **Closed loop input predictions**

$$\mathbf{u}_{\rightarrow OLP} = \begin{bmatrix} \mathbf{u}_{k|k} \\ \mathbf{u}_{k+1|k} \\ \vdots \\ \mathbf{u}_{k+n_c-1|k} \\ \hline -K\mathbf{x}_{k+n_c|k} \\ \vdots \\ -K\Phi^{n_y-n_c}\mathbf{x}_{k+n_c|k} \end{bmatrix} ; \quad \mathbf{u}_{\rightarrow CLP} = \begin{bmatrix} -K\mathbf{x}_k + \mathbf{c}_{k|k} \\ -K\mathbf{x}_{k+1|k} + \mathbf{c}_{k+1|k} \\ \vdots \\ -K\mathbf{x}_{k+n_c|k} + \mathbf{c}_{k+n_c-1|k} \\ \hline -K\mathbf{x}_{k+n_c|k} \\ \vdots \\ -K\Phi^{n_y-n_c}\mathbf{x}_{k+n_c|k} \end{bmatrix}$$

$$(2.105)$$

Firstly, note that the predictions are separated into two parts:

- Mode 1 is the first n_c steps. This includes any d.o.f. within the input predictions. From here it is clear that the d.o.f. in the OLP are the first n_c control moves, that is, $\mathbf{u}_{k+i}, i = 0, ..., n_c - 1$. In the CLP the d.o.f. are the perturbations (or inner loop inputs) $\mathbf{c}_{k+i}, i = 0, ..., n_c - 1$.

- Mode 2 is beyond the first n_c steps and has no d.o.f.. The inputs are determined solely by the fixed inner control law. The mode 2 predictions (mode 2 is where the inner feedback is assumed and is from the n_c^{th} sample onward) are the same, that is, based on the stabilising state feedback $\mathbf{u}_k = -K\mathbf{x}_k$.

Theorem 2.3 *The OLP and CLP paradigms give an identical prediction class.*

Proof: This is obvious, as during mode 1 the corresponding terms are

$$\mathbf{u}_{k+i|k} \quad : \quad -K\mathbf{x}_{k+i|k} + \mathbf{c}_{k+i|k} \qquad (2.106)$$

Clearly $\mathbf{u}_{k+i|k}$, $\mathbf{c}_{k+i|k}$ have the same dimension and hence can be selected to ensure equivalence. The mode 2 predictions are the same if $\mathbf{x}_{k+n_c|k}$ is the same which is true if the mode 1 predictions are the same.   □

Corollary 2.1 *The OLP and CLP paradigms parameterise the d.o.f. in a different way.*

Proof: This is obvious from the predictions. The significance will be seen later in Chapter 6 in the formulation of the cost function J using both prediction sets; in general the CLP predictions give a better conditioned optimisation because it avoids any ill-conditioning from using open-loop predictions in the first n_c steps and, if K is well chosen, embeds good dynamics into the nominal predictions.   □

Summary: The OLP and CLP are equivalent in the space covered but have different parameterisations of the d.o.f. In fact the CLP can be viewed as a means of normalising the space of optimisation variables which improves conditioning.

This section will derive the prediction equations for the CLP only, although readers should be able to manage the OLP themselves by combining the observations of the following subsection with the earlier Subsection 2.4.1.

2.9.4 Basic CLP predictions for state space models

It was noted in Section 2.4 that state space predictions could vary depending upon choices such as the computation of the offset term $\mathbf{d}_k$, the explicit integration of the target steady-state $\mathbf{u}_{ss}$ and expression in terms of $\Delta\mathbf{u}_k$. The reader will gather by the length of this chapter that including every possibility explicitly would take too long and at some point, readers have to be confident to work out fine details for slightly different scenarios by themselves.

Consequently, for simplicity the details of how to incorporate integral action are omitted and this section focuses on the principles of closed-loop prediction. The equations within the prestabilised loop (during prediction) are

$$\mathbf{x}_{k+i|k} = A\mathbf{x}_{k+i-1|k} + B\mathbf{u}_{k+i-1}; \quad \mathbf{u}_{k+i} = -K\mathbf{x}_{k+i|k} + \mathbf{c}_{k+i} \tag{2.107}$$

Removing the dependent variable $\mathbf{u}_{k+i}$ one gets:

$$\mathbf{x}_{k+i|k} = [A - BK]\mathbf{x}_{k+i-1|k} + B\mathbf{c}_{k+i-1}; \quad \mathbf{u}_{k+i} = -K\mathbf{x}_{k+i|k} + \mathbf{c}_{k+i} \tag{2.108}$$

Simulating these forward in time with $\Phi = A - BK$ one gets:

$$\underset{\rightarrow k+1}{\mathbf{x}} = \underbrace{\begin{bmatrix} \Phi \\ \Phi^2 \\ \Phi^3 \\ \vdots \end{bmatrix}}_{P_{cl}} \mathbf{x}_k + \underbrace{\begin{bmatrix} B & 0 & 0 & \dots \\ \Phi B & B & 0 & \dots \\ \Phi^2 B & \Phi B & B & \dots \\ \vdots & \vdots & \vdots & \vdots \end{bmatrix}}_{H_c} \underset{\rightarrow k}{\mathbf{c}} \tag{2.109}$$

or in more compact form

$$\underset{\rightarrow k}{\mathbf{x}} = P_{cl}\mathbf{x}_k + H_c \underset{\rightarrow k}{\mathbf{c}} \tag{2.110}$$

The corresponding input predictions can be written as

$$\underset{\rightarrow k}{\mathbf{u}} = \underbrace{\begin{bmatrix} -K \\ -K\Phi \\ -K\Phi^2 \\ \vdots \end{bmatrix}}_{P_{clu}} \mathbf{x}_k + \underbrace{\begin{bmatrix} I & 0 & 0 & \dots \\ -K & I & 0 & \dots \\ -K\Phi & -K & I & \dots \\ \vdots & \vdots & \vdots & \vdots \end{bmatrix}}_{H_c u} \underset{\rightarrow k}{\mathbf{c}} \tag{2.111}$$

or

$$\mathbf{u}_{\rightarrow k} = P_{clu}\mathbf{x}_k + H_{cu}\underset{\rightarrow k}{\mathbf{c}} \qquad (2.112)$$

The state after n_c steps will be denoted as

$$\mathbf{x}_{k+n_c|k} = P_{cl2}\mathbf{x}_k + H_{c2}\underset{\rightarrow k}{\mathbf{c}} \qquad (2.113)$$

Summary: Predictions are affine in the current state and the d.o.f. and hence have an equivalent form to (2.2).

STUDENT PROBLEM

Modify the algebra above to make use of the input predictions below which are based on deviation variables.

$$\begin{aligned} \mathbf{u}_{k+i} - \mathbf{u}_{ss} &= -K(\mathbf{x}_{k+i} - \mathbf{x}_{ss}) + \mathbf{c}_{k+i} & i &= 0, 1, \cdots, n_c - 1 \\ \mathbf{u}_{k+i} - \mathbf{u}_{ss} &= -K(\mathbf{x}_{k+i} - \mathbf{x}_{ss}) & i &\geq n_c \end{aligned}$$

HINT: Note analogy with Section 2.4.5.

2.9.5   Unbiased closed-loop prediction with autonomous models

Closed-loop prediction can get somewhat messy, algebraically, due to the number of interlinked equations that must be simulated forward in time and thus this section gives a proposal for how to handle this algebra in a neat and compact fashion.

Equations linked to unbiased closed-loop prediction

We need a model update equation, computation of expected steady-state inputs and states for given set point, disturbance estimate and the implied input prediction structure.

$$\mathbf{x}_{k+1} = A\mathbf{x}_k + B\mathbf{u}_k; \quad \mathbf{y}_k = C\mathbf{x}_k + \mathbf{d}_k; \quad \mathbf{d}_k = \mathbf{y}_p(k) - \mathbf{y}_m(k) \qquad (2.114)$$

$$\begin{bmatrix} \mathbf{x}_{ss} \\ \mathbf{u}_{ss} \end{bmatrix} = \begin{bmatrix} K_{xr} \\ K_{ur} \end{bmatrix} (\mathbf{r}_{k+1} - \mathbf{d}_k); \qquad \begin{array}{l} E[\mathbf{r}_{k+i}] = \mathbf{r}_{k+1} \\ E[\mathbf{d}_{k+i}] = \mathbf{d}_{k+1} \end{array} \; \forall i > 0$$

$$\begin{aligned} \mathbf{u}_{k+i} - \mathbf{u}_{ss} &= -K(\mathbf{x}_{k+i} - \mathbf{x}_{ss}) + \mathbf{c}_{k+i} & i &= 0, 1, \cdots, n_c - 1 \\ \mathbf{u}_{k+i} - \mathbf{u}_{ss} &= -K(\mathbf{x}_{k+i} - \mathbf{x}_{ss}) & i &\geq n_c \end{aligned} \qquad (2.115)$$

The proposal here is to combine all Equations (2.114,2.115) into a single equivalent state space model.

$$\mathbf{Z}_{k+1} = \underbrace{\begin{bmatrix} \Phi & (I-\Phi)K_{xr} & [B,0,\ldots,0] & 0 \\ 0 & I & 0 & 0 \\ 0 & 0 & I_L & 0 \\ K & -K.K_{xr}-K_{ur} & [I,0,0,\ldots] & 0 \end{bmatrix}}_{\Psi} \mathbf{Z}_k; \quad \mathbf{Z}_k = \begin{bmatrix} \mathbf{x}_k \\ \mathbf{r}_{k+1}-\mathbf{d}_k \\ \underset{\rightarrow k}{\mathbf{c}} \\ \mathbf{u}_{k-1} \end{bmatrix}$$

(2.116)

where $\Phi = A - BK$ and I_L is a block upper triangular matrix of identities.

> **Summary:** Closed-loop prediction can be performed using the simple update equation of
>
> $$\mathbf{Z}_{k+1} = \Psi \mathbf{Z}_k \quad \Rightarrow \quad \mathbf{Z}_{k+n} = \Psi^n \mathbf{Z}_k \qquad (2.117)$$
>
> where the different components of $\mathbf{Z}_k$ contain all the elements that may be of interest (state predictions, input predictions, d.o.f. and *expected target/disturbance*).

STUDENT PROBLEM

Write some code to validate that the predictions from the autonomous model (2.116) match those from simulating the Equations in (2.114)-(2.115) separately in parallel.

2.9.6 CLP predictions with transfer function models

In the case of transfer function models, a typical controller structure is given as follows:

$$D(z)\Delta\mathbf{u}_k = P(z)\mathbf{r}_k - N(z)\mathbf{y}_k \qquad (2.118)$$

One can then perturb the implied control action by a simple modification as follows:

$$D(z)\Delta\mathbf{u}_k = P(z)\mathbf{r}_k - N(z)\mathbf{y}_k + \mathbf{c}_k \qquad (2.119)$$

The implied structure is shown in Figure 2.6.

The following derives the dependence of the system predictions upon the d.o.f. $\mathbf{c}_k$, past inputs and outputs but omits the dependence on the future target $\mathbf{r}_k$. The derivations are analogous to those in Section 2.5 and thus are deliberately concise.
1. Let the model/controller equations be

$$A(z)\mathbf{y}_k = b(z)\Delta\mathbf{u}_k; \quad D(z)\Delta\mathbf{u}_k = -N(z)\mathbf{y}_k + \mathbf{c}_k \qquad (2.120)$$

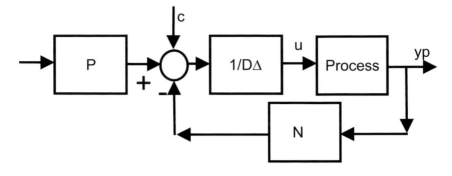

FIGURE 2.6
Example structure of prestabilised loop used using SISO transfer function models.

2. The model and controller equations over the prediction horizon are:

$$
\begin{aligned}
C_A \underset{\rightarrow k+1}{\mathbf{y}} + H_A \underset{\leftarrow k}{\mathbf{y}} &= C_{zb} \underset{\rightarrow k}{\Delta \mathbf{u}} + H_{zb} \underset{\leftarrow k-1}{\Delta \mathbf{u}} \\
C_{z^{-1}N} \underset{\rightarrow k+1}{\mathbf{y}} + H_{z^{-1}N} \underset{\leftarrow k}{\mathbf{y}} &= -C_D \underset{\rightarrow k}{\Delta \mathbf{u}} - H_D \underset{\leftarrow k-1}{\Delta \mathbf{u}} - \underset{\rightarrow k}{\mathbf{c}}
\end{aligned}
\tag{2.121}
$$

3. Eliminate $\underset{\rightarrow k}{\Delta \mathbf{u}}$ using the second of these equations:

$$
\underset{\rightarrow k}{\Delta \mathbf{u}} = C_D^{-1}\left[C_{z^{-1}N} \underset{\rightarrow k+1}{\mathbf{y}} + H_{z^{-1}N} \underset{\leftarrow k}{\mathbf{y}} - H_D \underset{\leftarrow k-1}{\Delta \mathbf{u}} - \underset{\rightarrow k}{\mathbf{c}}\right]
$$

4. Then substitute $\underset{\rightarrow k}{\Delta \mathbf{u}}$ into the first equation:

$$
\begin{aligned}
C_A \underset{\rightarrow k+1}{\mathbf{y}} + H_A \underset{\leftarrow k}{\mathbf{y}} &= C_{zb}C_D^{-1}\left[C_{z^{-1}N} \underset{\rightarrow k+1}{\mathbf{y}} + H_{z^{-1}N} \underset{\leftarrow k}{\mathbf{y}} - H_D \underset{\leftarrow k-1}{\Delta \mathbf{u}} - \underset{\rightarrow k}{\mathbf{c}}\right] \\
&\quad + H_{zb} \underset{\leftarrow}{\Delta \mathbf{u}} \\
[C_A - C_{zb}C_D^{-1}C_{z^{-1}N}] \underset{\rightarrow k+1}{\mathbf{y}} &= [C_{zb}C_D^{-1}H_{z^{-1}N} + H_A] \underset{\leftarrow k}{\mathbf{y}} + [C_{zb}C_D^{-1}H_{D_k} + H_{zb}] \underset{\leftarrow k-1}{\Delta \mathbf{u}} \\
&\quad - C_{zb}C_D^{-1} \underset{\rightarrow k}{\mathbf{c}} \\
\overbrace{[C_D C_A - C_{zb}C_{z^{-1}N}]}^{C_{P_c}} \underset{\rightarrow k+1}{\mathbf{y}} &= [C_{zb}H_{z^{-1}N} + C_D H_A] \underset{\leftarrow k}{\mathbf{y}} + [C_{zb}H_D + C_D H_{zb}] \underset{\leftarrow k-1}{\Delta \mathbf{u}} \\
&\quad - C_{zb} \underset{\rightarrow k}{\mathbf{c}} \\
\underset{\rightarrow k+1}{\mathbf{y}} &= C_{P_c}^{-1}\{[C_{zb}H_{z^{-1}N} + C_D H_A] \underset{\leftarrow k}{\mathbf{y}} \\
&\quad + [C_{zb}H_D + C_D H_{zb}] \underset{\leftarrow k-1}{\Delta \mathbf{u}} - C_{zb} \underset{\rightarrow k}{\mathbf{c}}\}
\end{aligned}
\tag{2.122}
$$

Note that P_c is the implied closed-loop polynomial derived with the given control law.

5. In a similar way one can derive that

$$
\underset{\rightarrow k}{\Delta \mathbf{u}} = C_{P_c}^{-1}\left[\underbrace{[C_{zb}H_{z^{-1}N} + C_D H_A]}_{P_y} \underset{\leftarrow k}{\mathbf{y}} + \underbrace{[C_{zb}H_D + C_D H_{zb}]}_{P_u} \underset{\leftarrow k-1}{\Delta \mathbf{u}} - C_A \underset{\rightarrow k}{\mathbf{c}}\right]
\tag{2.123}
$$

Remark 2.11 *The reader will notice in (2.122, 2.123) a nice separation between the part of the predictions dependent on past information and that dependent on the perturbations* $\underset{\rightarrow k}{\mathbf{c}}$. *This neat separation aids useful insight as will be seen in later chapters.*

> **Summary:** CLP predictions with transfer function models take the same neat form as given in Section 2.2 with the exception that the d.o.f. are expressed in terms of the inner loop perturbation $\mathbf{c}_k$ rather than in terms of input $\mathbf{u}_k$.

STUDENT PROBLEM

Modify the algebra in this section to include the dependence on the future target $\mathbf{r}_{k+1}$.

2.10 Chapter Summary

A brief summary of the key results is given next for reader convenience.

State space prediction is summarised in Section 2.4 and covers two main alternatives: one using the absolute value of the input and/or deviation variables, and the second using input increments. Which of these to use will be linked to the choice of algorithm and performance index as evident in later chapters. Both require a disturbance estimate to ensure unbiased prediction in the steady-state.

Transfer function prediction in Section 2.5 largely meets unbiased requirements by expressing predictions in terms of input increments. However, these predictions can be sensitive to noise and thus the use of T-filter is introduced as one mitigating approach.

Alternative prediction methods include the use of step response models, independent model structures and closed-loop structures but these would not normally be covered in an introductory course.

2.11   Summary of MATLAB code supporting prediction

This code is provided so that students can learn immediately by trial and error, entering their own parameters directly. Readers should note that this code is not *professional quality* and is deliberately simple, short and free of comprehensive *error catching* so that it is easy to understand and edit by users for their own scenarios. Please feel free to edit as you please.

mpc_predmat.m	Forms prediction matrices as given in (2.49). **imgpc_tf_predmat.m** is similar but adds the L needed for an independent model formulation.
mpc_predtfilt.m	Forms prediction matrices as given in (2.62).
caha.m	Forms Hankel and Toeplitz matrices as presented in (2.43).
convmat.m	Does multiplication of matrix polynomials, assuming a specified syntax.
tfpredexample*.m	Template files showing how to generate prediction matrices directly from the Toeplitz and Hankel matrices as in (2.45) or recursion as in Section 2.6.
sspredmat.m	Forms prediction matrices as in (2.14).
sspredmat_unbiased.m	Numerical example of unbiased prediction to support Section 2.4.3.
sssim.m	Utility file for doing open-loop prediction of a state space model with an arbitrary input.
ssmpc_predclp.m	Forms closed-loop predictions such as (2.112, 2.113) but also including dependence on $\mathbf{x}_{ss}, \mathbf{u}_{ss}$.

The code is available on an open website.
http://controleducation.group.shef.ac.uk/htmlformpc/introtoMPCbook.html

3

Predictive functional control

CONTENTS

3.1   Introduction

It is reasonable to begin a book on predictive control methods with the simplest approaches. Readers will appreciate that if a simple approach is good enough, then industry would generally prefer that to a more complicated and expensive alternative because simple implementations are easier and cheaper to install and maintain. The small losses in potential loop performance are regained through savings in downtime, upfront expenditure and so forth. Thus this chapter focuses on predictive functional control (PFC).

Readers may like to note that much like classical control, a core weakness of PFC is its simplicity which means that it is not as versatile as a more complicated alternative. This means that modifications are often needed to tackle more challenging scenarios such as unstable plant, output constraints, MIMO plant and so forth. Moreover, the modifications are different for each case and thus, in the literature [122, 123], there are a plethora of alternative formulations that are tailored to specific scenarios. It is not the intention of this book to go through each of those, but rather to communicate the attractiveness and ease of application of the core concepts which are sufficient to cover a wider range of real scenarios.

Section 3.3 will introduce the main concepts. Section 3.4 will focus on definitions and tuning for first-order models whereas the following section will demonstrate the tuning of PFC for higher-order models followed by sections on a range of specific issues.

Main observation: PFC has been very widely applied in industry across a huge range of industrial sectors [124], but in the main the application has been to SISO control loops and in a cost bracket that is competitive with PID approaches. This chapter will deal only with the SISO case.

Secondary observation: PFC is the only MPC strategy in the literature which has explicitly considered the case of tracking ramp and parabolic targets. A brief discussion of this is given at the end of the chapter as this is usually a special case.

3.2 Guidance for the lecturer/reader

The author would recommend that PFC is discussed in all introductory MPC courses because it is both widely used and also a simple entry point to the efficacy of the concepts. There is no need to dwell on details such as how to handle dead-time systems or uncertainty, or indeed constraint handling and challenging dynamics, in order to convey the core principles. In the author's experience one week of lectures is enough to cover the basics and enable students to design and test some controllers by themselves.

Students can engage with the basic concepts in Section 3.3 and also Section 3.4 which demonstrates the efficacy on first-order systems as this is straightforward and requires only elementary algebra. Thereafter the chapter comprises predominantly a number of illustrations giving insight into the design process which requires the selection of two parameters; students should be encouraged to read this but more importantly undertaking such investigations themselves as part of an assignment is more likely to lead to effective learning.

Typical learning outcomes for an examination assessment: Examination questions could focus on the core concepts and the use of first-order systems to demonstrate how the principles are converted into a control law. Using first-order systems enables a focus on concepts without the algebra becoming unduly messy.

Typical learning outcomes for an assignment/coursework: For an assignment students could be encouraged to investigate the efficacy of PFC and the ease of tuning for a number of case studies with differing dynamics and constraints. MATLAB® code could be provided so that students can focus on design and simulation.

3.3   Basic concepts in PFC

Predictive functional control (PFC) has been in use since the 1960s under various pseudonyms [57, 120, 122]. However, the purpose here is not to write history, but rather to summarise the core underlying principles and from that, to show how a PFC control law can be derived.

This section will focus on the core concepts and how, in principle, PFC incorporates integral action, deals with dead-time and facilitates constraint handling. Later sections will discuss algorithm tuning and how this is linked to model dynamics as well as give a number of numerical examples and some illustrative code.

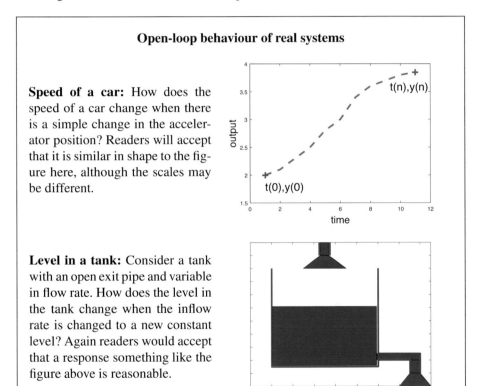

Open-loop behaviour of real systems

Speed of a car: How does the speed of a car change when there is a simple change in the accelerator position? Readers will accept that it is similar in shape to the figure here, although the scales may be different.

Level in a tank: Consider a tank with an open exit pipe and variable in flow rate. How does the level in the tank change when the inflow rate is changed to a new constant level? Again readers would accept that a response something like the figure above is reasonable.

3.3.1   PFC philosophy

PFC has an underlying principle that is closely linked to human behaviour, but critically depends upon the expectation that system behaviour moves smoothly from the current output to the new steady-state. In simple terms, it is assumed that with a constant input, it is reasonable to assume that the process output behaviour will follow a

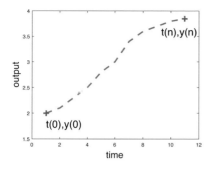

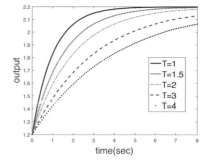

FIGURE 3.1
Underlying prediction assumption in PFC with constant input.

FIGURE 3.2
Desired output evolution shapes with initial/final target values of 1.2/2.2 respectively and for alternative choices of time constant T.

response curve which is close to a first-order response such as shown in Figures 3.1, 3.2.

Main observation: PFC design is premised on the output behaviour with a constant input moving relatively smoothly (that is, similar to a first-order response) from the current output to a new steady-state.

3.3.2 Desired responses

The next foundation stone in PFC is a description of *an ideal response*. In the aim of simplicity, ideal responses are represented solely as being equivalent to first-order dynamics, as shown in Figure 3.2. The speed of convergence is linked to the desired time constant T in that the underlying behaviour takes the form:

$$y(t) = y(\infty) - [y(\infty) - y(0)]e^{\frac{-t}{T}} \qquad (3.1)$$

The user is expected to have a strong viewpoint on what is a reasonable *time constant* T for the process of interest.

Remark 3.1 *Practitioners often refer to a target closed-loop settling time instead where it is assumed that the settling time is about three times the time constant.*

In practice PFC is often implemented in discrete time so, for sampling period T_s it is convenient to restate Eqn.(3.1) as follows:

$$\lambda = e^{\frac{-T_s}{T}}; \quad y_{k+n_y} = r_{k+n_y} = r_\infty - [r_\infty - y_k)]\lambda^{n_y}, \; n_y = 1, 2, \cdots \qquad (3.2)$$

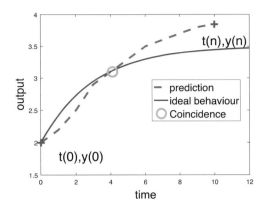

FIGURE 3.3
Coincidence between the system prediction and the ideal behaviour 4 seconds into
the future.

where here the term r_{k+n_y} is used to emphasise the fact that this is in essence a
target response trajectory (which thus varies with prediction horizon n_y) and with
asymptotic set point r_∞ (hereafter just r) used in place of y_∞ to emphasise that the
ultimate steady-state is also intended to be the desired asymptotic target.

> **Summary:** A key part of PFC design is the choice of the desired closed-
> loop time constant T. Within the technical literature, users are more likely to be
> comfortable with the choice of the implied closed-loop pole λ, $0 \le \lambda < 1$; the
> variables are linked through $\lambda = e^{-T_s/T}$, T_s the sampling time.

3.3.3   Combining predicted behaviour with desired behaviour: the coincidence point

A simple PFC algorithm combines the insights of Figure 3.1 with the desired behaviour of Figure 3.2, or in other words, there is an attempt to make these two match.
The key argument is that, if we can make *prediction* of Figure 3.1 match the desired response of 3.2 at some chosen point (the so-called coincidence point), then we
could be confident in expecting that the overall behaviour will be close to the desired
behaviour. An illustration of this is given in Figure 3.3.

> **Summary:** PFC has two key parameters which the user can select:
>
> 1.   The time constant T which defines the ideal behaviour of (3.2).

2. The prediction horizon n_y (seconds or samples) at which coincidence will be enforced between the system prediction and the ideal behaviour [hence so-called coincidence point].

3.3.4 Basic mathematical definition of a simple PFC law

The ideal response is given by Equation (3.2) where r is the target steady-state. The system predictions can be computed by a variety of methods as discussed in Chapter 2, for now we will use Equation (2.2), but noting that with the control horizon $n_u = 1$, it is assumed that $\Delta u_{k+i|k} = 0$, $i > 0$ and thus H has just one column. Then PFC is defined by solving the identity:

$$\mathbf{e}_{n_y}^T \underset{\rightarrow k+1|k}{y} = y_{k+n_y|k} = r - [r - y_k]\lambda^{n_y} \qquad (3.3)$$

where $\mathbf{e}_{n_y}$ is a standard basis vector with a 1 in the n_y^{th} position. Alternatively, for suitable H, P as in Equation (2.2):

$$\mathbf{e}_{n_y}^T [H\Delta u_k + P \underset{\leftarrow k}{\mathbf{x}}] = r - [r - y_k]\lambda^{n_y} \qquad (3.4)$$

Summary: A simple representation of a PFC law with prediction model (2.2) is determined by solving Equation (3.4) to determine Δu_k in terms of the current state and target. This depends on the selected λ and n_y.

Remark 3.2 *The vendors of PFC tend to use only a small subset of the prediction models from Chapter 2 and thus the remainder of this chapter will follow that practice. Of course, it is implicit that readers are not likewise restricted and can use whatever* unbiased *prediction model is convenient.*

3.3.5 Alternative PFC formulation using independent models

It is commonplace in PFC applications [123] to use the independent model format for prediction (Section 2.7). Moreover, a small sleight of hand is used to simplify the objective, that is, the definition of the basic PFC law.

Core observations

1. Within the predictions, the change in the model output y_m can be assumed to be the same as the change in the process output y_p, that is:

$$y_m(k + n_y|k) - y_m(k) = y_p(k + n_y|k) - y_p(k) \qquad (3.5)$$

2. The desired change in the process output, consistent with the objective (3.2) is

$$y_p(k+n_y|k) - y_p(k) = (r - y_p(k))(1 - \lambda^{n_y})$$

3. Hence, equivalent to PFC law (3.2) one can use:

$$y_m(k+n_y|k) - y_m(k) = (r - y_p(k))(1 - \lambda^{n_y}) \qquad (3.6)$$

where now, albeit implicit, the disturbance estimate $d(k) = y_p(k) - y_m(k)$ is not computed explicitly.

The user may, in principle, use any method to compute the model prediction $y_m(k+n_y|k)$ although PFC practitioners favour the independent model methods in Section 2.7.

For completeness, this section defines the basic PFC law for a high order system based on an independent model. Let the model predictions be given from Eqn.(2.79) (but with $n_u = 1$ and using brackets rather than subscripts for sample time), that is:

$$\underset{\rightarrow}{y_m}(k+1) = H\Delta u(k) + P\underset{\leftarrow}{\Delta u}(k-1) + Q\underset{\leftarrow}{y_m}(k) \qquad (3.7)$$

Combining this with the definition in (3.6) gives:

$$\mathbf{e}_{n_y}^T\{H\Delta u(k) + P\underset{\leftarrow}{\Delta u}(k-1) + Q\underset{\leftarrow}{y_m}(k)\} - y_m(k) = (r - y_p(k))(1 - \lambda^{n_y}) \qquad (3.8)$$

Rearranging to solve for the current input $u(k) = u(k-1) + \Delta u(k)$ gives:

$$u(k) = u(k-1) + \frac{-\mathbf{e}_{n_y}^T P\underset{\leftarrow}{\Delta u}(k-1) - \mathbf{e}_{n_y}^T Q\underset{\leftarrow}{y_m}(k) + y_m(k) + (r - y_p(k))(1 - \lambda^{n_y})}{\mathbf{e}_{n_y}^T H}$$

$$(3.9)$$

Remark 3.3 *The reader is of course free to use other forms of prediction (such as in Sections 2.7.3 and 2.8; this is commonplace where simpler coding is required which avoids the need to compute matrices P, H, Q. The choice of prediction method will have an impact on sensitivity and ease of model identification but will have no impact on nominal performance.*

Summary: A simple representation of a PFC law with a transfer function model is given by the following:

$$u(k) = u(k-1) + \frac{-\mathbf{e}_{n_y}^T P\underset{\leftarrow}{\Delta u}(k-1) - \mathbf{e}_{n_y}^T Q\underset{\leftarrow}{y_m}(k) + y_m(k) + (r - y_p(k))(1 - \lambda^{n_y})}{\mathbf{e}_{n_y}^T H}$$

$$(3.10)$$

where parameters H, P, Q are discussed in Section 2.7.

STUDENT PROBLEM

Derive the PFC control law equivalent to (3.9) for a number of the different prediction models covered in the previous chapter.

3.3.6 Integral action within PFC

The main point to note here is that, by definition, the setup of the PFC law (3.6) ensures integral action is implicit. Detailed algebra to prove the integral appears in the equivalent block diagram [150] is somewhat superfluous and indeed practitioners often claim that offset-free tracking is achieved without integral action which is misleading as the integral is implicit in the algebra, even if not explicit. Here we give a simple demonstration.

First it is necessary to assume that the system is closed-loop stable so that convergence occurs. Then it is sufficient to show that the process output y_p can only converge to the target r, irrespective of model uncertainty $G_m \neq G_p$ (see Figure 2.2) and disturbances d_k (e.g., see Eqn.(2.16)).

At steady-state, $u_k = u_{k-1} = u_{k-i}$, $\forall i > 0$. In consequence, recalling that future inputs are also assumed constant, it is obvious that:

$$y_m(k) = y_m(k+1|k) = y_m(k+i|k), \ \forall i > 0$$

This means that $y_m(k+n|k) - y_m(k) = 0$. Referring back to (3.6), this can only occur if $(r - y_p(k)) = 0$ as $\lambda^{n_y} \neq 0$!

> **Summary:** The proof of integral action, or offset-free tracking, given here has been done in general terms so that it clearly applies to any order of system and also in the presence of uncertainty. Indeed it could also be extended to MIMO systems. The proof is however premised on the closed-loop being stable.

3.3.7 Coping with large dead-times in PFC

As long as the dead time (say n samples) is known, then one can make a simple modification to (3.6) to determine the appropriate control law. The main premise is to explicitly account for the fact that the impact of any change in the input will be delayed by n samples and thus, to delay any expectation of the response accordingly.

The underlying thought process is to assume that the output response can begin to follow the ideal first-order response in n samples time, and thus the expected or reasonable target trajectory is defined as follows:

$$r(k+n+i|k) = r - [r - y(k+n|k)]\lambda^i, \quad i > 0 \tag{3.11}$$

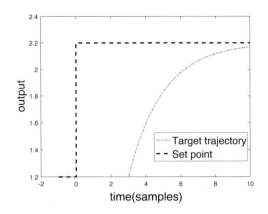

FIGURE 3.4
Typical target trajectory for a system with a delay of 3 samples.

where it is noted that the target is defined only for samples further than n samples into the future. A typical picture is given in Figure 3.4 where the current sample is taken as 0. The control law is determined by combining (3.11) with (3.6) and hence:

$$y_p(k+n_y+n|k) - y_p(k+n|k) = (r - y_p(k+n|k))(1 - \lambda^{n_y}) \qquad (3.12)$$

Numerical illustrations of the efficacy of this approach are given later in the chapter.

Remark 3.4 *In practice the independent or internal model G_m can be operated without a delay, thus the model in effect gives an estimate of the system output n steps ahead (i.e., $E[y_p(k+n|k)] = y_m(k) + y_p(k) - y_m(k-n)$). In this case, a modified law is defined as:*

$$y_m(k+n_y|k) - y_m(k) = (r - E[y_p(k+n|k)]))(1 - \lambda^{n_y}); \quad d_k = y_p(k) - y_m(k-n).$$
$$(3.13)$$

STUDENT PROBLEM

Derive the PFC control law equivalent to (3.12) for a number of the different prediction models covered in the previous chapter.

Summary: PFC deals with system delay in a natural way by simply modifying the objective (or coincident point) to be reasonable; in practice this means the coincidence horizon is in effect $n_y + n$ which must be larger than the delay n. The internal model is simulated without a delay, a policy akin to a Smith predictor, but with the obvious observation that the states are therefore n samples out of synchronisation with the equivalent plant states.

3.3.8 Coping with constraints in PFC

One particular nice thing about predictive based approaches is that the constraints can be built in at the outset rather than as an add-on. PFC achieves this benefit to some extent, although it is not as flexible or systematic as more advanced MPC approaches due to the focus on simple coding.

Where there are only input constraints and input rate constraints, PFC uses a form of saturation (or clamping).

Algorithm 3.1 *When solving the default PFC law, say (3.6), the user will arrive at a specific value of u_k. Before this is implemented, it is modified as follows:*

$$\begin{aligned} \Delta u_k > \overline{\Delta u_k} &\quad \Rightarrow \quad \Delta u_k = \overline{\Delta u_k} \\ \Delta u_k < \underline{\Delta u_k} &\quad \Rightarrow \quad \Delta u_k = \underline{\Delta u_k} \\ u_k > \overline{u_k} &\quad \Rightarrow \quad u_k = \overline{u_k} \\ u_k < \underline{u_k} &\quad \Rightarrow \quad u_k = \underline{u_k} \end{aligned} \qquad (3.14)$$

where $\overline{\Delta u_k}$, $\overline{u_k}$ are upper limits and $\underline{\Delta u_k}$, $\underline{u_k}$ are lower limits.

It should be noted however, that although such an approach is adequate for many scenarios with relatively benign dynamics, it is known from the general MPC literature (e.g., [133]) that this could be severely suboptimal or even lead to instability in more challenging cases. To be more specific, often when a default control law is taking a system up to its constraints, the optimal control action in fast transients needs to be backed well off constraints to give some slack in control movements in the future.

In the case of output or state constraints the approaches adopted by PFC are much less clear cut and begin to lose the transparency or ease of coding of eqns.(3.6, 3.14). A common example [123] is to form a sort of cascade loop where an outer loop, also based on PFC, modifies the set point to the inner loop where necessary to ensure the inner loop does not violate constraints; in fact this is somewhat analogous to reference governor approaches [43, 44, 46, 47]. The author's conjecture is that the many possible approaches to this [123] are less systematic than more modern alternatives in the literature and are premised on a form of block diagram implementation that is no longer essential. Consequently, these are pursued no further. Instead a relatively simple suggestion is made which is analogous to a reference governor approach [43]. The following algorithm is conceptual and explicit algebraic details are therefore not precise.

Algorithm 3.2 *Form the implied closed-loop predictions for a system using the current state* $\mathbf{x}$, *process output* y_p, *target* r *and expected unconstrained PFC control law. These predictions will take the following form (see Chapter 2):*

$$\underset{\rightarrow}{y_p}(k+1) = P\mathbf{x}(k) + Ly_p(k) + Qr; \quad \underset{\rightarrow}{u}(k) = P_u\mathbf{x}(k) + L_u y_p(k) + Q_u r \qquad (3.15)$$

Select the value of r *closest to the desired one such that these predictions satisfy all the constraints over some future prediction horizon.*

Remark 3.5 *In practice Algorithm 3.2 can be implemented online as single simple loop in code. However, it should be emphasised that it requires the computation of the prediction matrices* P, Q, P_u, L_u, Q_u *which is a complexity and computational demand that historical PFC practitioners have sought to avoid. Unsurprisingly, as the complexity of the problem increases, simple solutions are no longer available!*

> **Summary:** It is relatively straightforward to handle input constraints with PFC using a simple saturation approach and indeed, as known from the MPC literature, for benign cases this will often be optimal. However, where state and output constraints are important it is less straightforward to implement a computationally simple and transparent approach and, in the light of modern developments, the author suggests some form of reference governor wrapped around the PFC loop would be logical.

3.3.9 Open-loop unstable systems

This book is not going to discuss PFC design for systems with open-loop unstable poles or open-loop integrators separately and indeed this is covered more carefully in resources dedicated solely to PFC [122]. At one level, if the user chooses a CARIMA model for prediction as in Section 2.5.2 then the implementation is automatic via control law (3.3) and nothing needs to change, although tuning is not always easy, as observed in the numerical illustrations later in this chapter.

However, the use of independent models is not straightforward as *an input which ensures convergence of process output* y_p *would almost inevitably imply divergence of model output* y_m *due to the small difference in pole positions between* G_p *and* G_m. In consequence, independent model prediction is not advisable or indeed often not possible, for such cases. In the author's view the most preferable option is now popular in mainstream MPC which is to pre-stabilise [138] and thus use a form of cascade where the MPC design is applied to an already stable system. This does of course complicate constraint handling.

3.4 PFC with first-order models

This section will show that for the special case of first-order models it is possible to get neat and strong stability and performance results with PFC. This means that tuning is straightforward and reduces to the conventional trade-off between input activity and loop bandwidth.

3.4.1 Analysis of PFC for a first-order system

For the sake of simplicity, this section will assume that $y_m = y_p$ in the loop analysis as it is conventional to base such analysis on the nominal case. Obviously, in the case of parameter uncertainty, the actual loop behaviour will vary but a study of loop sensitivity [13, 50, 63, 74] is non-trivial in general and beyond the remit of this book.

Closed-loop poles with PFC of a first-order model

1. The model/process has an update equation and associated predictions (with constant input):

$$y_{k+1} = bu_k - ay_k; \quad y_{k+n_y|k} = K_n u_k + (-a)^{n_y} y_k \quad (3.16)$$

where $K_n = b - ab + \cdots + (-a)^{n_y-1}b$.

2. The control law (3.6) is summarised as:

$$y_{k+n_y|k} = (r - y_k)(1 - \lambda^{n_y}) + y_k \quad (3.17)$$

3. Combining (3.16, 3.17) gives an implied control law of:

$$K_n u_k + (-a)^{n_y} y_k = (r - y_k)(1 - \lambda^{n_y}) + y_k \quad (3.18)$$

4. Substituting (3.18) back into the model equation

$$y_{k+1} + ay_k = b\frac{(r - y_k)(1 - \lambda^{n_y}) + y_k - (-a)^{n_y} y_k}{K_n} \quad (3.19)$$

which gives an implied closed-loop pole polynomial of:

$$
\begin{aligned}
p_c &= 1 + z^{-1}[a + b\frac{(1-\lambda_y^n)-1+(-a)^{n_y}}{K_n}] \\
&= 1 + z^{-1}[a + \frac{-\lambda_y^n+(-a)^{n_y}}{1-a+\cdots+(-a)^{n_y-1}}]
\end{aligned} \quad (3.20)
$$

Lemma 3.1 *For a first-order model with a coincidence horizon of $n_y = 1$, the desired closed-loop pole is achieved exactly in the nominal case.*

Proof: This follows automatically from (3.20). Setting $n_y = 1$ gives:

$$p_c = 1 + z^{-1}[a + \frac{-\lambda - a}{1}] = 1 - \lambda z^{-1} \quad \Box$$

Lemma 3.2 *For a first-order order model with a large coincidence horizon, the actual closed-loop pole tends towards the open-loop pole and thus the tuning parameter of λ is superfluous.*

Proof: This follows automatically from (3.20). Let $n_y \to \infty$, then:

$$\lim_{n_y \to \infty} [\frac{-\lambda^{n_y} + (-a)^{n_y}}{1 - a + \cdots + (-a)^{n_y - 1}}] = \lim_{n_y \to \infty} (-\lambda^{n_y} + (-a)^{n_y})(1 + a) = 0$$

It is implicit that $|a| < 1, |\lambda| < 1$. $\quad \Box$

Theorem 3.1 *With a first-order model, the tuning parameter of λ is most meaningful when $n_y = 1$ and moreover, in this case, the closed-loop dynamics are defined explicitly. The proof is an obvious consequence of the above two lemmata.*

Remark 3.6 *It is relatively straightforward to show that the same algebra can be applied to first-order systems with a delay with the minor adjustment that the control law is given as in Eqn.(3.13). Obviously however, the corresponding feedback loop may be sensitive to parameter and dead-time uncertainty.*

STUDENT PROBLEMS

1. Show that if the user chooses $\lambda = -a$ (for a first-order system) then the desired closed-loop dynamic (equivalent to the open-loop response) is achieved irrespective of the choice of n_y.
2. Write some code to implement PFC for a first-order model and verify that, with a choice of $n_y = 1$, the closed-loop behaviour in the nominal case achieves the desired target dynamic. Verify this for a variety of models and choices of λ.
3. Also demonstrate that, for higher choices of n_y, the closed-loop behaviour differs from the target dynamic.

Summary: When using PFC with a first-order model you should always choose the coincidence horizon n_y to be one and λ to be the desired closed-loop pole, $0 \le \lambda < 1$.

3.4.2 Numerical examples of PFC on first-order systems

The following section comprises a number of simple illustrations of the application of PFC on first-order systems and also the efficacy of the tuning parameters:

1. Coincidence horizon n_y.

2. Desired closed-loop time constant T or equivalently desired closed-loop pole λ, $0 \leq \lambda < 1$ (where $\lambda = e^{-T_s/T}$), T_s the sampling time.

3.4.2.1 Dependence on choice of desired closed-loop pole λ with $n_y = 1$

Consider the following system.

$$y(z) = \frac{0.25z^{-1}}{1 - 0.8z^{-1}} u(z) \qquad (3.21)$$

Overlay the closed-loop responses for $n_y = 1$ and a variety of choices of λ. Also include the implied target trajectory of (3.2) for $\lambda = 0.8$.
It is clear that:

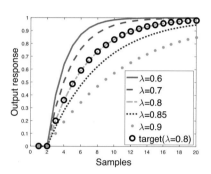

- The closed-loop behaviour with $\lambda = 0.8$ matches the target behaviour; these two curves are overlaid to emphasise this point.

- In fact, irrespective of λ the closed-loop and target trajectories are matched; it is clear that the closed-loop trajectories change as expected as λ changes.

- This means tuning is intuitive for practitioners who can simply decide on what closed-loop time constant/pole is desired.

3.4.2.2 Dependence on choice of coincidence horizon n_y with fixed λ

Consider the following system.

$$y(z) = \frac{0.2z^{-1}}{1 - 0.9z^{-1}} u(z) \qquad (3.22)$$

Overlay the closed-loop responses for a variety of choices of n_y for a single choice of $\lambda = 0.75$. It is clear that:

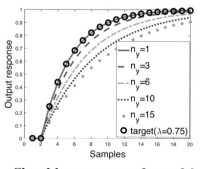

Closed-loop responses for model (3.22) with various n_y.

- The closed-loop behaviour matches the desired target behaviour only when $n_y = 1$.

- As n_y is increased there is a bigger and bigger mismatch between the target behaviour(circles) and the actual behaviour.

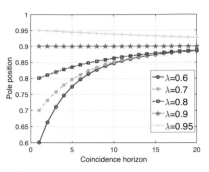

For completeness here we show how the closed-loop pole position is linked to the choice of both λ and n_y. Ideally the closed-loop pole and λ should be the same but this only occurs when $n_y = 1$.

Closed-loop pole dependence on n_y and λ for model (3.22).

3.4.2.3   Effectiveness at handling plants with delays

Consider the following system with a delay of $n = 5$ samples.

$$y(z) = \frac{0.25z^{-6}}{1 - 0.8z^{-1}} u(z) \qquad (3.23)$$

The figure to the right overlays the closed-loop responses for the input u, model output y_m, process output y_p and target trajectory based on r.

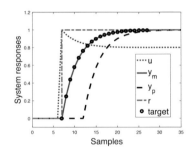

- The closed-loop behaviour matches the desired target behaviour (here $\lambda = 0.75$ so a slight speed up compared to open-loop), but delayed by n samples.

- The model output y_m is 5 samples in front of the process response y_p and thus is a good predictor.

3.4.2.4   Effectiveness at handling plants with uncertainty and delays

Consider the following model and process with a delay of $n = 3$ samples.

$$y_m(z) = \frac{0.25z^{-4}}{1 - 0.8z^{-1}} u(z); \qquad (3.24)$$

$$y_p(z) = \frac{0.26z^{-4}}{1 - 0.83z^{-1}} u(z)$$

The figure to the right overlays the closed-loop responses for the input u, model output y_m, process output y_p and set point r.

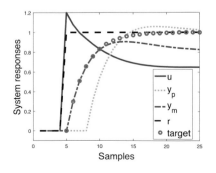

- The closed-loop behaviour matches the desired target behaviour quite well (here $\lambda = 0.7$ so a slight speed up compared to open-loop), but delayed by $n = 3$ samples. The difference later on is due to the model uncertainty.

- Despite the parameter uncertainty and therefore difference between the model and process ouputs y_m, y_p, and moreover despite the delay, the PFC control law has behaved well and achieved good closed-loop responses with no steady-state offset.

3.4.2.5 Effectiveness at handling plants with uncertainty, constraints and delays

Consider the model and process of (3.24) and add in input limits.

$$-0.2 \leq u_k \leq 0.8;$$

$$-0.3 \leq \Delta u_k \leq 0.2$$

Also add in an output disturbance d and an input disturbance d_u. The figure here overlays the closed-loop responses for the input u, model output y_m, process output y_p and set point r.

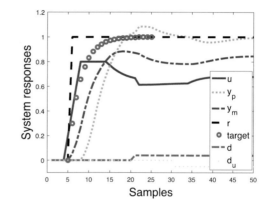

- The closed-loop behaviour no longer appears to match the desired target behaviour and lags behind significantly although this is clearly because of the input saturation.

- Despite the parameter uncertainty, delay, non-zero disturbances and constraints, PFC has delivered good closed-loop responses with no steady-state offset.

STUDENT PROBLEMS

1. Write some code to implement PFC for a first-order model with a delay and demonstrate that one can still achieve a tight link between the tuning parameter λ and the corresponding closed-loop responses, for the nominal case.

2. Using the provided code, or otherwise, explore the efficacy of PFC for first-order systems when a coincidence horizon n_y is set to be greater than 1.

3. Explore the impact of changing the tuning parameter λ on the resulting system behaviour.

4. Generate a variety of first examples of your own and explore the extent to which PFC can cope with delays, parameter uncertainty, input and output disturbances and input constraints.

MATLAB code to help with these tasks is summarised in Section 3.9.

3.5   PFC with higher-order models

While the previous section demonstrated that PFC is very effective with first-order models, unfortunately the same does not automatically follow with higher-order systems. This should not be unsurprising given the simplicity of the underlying philosophy, to match predictions and desired behaviour at a single point. This section considers the extent to which PFC is still an effective approach for higher-order systems, and should be tempered by the observation that it is very widely used in industry, so notwithstanding some theoretical weaknesses, it is popular for a reason!

Over damped second-order system with input constraints

Consider the following system.

$$y(z) = \frac{0.1z^{-1} + 0.4z^{-2}}{1 - 1.4z^{-1} + 0.45z^{-2}} u(z)$$
(3.25)

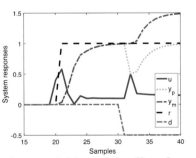

Plot the closed-loop responses for $\lambda = 0.6, n_y = 3$, no parameter uncertainty and a non-zero output disturbance after 30 samples. An input rate constraint of 0.4 is also included.

It is clear that the closed-loop behaviour is acceptable and the disturbance has been rejected. In other words, PFC can work quite well in relatively challenging scenarios!

This section will give a number of case studies looking at the efficacy of the PFC tuning parameters, that is the coincidence horizon n_y and the target closed-loop pole λ. This set of studies is intended to help the reader understand how the tuning can be performed in practice. The reader is reminded that, for open-loop unstable systems, an independent model form of prediction is inappropriate and hence another form must be used.

To give a good coverage of possible issues, it is useful to have a set of example problems with a range of characteristics. These are listed in Table 3.1.

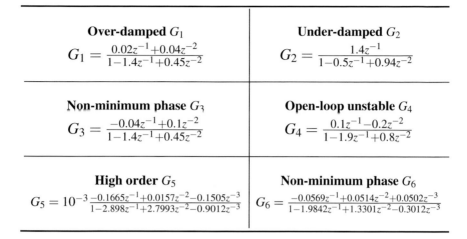

Over-damped G_1	Under-damped G_2
$G_1 = \dfrac{0.02z^{-1}+0.04z^{-2}}{1-1.4z^{-1}+0.45z^{-2}}$	$G_2 = \dfrac{1.4z^{-1}}{1-0.5z^{-1}+0.94z^{-2}}$
Non-minimum phase G_3	**Open-loop unstable G_4**
$G_3 = \dfrac{-0.04z^{-1}+0.1z^{-2}}{1-1.4z^{-1}+0.45z^{-2}}$	$G_4 = \dfrac{0.1z^{-1}-0.2z^{-2}}{1-1.9z^{-1}+0.8z^{-2}}$
High order G_5	**Non-minimum phase G_6**
$G_5 = 10^{-3}\dfrac{-0.1665z^{-1}+0.0157z^{-2}-0.1505z^{-3}}{1-2.898z^{-1}+2.7993z^{-2}-0.9012z^{-3}}$	$G_6 = \dfrac{-0.0569z^{-1}+0.0514z^{-2}+0.0502z^{-3}}{1-1.9842z^{-1}+1.3301z^{-2}-0.3012z^{-3}}$

TABLE 3.1
Example systems for use with PFC.

3.5.1 Is a coincidence horizon of 1 a good choice in general?

In this section an illustration is given of the closed-loop behaviour of the models in Table 3.1 when the coincidence horizon is fixed at one. For first-order models this was an ideal solution so one might naively think this would carry across to other examples. However, simulations show a very worrying contradiction to this.

Closed-loop simulations of systems in Table 3.1 with $n_y = 1, \lambda = 0.8$.

- In the initial transients, all the output responses seem to be ideal and following the desired behaviour.

- However, later on some of the output responses for G_1, G_3, G_4 seem to rather suddenly go unstable.

- The input responses appear chaotic throughout with the exception of G_2.

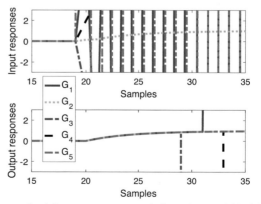

In summary, while $n_y = 1$ has worked for some systems, it has clearly failed for others.

3.5.1.1 Nominal performance analysis with $n_y = 1$

This section will analyse the closed-loop performance and stability a little more carefully to develop more insight into PFC tuning.

Lemma 3.3 *For a coincidence horizon of $n_y = 1$ and any choice of λ $(0 < \lambda < 1)$, in the nominal case the closed-loop system output should exactly match the target behaviour of (3.2).*

Proof: The control law is defined from:

$$y(k+1|k) - y(k) = (R - y(k))(1 - \lambda) \tag{3.26}$$

Moreover, the input is chosen to enforce (3.26) which means that, in the nominal case, the output will follow the target precisely as this law is used recursively! The error is monotonically decreasing in magnitude, which implies convergence of the output and with precisely the desired first-order response! That is:

$$e(k+i) = \lambda e(k+i-1); \quad e(k+i) = R - y(k+i) \tag{3.27}$$

$\square$

This lemma explains the observation from the figure that, in early transients, all the output responses appear to be behaving as expected.

3.5.1.2 Stability analysis with $n_y = 1$

The reader will note that the previous lemma was careful not to assert stability. In order for the closed-loop to be stable, it is also necessary that the corresponding input is convergent.

Theorem 3.2 *PFC with a coincidence horizon of $n_y = 1$ is guaranteed closed-loop stable in the nominal case iff the system has no unstable zeros.*

Proof: Combining (3.26) with a generic first-order model:

$$\left\{ \begin{array}{l} y(z) = \frac{1-\lambda}{1-\lambda z^{-1}} \frac{R}{1-z^{-1}} \\ a(z)y(z) = b(z)u(z) \end{array} \right\} \Rightarrow u(z) = \frac{a(z)}{b(z)} \frac{1-\lambda}{1-\lambda z^{-1}} \frac{R}{1-z^{-1}} \tag{3.28}$$

Hence $u(z)$ is stable iff $b(z)$ has no unstable roots. $\square$

So, if the system numerator has any unstable roots, then the use of $n_y = 1$ is guaranteed to give closed-loop divergence of the inputs and this explains the sudden divergence of the outputs for some systems; at some point the mathematical precision of the software breaks down so that *expected* behaviour of the output can no longer be retained. Some readers will recognise the parallels between this result and the minimum variance literature [24]; using $n_y = 1$ is equivalent to inverting the plant which is ill-advised in general.

The reader will note that examples G_1, G_3, G_4 have unstable zeros, and hence cannot deploy a coincidence horizon of 1.

Remark 3.7 *In practice, with $n_y = 1$, input constraints would cause a breakdown of relationship (3.27) much more quickly with unpredictable consequences in general.*

In summary, a choice of $n_y = 1$ will give the desired output dynamic exactly, as long as the system has stable zeros. This is at the expense of whatever input activity is required (implied by (3.28)) and thus may not be a good design! Where the implied input is over-active, the use of $n_y = 1$ is inadvisable.

STUDENT PROBLEMS

Develop some numerical examples to demonstrate that PFC may be ineffective or may fail entirely with $n_y = 1$. Also investigate the extent to which the resulting closed-loop dynamics (input and output) are linked to the choice of λ.

Summary: When using PFC with a high order model, it is generally inadvisable to use $n_y = 1$ as this implies plant inversion which in turn implies input signals that often cannot be implemented, or in the worst case, are divergent.

3.5.2 The efficacy of λ as a tuning parameter

A key selling point of PFC is the intuitive nature of the main tuning parameter, that is λ, as in effect this means the user can choose the desired closed-loop time constant (see Eqn.(3.2)). However, such an assertion only has value if the closed-loop behaviour does indeed acquire the target pole. This section will investigate the link between the chosen λ and the actual closed-loop poles which result when PFC is used, again using the examples of Table 3.1.

In the following illustrations, each line plot indicates how the dominate/slowest closed-loop pole changes as n_y is changed, for a specified choice of λ. An effective design would imply that the desired λ becomes a closed-loop pole; horizons where PFC gives wholly unsatisfactory behaviour are marked with a pole of 1.

3.5.2.1   Closed-loop poles/behaviour for G_1 with various choices of n_y

Consider system G_1 and plot the slowest
closed-loop pole over a range of n_y and λ,
but only where system is closed-loop stable. A value of 1 is used to denote closed-
loop instability. It is clear that, in this case
the closed-loop behaviour is acceptable as
long as $n_y > 1$. Moreover, the link between
the desired λ and the actual closed-loop
pole is good if and only if $n_y < 4$.

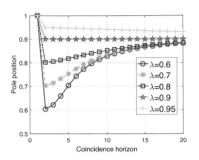

3.5.2.2   Closed-loop poles/behaviour for G_3 with various choices of n_y

Consider system G_3 and plot the slowest
closed-loop pole over a range of n_y and λ,
but only where system is closed-loop stable. A value of 1 is used to denote closed-
loop instability.
In this case, the closed-loop behaviour dependence upon both $n_y > 1$ and λ and the
link between λ and the actual dominant
closed-loop pole is useful only for $n_y \approx 3$
and also the system is closed-loop unstable for small n_y.

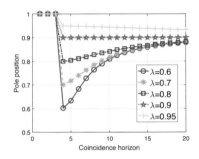

WARNING: *What is not apparent from the pole positions is that the closed-loop
responses are poor with small values of n_y, say less than 5. If one notes that it is
necessary to use $n_y \geq 5$ to get reasonable behaviour, then the connection between λ
and closed-loop behaviour is lost!*

Consider system G_3 and plot the
closed-loop step responses for
various n_y but the same $\lambda = 0.8$.
In all cases the expected *poles* are
OK (although slower than the target λ), but this is not the same
as the resulting behaviour being
good!

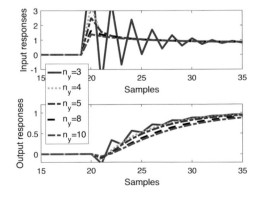

3.5.2.3   Closed-loop poles/behaviour for G_2 with various choices of n_y

Consider system G_2 and plot the slowest
closed-loop pole over a range of n_y and λ,
but only where system is closed-loop stable. A value of 1 is used to denote closed-
loop instability.

In this case the closed-loop behaviour dependence upon both $n_y > 1$ and λ is er-
ratic and unpredictable. [Readers should
also note that the plot shows the absolute
value of the poles as these are in fact complex poles].

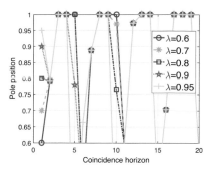

3.5.2.4   Closed-loop poles/behaviour for G_4 with various choices of n_y

Consider system G_4 and plot the slowest
closed-loop pole over a range of n_y and λ,
but only where system is closed-loop stable. A value of 1 is used to denote closed-
loop instability.

The closed-loop behaviour dependence
upon both $n_y > 1$ and λ is fairly pre-
dictable, but again the link between λ and
the actual dominant closed-loop pole is
useful only for $n_y \approx 4$ and also the system
is closed-loop unstable for $n_y < 4$.

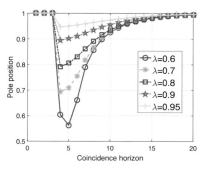

REPEAT WARNING: *What is not apparent from the pole positions is that the
closed-loop responses may be poor, for smaller n_y even when the dominant pole
seems to be in a sensible position, so often one needs to resort to a higher n_y!*

STUDENT PROBLEMS

Develop some numerical examples to demonstrate that the tuning parameter λ has limited efficacy with higher-order models when it is
necessary that $n_y > 1$.

Summary: When using PFC with a higher-order model, the link between
the parameter λ and the resulting closed-loop poles can be poor and moreover,
finding a suitable pairing of n_y, λ to give reasonable closed-loop behaviour may
take significant trial and error; indeed in some cases this may fail altogether.

3.5.3   Practical tuning guidance

It is clear from the examples above that the tuning parameters deployed by PFC may or may not be effective for higher-order models, that is when the expectation of a closed-loop behaviour close to a first-order response is invalid or not realistic.

3.5.3.1   Closed-loop poles/behaviour for G_5 with various choices of n_y

Consider system G_5 and plot the slowest closed-loop pole over a range of n_y and λ, but only where system is closed-loop stable. A value of 1 is used to denote closed-loop instability.

The closed-loop behaviour dependence upon both $n_y > 1$ and λ is fairly constant here for $n_y < 10$, although the link between λ and the actual dominant closed-loop pole is still only useful only if $\lambda > 0.9$, and for $n_y > 2$ ($n_y = 1$ gives instability). Moreover, consideration of the closed-loop behaviour shows that $n_y < 20$ gives rather undesirable behaviour although the plots are not shown here.

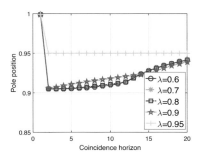

- Where a small coincidence horizon is possible, there is a good link between the chosen λ and the dominant closed-loop pole.

- However, for higher-order systems, a small coincidence horizon often leads to poor transient behaviour.

- Practical PFC tuning guidance often requires a form of global search over both n_y and λ to find a pairing which results in suitable behaviour, but this is clearly an unacceptable tuning methodology in practice as it is not systematic and may fail. However, a small scale search may be acceptable for fine tuning.

Consequently, there is a need for some more insightful, but simple, guidance which will enable users to get a *right first time* approach to PFC design.

This section uses case study illustrations to argue for a common sense approach which hence is easy to implement for non-expert users and will also accord with engineering intuition. There is no suggestion that this approach has any mathematical rigour.

3.5.3.2   Mean-level approach to tuning

An alternative and simple approach to tuning is the so-called mean-level approach. This assumes that the open-loop dynamics are satisfactory (implicitly stable), and therefore the only requirement is to ensure offset-free tracking and constraint satisfaction. One can view this like a PFC control law with a coincidence horizon of infinity, that is:

$$\lim_{n_y \to \infty} \{y_p(k+n_y|k) = r - [r - y_p(k)]\lambda^{n_y}\} \quad \Rightarrow \quad \lim_{n_y \to \infty} y_p(k+n_y|k) = r \qquad (3.29)$$

Such a control law reduces to estimating the expected steady-state value for the input:

$$u(k) = u_{ss} = E[u(k)] \text{ s.t. (3.29)} \qquad (3.30)$$

This can be calculated from the model steady-state $G(1)$ as:

$$\lim_{n_y \to \infty} y_p(k+n_y|k) = G(1)u_{ss} + y_p(k) - y_m(k) = r \qquad (3.31)$$

Remark 3.8 *A mean-level approach is guaranteed convergent in the nominal case for stable open-loop processes. The proof of this is obvious in that, in the absence of a change in r, the estimated value of u_{ss} would not change and a constant input must necessarily give a convergent output.*

Remark 3.9 *Using a mean-level algorithm as a base, one can superimpose a closed-loop response dynamic (pole of $1 - \beta$) over the open-loop dynamics by using a control law such as the following:*

$$u(k) - u(k-1) = \beta(u_{ss} - u(k-1)) \qquad (3.32)$$

The proof of convergence and pole positioning is obvious.

> **Summary:** When the open-loop dynamic is satisfactory both in speed of response and the transient behaviour, then a mean-level control law is very simple to design and implement and requires negligible coding.

STUDENT PROBLEM

For a number of high-order models of your choice compare a mean-level approach with PFC using a low coincidence horizon. Make some conclusions on the efficacy of a mean-level approach in general. [To save time writing new code, standard PFC code supplied in Section 3.9, will replicate mean-level if n_y is chosen to be large enough.]

3.5.3.3 Intuitive choices of coincidence horizon to ensure good behaviour

In cases where the open-loop response does not include significant oscillation or instability and moves smoothly to the steady-state, it is possible to demonstrate that matching the step response to a first-order response at some point in the future (Figure 3.2) still makes good intuitive sense and is likely to lead to a well-posed decision making or control law; of course, this is notwithstanding the fact that a rigorous proof or analysis may not be possible in many cases. This section will give some demonstrations of how one can be systematic at designing a PFC control law that works reasonably well, that is choosing a coincidence horizon that works reasonably well and achieves close to the target time constants.

The proposal here is linked to the underlying common sense or intuitive nature of the PFC approach, that is the argument that the part of the prediction which is ignored (further into the future than the coincidence point) is implicitly moving closer to the desired steady-state, and hence, iterative decision making will be well-posed if it gradually moves the coincidence point nearer and nearer to the target (in absolute terms). This will happen automatically if n_y is based on a point where the step response is increasing, and changes relative to the steady-state are close to monotonic thereafter.

Remark 3.10 *One might wonder whether a PFC based on the delay format of (3.12) would be a more logical start point for systems with a noticeable lag or non-minimum phase characteristic, but this approach has not yet been considered in the literature and is left for the reader to explore.*

3.5.3.4   Coincidence horizon of $n_y = 1$ will not work in general

Typical higher-order systems have some lag in their step response as seen in this figure for both models 1 and 2. Consequently, as these systems do not respond

quickly to a change in the input, trying to match such a response to *the target* of (3.2) just one step ahead will clearly imply an input that is either the wrong sign or far too large in magnitude. In the figure, $n_y = 1$ is marked with the vertical dashed line and it is clear that for any input magnitude, matching models 1 and 2 with the target at this horizon is unreasonable.

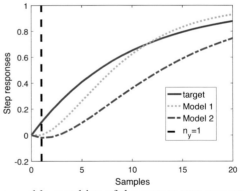

In these cases, one can only get a reasonable matching of the system step responses to a *target* if one uses a coincidence horizon of the order of $10 \le n_y \le 15$; this of course also has repercussions on what constitutes an achievable λ, i.e., $\lambda^{10} \gg 0$!.

KEY OBSERVATION: It is unsurprising therefore that for many high order systems a choice of $n_y = 1$ leads to closed-loop instability as the corresponding choice of input is entirely misconceived. (Also see Theorem 3.2.)

3.5.3.5   Small coincidence horizons often imply over-actuation

One can only force coincidence
with faster poles/targets (and implicitly therefore low n_y) by having significant over-actuation during transients. For models 1 and 2
respectively, even to ensure coincidence at $n_y = 4$, the implied inputs are $2\times, 6\times$ over the required
steady-state.

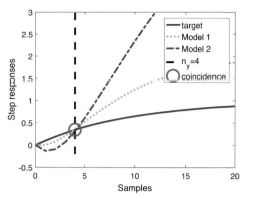

Conjecture 3.1 *For systems with close to monotonically convergent behaviour after immediate transients, a good choice of coincidence horizon is one where the
open-loop step response has risen to around 40% to 80% of the steady-state and the
gradient is significant.*

3.5.3.6   Example 1 of intuitive choice of coincidence horizon

Consider the over-damped system
G_1 (normalised to give a steady-
state of unity) of Table 3.1 with
the following open-loop step response. The target is based on
$\lambda = 0.8$.
Here a value of 40-80% implies a
horizon $6 \leq n_y \leq 15$.

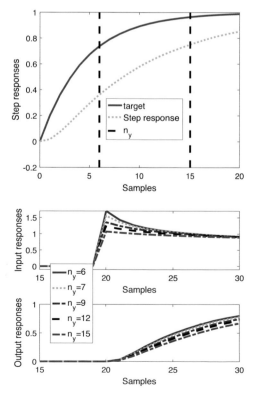

The closed-loop behaviour shown
here with $\lambda = 0.8$ and various n_y
within the recommended range is
good in all cases, although as expected, the desired speed-up of
G_1 is not achieved due to the need
to use high n_y; that is, the closed-
loop pole is not close to λ.

Summary: For many systems, the tuning of PFC is not obvious, and the designer is left with, in effect, a global search across λ, n_y, but without a clear expectation of whether this will give a good result or not. However, in practice one can determine a logical n_y from the open-loop step response (conjecture 3.1). Likewise, the choice of λ should be consistent with the open-loop behaviour, an achievable settling time and also the magnitude of n_y as readers will note that the influence of λ decreases as n_y increases.

Remark 3.11 *This book does not give examples for open-loop unstable [167] or oscillatory systems [168] because for those cases it is far more difficult to come up with generic guidance. In practice the author would not recommend a default PFC algorithm for either case and modified approaches are beyond the remit of this book. Intuitively, the core failing is the desire to follow a first-order trajectory, and using a constant future input as this is clearly unrealistic for systems with complex dynamics. Therefore, to extend the benefits of PFC, one needs to extend the target behaviours and/or input parametrisations (e.g., [1, 159, 167]).*

3.5.3.7 Example 6 of intuitive choice of coincidence horizon

Consider the non-minimum phase system G_6 of Table 3.1 (normalised to give a steady-state of unity) with the following open-loop step response. The target is based on $\lambda = 0.8$.

Here a value of 40-80% implies a horizon $7 \leq n_y \leq 10$.

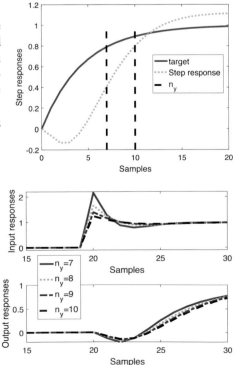

The figure here shows that the closed-loop behaviour with $\lambda = 0.8$ and various n_y is good in all cases, although as expected, the desired pole λ is not achieved and the actual response is notably slower than the target; the fastest corresponds to the smallest n_y. One might argue that the responses are best with the largest n_y.

STUDENT PROBLEMS

For a number of high order models of your choice and using the provided code, or otherwise, explore the efficacy of PFC as a control design method and in particular explore the efficacy of the tuning parameters.

1. Experiment with Conjecture 3.1 and hence explore the extent to which PFC can be used to obtain good closed-loop behaviour using a minimum of high level design decisions. For simplicity choose λ to be close to the dominant open-loop dynamic or slightly faster.

2. Explore the impact of changing the tuning parameter λ on the resulting system behaviour. How do you go about choosing an appropriate coincidence horizon n_y?

3. Explore the sensitivity of PFC to output disturbances and parameter uncertainty.

4. Explore the efficacy of PFC for open-loop unstable processes and processes with significant under-damping or significant non-minimum phase characteristics.

3.6 Stability results for PFC

As will become apparent in the later chapters, there is a lot of interest in the academic community in so-called *a priori* stability results. That is, before doing any analysis such as closed-loop root computations, can we determine whether a MPC law will be closed-loop stabilising, either for the nominal or uncertain case?

What is clear however, is that strong stability results come with a price [100]. The problem needs to be tied down more carefully and the associated computations may be more demanding; both of these observations are contrary to the underlying principles of PFC which are computational simplicity and transparency. Consequently it is unsurprising that stability results for PFC are scarce [169] and quite weak. In fact the strongest argument is that the underlying control law of (3.2) makes good sense and is in accord with the sort of strategies that humans use, successfully, all the time. A caviat would be that humans use these strategies for systems with relatively benign dynamics and hence the difficulties when considering the use of PFC for more challenging dynamics such as with open-loop unstable or highly under damped processes.

Strong stability results can be given for two cases:

1. For stable first-order systems it was shown that irrespective of the choice of n_y and λ, the implied closed-loop is always stable. However a choice of $n_y = 1$ is advised.

2. The mean level approach of Section 3.5.3.2 is guaranteed stabilising, for stable open-loop systems although of course this implies that one is happy with open-loop behaviours.

The latter observation has a corollary that, in general, for stable open-loop systems, a large n_y will lead to a stable closed-loop, but not necessarily good behaviour.

> **Summary:** With the exception of trivial cases, a priori stability analysis is difficult with PFC so one can only determine the efficacy a posteriori, that is, build the control law and try it!

3.7 PFC with ramp targets

The popular MPC literature in the main assumes that the target signals are constant and indeed this is common in the petrochemical industry and processing industry where multivariable MPC implementations have become very successful. Nevertheless, there exists a subset of problems where the target is not constant and one facet of PFC is that the authors have deliberately targeted this scenario and asked how one can form a predictive control approach which has both no offset or lag.

The underlying approach is analogous to ideas in [116, 138, 171, 192], that is, one should parameterise the input trajectory in such a way that it automatically gives output predictions close to those which are most desired. In general this means that the predicted input trajectory cannot be just a constant, rather it must include some appropriate dynamics. In the case of ramp targets, for systems which do not already contain an integrator, it is clear that the asymptotic input must also be a ramp. Hence given a linear model $y(s) = G(s)u(s)$, then

$$\lim_{t\to\infty} y(t) = a + bt \quad \Rightarrow \quad \lim_{t\to\infty} u(t) = c + dt \tag{3.33}$$

for appropriate a, b, c, d. Given this observation, PFC practitioners made the proposal that in cases where the asymptotic target is a ramp, then the future input should be parameterised as follows:

$$\underset{\to k}{u} = \begin{bmatrix} \gamma \\ \gamma \\ \gamma \\ \vdots \end{bmatrix} + \begin{bmatrix} \beta \\ 2\beta \\ 3\beta \\ \vdots \end{bmatrix} \quad \text{or} \quad \Delta\underset{\to k}{u} = \begin{bmatrix} \gamma \\ 0 \\ 0 \\ \vdots \end{bmatrix} + \begin{bmatrix} \beta \\ \beta \\ \beta \\ \vdots \end{bmatrix} \tag{3.34}$$

where γ, β are two d.o.f. to be determined at each sampling instant.

To be consistent with the underlying concepts of PFC outlined in Section 3.3, two d.o.f. require two coincident points (with horizons n_{y1} and n_{y2}), and hence instead of (3.6), the ramp PFC law is defined by solving the following simultaneous equations.

$$
\begin{aligned}
y_{k+n_{y1}|k} - y_k &= [r_{k+n_{y1}} - r_k] - [r_k - y_k](1 - \lambda^{n_{y1}}) \\
y_{k+n_{y2}|k} - y_k &= \underbrace{[r_{k+n_{y2}} - r_k]} _{\delta r_{n_{y2}}} - [r_k - y_k](1 - \lambda^{n_{y2}})
\end{aligned}
\tag{3.35}
$$

KEY POINT: To ensure no lag in the response, readers will note the addition to the right-hand side of the expected change $\delta r_{n_{y2}}$ in the target. Hence the change in the output is based upon two components: (i) current distance from the current target can decay with the desired time constant or dynamic given by λ and (ii) expected change in the target value which must be followed without any lag.

Indicative PFC for following ramp targets

---

1. Let the model predictions be given by (2.49)

$$
\underset{\rightarrow}{y}_{k+1} = H\Delta \underset{\rightarrow}{u}_k + P\Delta \underset{\leftarrow}{u}_{k-1} + Q\underset{\leftarrow}{y}_k
\tag{3.36}
$$

2. Adding in model and process outputs information to ensure robust behaviour, the coincident points of (3.35) can be represented as follows:

$$
\begin{aligned}
y_m(k+n_{y1}|k) - y_m(k) &= [r_{k+n_{y1}} - r_k] - [r_k - y_p(k)](1 - \lambda^{n_{y1}}) \\
y_m(k+n_{y2}|k) - y_m(k) &= [r_{k+n_{y2}} - r_k] - [r_k - y_p(k))](1 - \lambda^{n_{y2}})
\end{aligned}
\tag{3.37}
$$

3. Substituting in from (3.34,3.36) and using $L = [1, 1, \cdots]^T$ the parameterised predictions become:

$$
\begin{aligned}
y_m(k+n_{y1}|k) &= \mathbf{e}_{n_{y1}}^T [H\mathbf{e}_1\gamma + HL\beta + P\Delta\underset{\leftarrow}{u}_{k-1} + Q\underset{\leftarrow}{y}_k] \\
y_m(k+n_{y2}|k) &= \mathbf{e}_{n_{y2}}^T [H\mathbf{e}_1\gamma + HL\beta + P\Delta\underset{\leftarrow}{u}_{k-1} + Q\underset{\leftarrow}{y}_k]
\end{aligned}
\tag{3.38}
$$

4. Simple rearrangement of (3.38) leads to:

$$
\begin{bmatrix} y_m(k+n_{y1}|k) \\ y_m(k+n_{y2}|k) \end{bmatrix} = \underbrace{\begin{bmatrix} \mathbf{e}_{n_{y1}}^T H\mathbf{e}_1 & \mathbf{e}_{n_{y1}}^T HL \\ \mathbf{e}_{n_{y2}}^T H\mathbf{e}_1 & \mathbf{e}_{n_{y2}}^T HL \end{bmatrix}}_{H_r} \begin{bmatrix} \gamma \\ \beta \end{bmatrix} + \underbrace{\begin{bmatrix} \mathbf{e}_{n_{y1}}^T (P\Delta\underset{\leftarrow}{u}_{k-1} + Q\underset{\leftarrow}{y}_k) \\ \mathbf{e}_{n_{y2}}^T (P\Delta\underset{\leftarrow}{u}_{k-1} + Q\underset{\leftarrow}{y}_k) \end{bmatrix}}_{y_f}
\tag{3.39}
$$

5. As H_r is square, substitution of (3.39) into (3.37) allows the solution of γ, β and hence definition of the control increment $\Delta u_k = \gamma + \beta$.

$$H_r \begin{bmatrix} \gamma \\ \beta \end{bmatrix} + y_f - \begin{bmatrix} y_m(k) \\ y_m(k) \end{bmatrix} = \begin{bmatrix} (r_{k+n_{y1}} - r_k) - (r_k - y_p(k))(1 - \lambda^{n_{y1}}) \\ (r_{k+n_{y1}} - r_k) - (r_k - y_p(k)))(1 - \lambda^{n_{y2}}) \end{bmatrix}$$

(3.40)

Remark 3.12 *One might argue that this is no longer simple to code and nor is it transparent and obvious to a non-academic user. However, the counter argument is that more demanding scenarios inevitably lead to more involved control structures and the coding of (3.40) is still moderately straightforward in practice, especially for low-order models where H, P, Q can be defined by inspection.*

<hr>

Example of PFC applied to ramp target

Consider the system G_1 from Table 3.1 and implement ramp PFC with coincidence horizons of $n_{y1} = 5, n_{y2} = 10$. The resulting closed-loop responses show that the target is tracked with no lag and rejects disturbances (a step disturbance d is introduced around sample 25). Also, it is seen from the input signal that the control law has anticipative (or feedforward) action due to the incorporation of future target values into the control law.

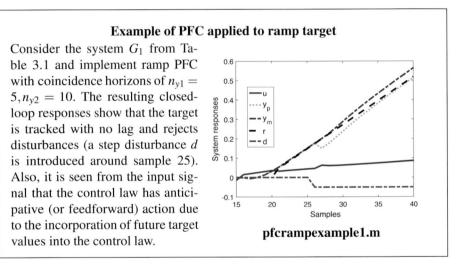

pfcrampexample1.m

<hr>

Summary: In principle the extension of PFC to track ramps is straightforward and uses identical concepts to step targets. The main additional insights are the need for two coincident points as there are now two d.o.f. to solve and a careful definition of the desired output change (3.35) to ensure the target is tracked without lag.

3.8 Chapter summary

PFC is very widely used in industry, largely as a competitor with PID technology. Being based on prediction it allows the handling of constraints and dead times far more easily than classical methods and moreover, being model-based, automatically modifies the controller order to deal, to some extent, with high-order dynamics. The key selling point is computational simplicity and transparency (intuitive design) so the technology is cheap and relatively easy to tune. Indeed, where publicised to technical staff, it is widely adopted [123].

However, the downside is the same as the strength. Its very simplicity means that tuning or implementation for systems with more challenging dynamics or constraints is no longer as straightforward and indeed, a good solution often may not exist. In such cases readers are recommended to consider more advanced MPC strategies which are discussed in the remainder of this book.

3.9 MATLAB code available for readers

The following code is available on the website for readers who wish to explore the capabilities of PFC with first-order models and systems.

- **pfc1stordermodel.m** This function file allows the user to enter model and process parameters, non-zero dead-time and input constraints and performs the simulation for a unit target. It is assumed that the model poles are stable.

- **pfc1storderdemo.m** A script file to demonstrate the use of **pfc1stordermodel.m**. Readers can edit the system parameters directly in this script file to simulate examples and scenarios of their choice.

The following code is available on the website for readers who wish to explore the capabilities of PFC with higher-order models and systems.

- **pfcsimulateim.m** This function file allows the user to enter model and process parameters, non-zero dead-time and input constraints and performs the simulation for a unit target. A simple plot is produced.

- **pfcclosedlooppoles.m** Computes the implied closed-loop poles (nominal case) for a PFC control law. Useful to explore the links between the tuning parameters and the actual closed-loop behaviour. The file **pfchighorderpoles1.m** use this to produce a plot of poles for a range of tuning parameters.

- **pfchighorderexample1.m** Gives an exemplar script file calling the above two files and producing neat figures. Shows how to enter model, process, simulation and tuning data.

- **pfcsimulateimramp.m** Simulate a PFC law for a ramp target.

- **pfcrampexample1.m** Used to produce the figure in Section 3.7. Also shows how to run PFC ramp examples with your own models.

- The files above make use of utility m-files **caha.m, findpfccontrollaw.m, bodechange.m**.

- Miscellaneous files used to produce figures within this chapter: **pfc_1storderex1.m, pfc_1storderex2.m, pfc_1storderex1_dead.m, pfc_1storder_uncertainty.m, pfc_1storder_demo.m, fixlambdavaryny_ex1,3,6.m**.

Readers should note that the code is utilitarian rather than professional, avoids any complicated coding notation or functionality and hence should be transparent to anyone familiar with MATLAB and very easy to edit. The code is available on an open website.

> http://controleducation.group.shef.ac.uk/htmlformpc/introtoMPCbook.html

4

Predictive control – the basic algorithm

CONTENTS

4.1 Introduction

The purpose of this chapter is to introduce a standard 2-norm predictive control algorithm. Most emphasis will be placed on generalised predictive control (GPC [25]), as the majority of standard algorithms are very similar in principle, the minor differences are mainly due to modelling/prediction assumptions. Brief sections will illustrate the changes required for other variants such as the popular industrial choice dynamic matrix control (DMC) [28].

There is no attempt to revisit algorithms such as minimum variance which some staff might like to include in a predictive control course, although the argument could be that this is subsumed in the previous chapter on PFC. Also, there is no discussion of algorithms which use 1-norms or ∞-norms; these are available in the literature but rarely used in practice for feedback and in fact are more likely to be used higher up in a control hierarchy, such as for set point selection.

4.2 Guidance for the lecturer/reader

This chapter focuses on the use of a finite horizon quadratic performance index which is by far the most popular and widely adopted approach to MPC. For completeness, a number of variants are presented, but for a basic course in predictive control, I would suggest a focus on the following sections and doing just one of transfer function or state space approaches:

1. Section 4.3 summarises the main components of MPC.

2. Section 4.4 summarises typical choices of performance index (often denoted cost function).

3. Section 4.5 for transfer function models (you could focus on the sections before the T-filter subsection; the T-filter is best done by illustration and focusing on concepts rather than with all the algebra).

4. Section 4.8 for state space models.

The other sections (such as those on using step response functions or inclusion of the T-filter) will be useful for those who wish to broaden their understanding but are likely to include too much algebra/variation and/or too many *issues*, which consequently could be confusing for students doing a single module in this topic.

As throughout this book, the author would favour supplying students with basic code to create and simulate designs, as writing and validating code for a whole MPC algorithm will take too long to incorporate into most modules at the expense of other learning outcomes. The relevant MATLAB® code is summarised in Section 4.10 and related tutorial exercises encouraging students to learn by doing are spread throughout the chapter. More comprehensive tutorial, examination and assignment questions are in Appendix A.

Typical learning outcomes for an examination assessment: In terms of examinations, the author would focus on students demonstrating the key algorithmic steps: (i) defining the performance index; (ii) substituting in the predictions and (iii) deriving the optimal unconstrained law using a gradient operation. However, I would tend to expect this with algebra and not *numbers*.

Typical learning outcomes for an assignment/coursework: Assignment briefings are more likely to call on the following chapter as the author would recommend providing students with basic working code which thus largely negates the assessment of the current chapter content (within an assignment).

4.3 Summary of main results

For those readers who are interested in the main results and do not want to read through the detail, these are given now.

4.3.1 GPC control structure

The details of how to compute the controller parameters follow in this chapter.

1. In the absence of constraints, a GPC control law reduces to a fixed linear feedback.

   (a) For transfer function based predictions, the control law takes the form (see Figure 4.3):

   $$D_k(z)\Delta\mathbf{u}_k = P_r(z)\mathbf{r}_{k+1} - N_k(z)\mathbf{y}_k \qquad (4.1)$$

   (b) For state space based predictions, the control law takes the form:

   $$\mathbf{u}_k - \mathbf{u}_{ss} = -K(\mathbf{x}_k - \mathbf{x}_{ss}) \qquad (4.2)$$

   where K is the GPC state feedback and $\mathbf{u}_{ss}, \mathbf{x}_{ss}$ are the expected steady-states.

2. In the presence of constraints (see Chapter 7) the optimum predicted control trajectory is defined through the on-line solution of a quadratic programming problem which takes the form:

   $$\min_{\Delta\underset{\rightarrow k}{\mathbf{u}}} \Delta\underset{\rightarrow k}{\mathbf{u}}^T S \Delta\underset{\rightarrow k}{\mathbf{u}} + \Delta\underset{\rightarrow k}{\mathbf{u}}^T \mathbf{q}_k \quad \text{s.t.} \quad C\Delta\underset{\rightarrow k}{\mathbf{u}} - \mathbf{f}_k \leq 0 \qquad (4.3)$$

   where S is positive definite and $\mathbf{q}_k$, $\mathbf{f}_k$ are time varying (dependent on the current state and target).

4.3.2 Main components of an MPC law

This is a brief reminder of the discussion in Chapter 1, culminating in Section 1.7. To form a predictive control law, the following components are needed:

1. A definition of system predictions and their dependence on d.o.f. (typically the future control moves). This was covered in Chapter 2.

2. A performance index which gives a numerical measure for comparing alternative control strategies.

3. An optimisation which selects the control strategy giving the best/optimum predicted value for the performance index.

This chapter will consider the latter two aspects in turn and thus give a systematic development of the GPC control law.

4.4 The GPC performance index

This section introduces the performance index. This is a numerical measure of expected future performance.

4.4.1 Concepts of good and bad performance

Often humans use rather vague performance indices when making choices and we trade off between different aspects such as rise-time, oscillation, overshoot and settling time in somewhat arbitrary ways.

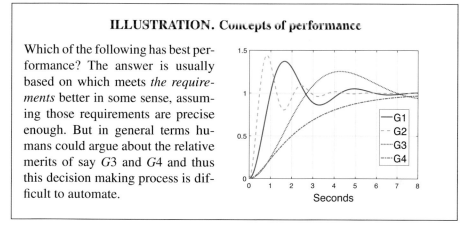

ILLUSTRATION. Concepts of performance

Which of the following has best performance? The answer is usually based on which meets *the requirements* better in some sense, assuming those requirements are precise enough. But in general terms humans could argue about the relative merits of say *G*3 and *G*4 and thus this decision making process is difficult to automate.

Possible criteria one could use to define best include:

- Fastest settling time (to within, say 5%).

- Smallest actuation energy (to avoid fatigue and minimise fuel costs).

- Smallest error on average, say $\int_o^t |e(t)| dt$.

- Fastest rise time (so operating close to target).

- Minimise the maximum deviation when subject to a disturbance (ensures quality remains within given tolerance).

However, the problem with many of these criteria is that they are non-linear, and possibly discontinuous, in the parameters we can choose and thus do not lead to tractable optimisations in general.

> **Summary:** MPC is typically based on a numerical measure of performance as this allows a precise definition of "optimum". However, this definition can be viewed as being equally arbitrary and is simply a tool to enable a unique solution.

4.4.2 Properties of a convenient performance index

For convenience define the performance index as follows:

$$J = J(\mathbf{x}_k, \underset{\rightarrow k}{\Delta \mathbf{u}}, \underset{\leftarrow k-1}{\Delta \mathbf{u}}, \underset{\leftarrow k}{\mathbf{y}}, \underset{\rightarrow k+1}{\mathbf{r}}) \tag{4.4}$$

That is, it depends upon the current state, future predictions and also future targets. In order for a performance index to be a reliable indicator of good performance, it needs some properties which are listed next.

1. Behaviour with poor performance should give a notably higher value of J as compared to behaviour with good performance; good and poor need to be defined.

2. The values of J should be continuously differentiable with respect of any degrees of freedom in the future behaviour so that one can easily optimise J.

3. Ideally, J should have a unique minimum to avoid the possibility of finding a local as opposed to a global minimum and thus ease on-line computation.

4. The performance index should be unbiased in the sense that it must not drive a process away from the ideal steady-state.

Remark 4.1 *The desire for continuous differentiability and a unique minimum has led to the dominance of 2-norm measures [25, 40] analogous to those used for optimal control. This preference is so strongly established in the community that this book will not consider alternatives although readers can find some discussion in the academic literature, e.g., [2, 182, 199].*

STUDENT PROBLEM

Suggest different measures that could be included in a performance index. Which of these have a simple dependence on the d.o.f. within system predictions and moreover, lend themselves to simple optimisation with unique minima.

Summary: Although 1-norms, ∞-norms and other measures of performance are possible in general and indeed may make good engineering sense, MPC is typically based on quadratic measures because these are *more convenient and easier to work with*. The 2-norm measure has the advantage of not distinguishing as much between small errors, which may be unimportant in practice, but emphasising much more the need to drive down high errors. Of course, this emphasises the point that J is, to some extent, arbitrary.

4.4.3   Possible components to include in a performance index

The performance index (often denoted as cost function) is a measure of future predictive performance used to distinguish between good and bad predictions. Conse-

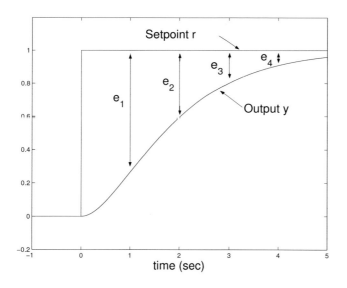

FIGURE 4.1
Illustration of tracking errors for a horizon $n_y = 4$.

quently, an obvious measure to use is the predicted error between the target and the predicted output; for example see Figure 4.1. Hence, with W_y some weighting matrix:

$$J_e = \sum_{i=1}^{n_y} \mathbf{e}_{k+i}^T \mathbf{e}_{k+i} \ \text{ or } \ J_e = \sum_{i=1}^{n_y} \mathbf{e}_{k+i}^T W_y \mathbf{e}_{k+i} \tag{4.5}$$

The reader should note that, as mentioned earlier, there is a marked preference for using quadratic terms in J, and this is clearly seen in J_e.

However, as is known from the minimum variance literature [24], a focus solely on tracking errors can lead to over active inputs and even *inversion of the model*. Consequently, it is important to also have some measure of input activity such as:

$$J_{\Delta u} = \sum_{i=0}^{n_u-1} \Delta \mathbf{u}_{k+i}^T \Delta \mathbf{u}_{k+i} \ \text{ or } \ \sum_{i=0}^{n_u-1} \Delta \mathbf{u}_{k+i}^T W_{\Delta u} \Delta \mathbf{u}_{k+i} \tag{4.6}$$

$$J_u = \sum_{i=0}^{n_u-1} [\mathbf{u}_{k+i}^T - \mathbf{u}_{ss}][\mathbf{u}_{k+i} - \mathbf{u}_{ss}] \ \text{ or } \ \sum_{i=0}^{n_u-1} [\mathbf{u}_{k+i}^T - \mathbf{u}_{ss}]W_u[\mathbf{u}_{k+i} - \mathbf{u}_{ss}] \tag{4.7}$$

where $W_u, W_{\Delta u}$ are weighting matrices to be chosen. Readers will note that the term J_u uses distances from the expected steady-state $\mathbf{u}_{ss}$. For computational reasons, it is normal to restrict the predicted input trajectories to those with n_u changes, as illustrated in Figure 4.2 where $n_u = 3$.

$$\Delta \mathbf{u}_{k+i|k} = 0, \ i \geq n_u \tag{4.8}$$

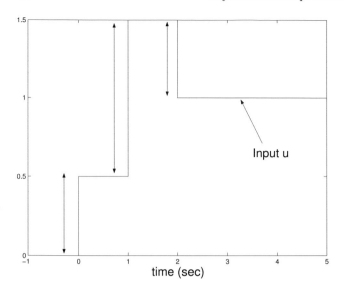

FIGURE 4.2
Illustration of input prediction with horizon $n_u = 3$. The double arrows indicate the input increments Δu.

In summary, a typical performance index would take the following form:

$$J = J_e + J_u + J_{\Delta u} \tag{4.9}$$

$$J = \sum_{i=1}^{n_y} \mathbf{e}_{k+i}^T W_y \mathbf{e}_{k+i} + \sum_{i=0}^{n_u-1} [\mathbf{u}_{k+i}^T - \mathbf{u}_{ss}] W_u [\mathbf{u}_{k+i} - \mathbf{u}_{ss}] + \sum_{i=0}^{n_u-1} \Delta \mathbf{u}_{k+i}^T W_{\Delta u} \Delta \mathbf{u}_{k+i}$$

STUDENT PROBLEM

Suggest different quadratic measures of performance that could be included in a performance index. Can these be easily subsumed in the cost function of (4.9)?

Summary: A typical MPC performance index has sums of squares of terms linked to the weighted output (or state), input and input increment.

4.4.4 Concepts of biased and unbiased performance indices

The use of an unbiased performance index is essential to ensure offset-free tracking. The concept of bias with regard to the performance index is introduced here through a few illustrations.

ILLUSTRATION: Concepts of an unblased performance index

An unbiased performance index is one which allows the system to reach and remain at the correct steady-state, irrespective of model uncertainty. Imagine the system is already at the correct steady-state so that:

$$\mathbf{y}_k = \mathbf{r_k}, \quad \mathbf{u}_k = \mathbf{u}_{ss}, \quad \forall k \leq 0$$

What choice of future input is required so that the predictions remain at the correct steady-state?

It should apparent that the predictions must be:

$$\mathbf{u}_{k+i} = \mathbf{u}_{ss}, \quad \Delta\mathbf{u}_{k+i} = 0, \quad \forall i \geq 0$$

Such a choice is consistent with $J = 0$, where J is defined in (4.9), and given J comprises a sum of squares, zero is the unique minimum. In summary, an unbiased performance index J is such that minimising J will give an input prediction that keeps the system at the correct steady-state!

Remark 4.2 *It is also assumed that the estimation of $\mathbf{u}_{ss}$ is unbiased and the predictions themselves are unbiased, as discussed in Sections 2.4.3 and 2.5.1.*

ILLUSTRATION: Concept of a biased performance index

One might ask whether using a performance index which weighted the inputs rather than the input deviations would cause bias. For example:

$$J_{u2} = \sum_{i=0}^{n_u-1} \mathbf{u}_{k+i}^T \mathbf{u}_{k+i} \qquad (4.10)$$

Note that (ignoring uncertainty for simplicity), in steady-state a relationship of the following form must hold:

$$\mathbf{y}_{ss} = G(1)\mathbf{u}_{ss} \qquad (4.11)$$

It should now be apparent that the following minimisation

$$\min_{\underset{\rightarrow k}{\mathbf{u}}} J = J_e + J_{u2} \quad \text{s.t. } (4.11)$$

will inevitably lead to a compromise as one cannot simultaneously achieve $J_e = 0$ and $J_{u2} = 0$ and thus, in practice, the system will settle where $J_e \neq 0$; that is there is bias or offset.

In summary, for most systems, the component J_{u2} cannot be used in the performance index as it would cause bias.

It should be obvious that the components $J_e, J_u, J_{\Delta u}$ are very carefully chosen to ensure the performance index is unbiased in the steady-state. While other choices of unbiased components are possible, there is little evidence in the literature of a need.

ILLUSTRATION: Typical performance indices

It is commonplace to use even simpler performance indices than (4.9).

1. When using transfer function models and incremental predictions such as (2.49), the term weighting $\mathbf{u}_{k+i} - \mathbf{u}_{ss}$ is partially redundant as it overlaps with the tracking error; consequently this is usually omitted. Thus:

$$J = \sum_{i=1}^{n_y} \mathbf{e}_{k+i}^T W_y \mathbf{e}_{k+i} + \sum_{i=0}^{n_u-1} \Delta \mathbf{u}_{k+i}^T W_{\Delta u} \Delta \mathbf{u}_{k+i} \qquad (4.12)$$

2. When using state space models, links to LQR design and the unbiased prediction structure such as (2.10) suggest a preference for ignoring the term based on $\Delta \mathbf{u}_{k+i}$ and linking the error term to the state $\mathbf{x}$. Also it is more conventional to use matrices Q, R for the weighting matrices. Hence:

$$J = \sum_{i=1}^{n_y} [\mathbf{x}_{k+i} - \mathbf{x}_{ss}]^T Q [\mathbf{x}_{k+i} - \mathbf{x}_{ss}] + \sum_{i=0}^{n_u-1} [\mathbf{u}_{k+i}^T - \mathbf{u}_{ss}] R [\mathbf{u}_{k+i} - \mathbf{u}_{ss}]$$
$$(4.13)$$

Hereafter, for simplicity of algebra, we will assume that the performance index takes the form of either (4.12) or (4.13). Nevertheless, the reader can use all three terms $J_e, J_u, J_{\Delta u}$ where the context suggests this is reasonable.

Summary: A suitable performance index has the following properties in general:

1. It comprises solely quadratic terms, or more precisely sums of squares, and thus has a unique minimum.

2. The terms in the performance index are such that they are expected to be zero when the system has reached the correct steady-state.

Thus the components $J_e, J_u, J_{\Delta u}$ ensure the performance index is unbiased in the steady-state. While other choices of unbiased components are possible and at times may make good engineering sense, there is little evidence in the literature of a need.

STUDENT PROBLEM

Suggest different quadratic measures of performance that could be included in a performance index and would be unbiased. Is there any reason why these could not be used and lead to sensible solutions?

4.4.5 The dangers of making the performance index too simple

A reader may wonder whether one could use a performance index comprising J_e on its own, which would thus put all the emphasis on the tracking behaviour. However, this is generally accepted to be unwise.

The dangers of using $J = J_e$

If one uses a performance index with just the predicted errors, then there must exist a future input trajectory which sets all these errors to zero, which is thus the minimum. Consider the impact of this in general for a system $y(z) = G(z)u(z)$:

$$e(z) = 0 \quad \Rightarrow \quad y(z) = r(z) \quad \Rightarrow \quad u(z) = G(z)^{-1}r(z) \qquad (4.14)$$

The corresponding input is determined by inverting $G(z)$! For minimum phase $G(z)$ this is safe even if somewhat sensitive and poorly shaped, but for non-minimum phase $G(z)$ the input is divergent!

ILLUSTRATION: Illustration of performance associated to $J = J_e$

Consider the system

$$G = \frac{z^{-1} - 0.05z^{-2} - 0.765z^{-3}}{1 - 1.6z^{-1} + 0.64z^{-2}}$$

and compute the input signal to track a step target exactly. This input signal is far from acceptable in general and one would prefer to compromise on tracking to improve the input activity.

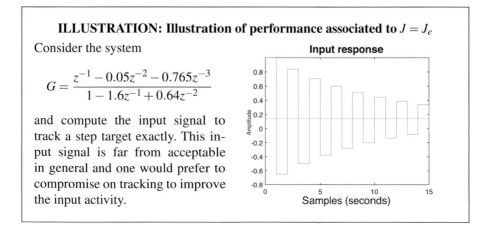

Summary: In general the choice $J = J_e$ is not recommended and is definitely not advisable for non-minimum phase processes.

4.4.6   Integral action in predictive control

A typical GPC control law implicitly includes integral action. But, as illustrated in the previous chapter on PFC, it is rare to see analysis which demonstrates the integral as an explicit term. Rather, demonstration of this fact is usually done by a *common sense* type of approach. For simplicity take the constraint-free case as constraints can cause unreachable targets and thus offset.

1. Assume that you are using unbiased predictions, that is predictions which have no offset in the steady-state, irrespective of model uncertainty.

2. Assume that the performance index is unbiased, that is comprises terms such as $J_e, J_u, J_{\Delta u}$.

3. If the system settles to a steady-state, this must be the correct steady-state because any other outcome is a contradiction.

    • If at the correct steady-state, because the predictions are unbiased a choice of $\mathbf{u} = \mathbf{u}_{ss}, \Delta\mathbf{u} = 0$ will imply the system remains at the correct steady-state and this is also consistent with $J = 0$ and thus corresponds to minimising J.

    • If the system is at an incorrect steady-state, the choice $\mathbf{u} = \mathbf{u}_{ss}, \Delta\mathbf{u} = 0$ is not consistent with $\mathbf{e} = 0$ and thus, the input must change to achieve a compromise between the terms in $J_e, J_u, J_{\Delta u}$ in the minimisation of J; any change in the input means a move away from the current steady-state.

Summary: The combination of unbiased predictions with an unbiased performance index ensures that, **assuming stability**, predictive control gives offset-free tracking in the steady-state *for constant targets*.

Remark 4.3 *Consideration of non-constant targets is rare in the predictive control literature and would require some adjustments to the choice of J and the choice of input d.o.f. (e.g., [191]), but such special cases are beyond the remit of the current book.*

4.4.7 Compact representation of the performance index and choice of weights

The cost function is often better represented in more compact notation (e.g., see Equations 2.1) using vectors and matrices. Hence J from (4.9) can be written as:

$$J = \|W_y^{0.5}(\underset{\rightarrow k+1}{\mathbf{r}} - \underset{\rightarrow k+1}{\mathbf{y}})\|_2^2 + \|W_{\Delta u}^{0.5} \underset{\rightarrow k}{\Delta \mathbf{u}}\|_2^2 + \|W_u^{0.5}(\underset{\rightarrow k}{\mathbf{u}} - L\mathbf{u}_{ss})\|_2^2 \tag{4.15}$$

or

$$J = (\underset{\rightarrow k+1}{\mathbf{r}} - \underset{\rightarrow k+1}{\mathbf{y}})^T W_y (\underset{\rightarrow k+1}{\mathbf{r}} - \underset{\rightarrow k+1}{\mathbf{y}}) + \underset{\rightarrow k}{\Delta \mathbf{u}}^T W_{\Delta u} \underset{\rightarrow k}{\Delta \mathbf{u}} + (\underset{\rightarrow k}{\mathbf{u}} - L\mathbf{u}_{ss})^T W_u (\underset{\rightarrow k}{\mathbf{u}} - L\mathbf{u}_{ss}) \tag{4.16}$$

for suitable weights $W_y, W_{\Delta u}, W_u$. Readers will note how using this compact notation will be invaluable when we move to optimisation in the following sections.

The weights can be used to decide how much emphasise is placed on each term and thus, for example, whether to emphasis tracking errors or control activity. However, in practice the literature gives very little attention to the weights in this performance index and minimal if any guidance on how they might be selected.

A typical design procedure, performance off-line, is as follows:

1. Normalise all the input and output signals to an equivalent range, say 0 to 1.

2. Set all the weights to be identity matrices.

3. Design the control law and assess the balance between input activity and tracking performance.

4. Increase weights according to design requirements: (i) increase W_y to improve tracking at the expense of increased input activity; (ii) increase $W_{\Delta u}$ to reduce input activity at the expense of tracking performance. Iterate over steps 3 and 4.

Readers will note that this is not particularly systematic and indeed this could be argued as a major failing of MPC. The design of the weights in the performance index has a potentially poor link to the actual performance. Nevertheless, in practice for many cases, the procedure works well enough and performance is often good after just step 2 above.

SHORTCUT: In many cases users assume $W_y = I$ and change only the input weight. Moreover, it is common to assume that $W_{\Delta u} = \lambda I$ or $W_U = \lambda I$, thus reducing the number of weights to just one, that is λ.

Remark 4.4 *Although not included in (4.15), one can also make the weights time varying and thus, place a different emphasis on terms at different sampling instants; this is done very rarely [192]. Moreover, readers should note that, the weights should increase further into the future or you may get a conflict with stability assurances. However, this additional complexity is rarely justified or used.*

Summary: Although other choices are possible, the most common performance index takes the form with a single weight λ:

$$J = \|(\underset{\rightarrow k+1}{\mathbf{r}} - \underset{\rightarrow k+1}{\mathbf{y}})\|_2^2 + \lambda \|\underset{\rightarrow k}{\Delta \mathbf{u}}\|_2^2 \qquad (4.17)$$

It is implicit here that the signals are normalised to a similar magnitude range.

4.5 GPC algorithm formulation for transfer function models

Much of this section is written assuming an MFD model and hence the SISO case is automatically included. The constraint-free case only is considered next; constraints are introduced in Chapter 7. First an outline is given of how to compute the GPC [25] control law.

The reader is reminded of the key stages in an MPC algorithm formulation.

1. A definition of system predictions and their dependence on d.o.f.

2. Definition of a suitable performance index.

3. An optimisation which selects the control strategy minimising the performance index.

Chapter 2 demonstrated how to form predictions for an MFD model, neatly summarised in Equation (2.49). The previous section developed a suitable performance in index, that is Equation (4.12) or (4.17). This section will summarise the key points from those and then add the optimisation component.

4.5.1 Selecting the degrees of freedom for GPC

While earlier sections have outlined the basic prediction structure of (2.49), it is necessary to clarify precisely the d.o.f. within this prediction. As indicated in Figure 4.2 and Eqn.(4.8), it is normal to assume that the future *predicted* control increments become zero after n_u steps. This has significant repercussions on the structure of the prediction equations. It is noted that the matrix H multiplies upon $\underset{\rightarrow k}{\Delta \mathbf{u}}$. Hence, the prediction Equations (2.49) assume that

$$
H\Delta \mathbf{u}_{\rightarrow k} =
\begin{bmatrix}
h_0 & 0 & 0 & \cdots & 0 & \cdots & 0 \\
h_1 & h_0 & 0 & \cdots & 0 & \cdots & 0 \\
\vdots & \vdots & \vdots & \vdots & \vdots & \vdots & \vdots \\
h_{n_y-1} & h_{n_y-2} & h_{n_y-3} & \cdots & h_{n_y-n_u} & \cdots & h_0
\end{bmatrix}
\begin{bmatrix}
\Delta \mathbf{u}_{k|k} \\
\Delta \mathbf{u}_{k+1|k} \\
\vdots \\
\hline
\Delta \mathbf{u}_{k+n_u-1|k} \\
\vdots \\
\Delta \mathbf{u}_{k+n_y-1|k}
\end{bmatrix}
\tag{4.18}
$$

However, the restriction (4.8), that is $\Delta \mathbf{u}_{k+n_u+i|k} = 0, i \geq 0$, implies that:

$$
H\Delta \mathbf{u}_{\rightarrow k} =
\underbrace{
\begin{bmatrix}
h_0 & 0 & 0 & \cdots & 0 \\
h_1 & h_0 & 0 & \cdots & 0 \\
\vdots & \vdots & \vdots & \vdots & \vdots \\
h_{n_y-1} & h_{n_y-2} & h_{n_y-3} & \cdots & h_{n_y-n_u}
\end{bmatrix}
}_{H_{n_u}}
\begin{bmatrix}
\Delta \mathbf{u}_{k|k} \\
\Delta \mathbf{u}_{k+1|k} \\
\vdots \\
\Delta \mathbf{u}_{k+n_u-1|k}
\end{bmatrix}
\tag{4.19}
$$

where H_{n_u} constitutes only the first n_u columns of H. Hereafter, for convenience, wherever the reader sees the prediction equation $H\Delta \mathbf{u}_{\rightarrow k}$, it can be assumed that H is tall and thin, that is $H = H_{n_u}$.

> **Summary:** In the term $H\Delta \mathbf{u}_{\rightarrow k}$ of the prediction equations (e.g., (2.49)) it is assumed that H is block dimension $n_y \times n_u$ and the vector $\Delta \mathbf{u}_{\rightarrow k}$ comprises block n_u terms and thus in effect one is using (4.19).

4.5.2 Performance index for a GPC control law

The second step in a GPC law formulation is to substitute the prediction equation of (2.49) into the performance index of (4.17). The predictions are given as:

$$
\mathbf{y}_{\rightarrow k+1} = H\Delta \mathbf{u}_{\rightarrow k} + P\Delta \mathbf{u}_{\leftarrow k-1} + Q\mathbf{y}_{\leftarrow k}
$$

Hence, substituting these into (4.17) gives:

$$
J = \| \mathbf{r}_{\rightarrow k+1} - H\Delta \mathbf{u}_{\rightarrow k} - P\Delta \mathbf{u}_{\leftarrow k-1} - Q\mathbf{y}_{\leftarrow k} \|_2^2 + \lambda \|\Delta \mathbf{u}_{\rightarrow k}\|_2^2
\tag{4.20}
$$

The performance index can be put in a more obvious quadratic form by expanding each term and hence:

$$
J = \Delta \mathbf{u}_{\rightarrow k}^T \underbrace{(H^T H + \lambda I)}_{S} \Delta \mathbf{u}_{\rightarrow k} + \Delta \mathbf{u}_{\rightarrow k}^T \underbrace{2H^T [P\Delta \mathbf{u}_{\leftarrow k-1} + Q\mathbf{y}_{\leftarrow k} - \mathbf{r}_{\rightarrow k+1}]}_{p} + q
\tag{4.21}
$$

$q = \| \mathbf{r}_{\rightarrow k+1} - P\Delta \mathbf{u}_{\leftarrow k-1} - Q\mathbf{y}_{\leftarrow k} \|_2^2$ contains terms that do not depend upon $\Delta \mathbf{u}_{\rightarrow k}$ and hence can be ignored in any optimisation.

Remark 4.5 *Readers will note that the performance index reduces to the form given in Appendix C.3.2.*

4.5.3   Optimising the GPC performance index

The d.o.f. are the future control increments, that is $\Delta\mathbf{u}_{\rightarrow k}$ and hence the optimum strategy is obtained by minimising J w.r.t to these d.o.f. It is not the role of this book to *teach* differentiation and hence the result is stated by inspection, noting that due to the quadratic form, there is a single stationary point (a minimum) and hence this is achieved by setting the gradient to zero. See Appendix C.3.2 for some elaboration.

The aim is to perform the minimisation:

$$\min_{\Delta\mathbf{u}_{\rightarrow k}} J \qquad (4.22)$$

The gradient (in effect differentiation of (4.21)) gives:

$$\frac{dJ}{d\Delta\mathbf{u}_{\rightarrow k}} = 2(H^T H + \lambda I)\Delta\mathbf{u}_{\rightarrow k} + 2H^T[P\Delta\mathbf{u}_{\leftarrow k-1} + Q\mathbf{y}_{\leftarrow k} - \mathbf{r}_{\rightarrow k+1}] \qquad (4.23)$$

Setting this derivative to zero gives:

$$\frac{dJ}{d\Delta\mathbf{u}_{\rightarrow}} = 0 \;\Rightarrow\; \Delta\mathbf{u}_{\rightarrow k} = (H^T H + \lambda I)^{-1} H^T [\mathbf{r}_{\rightarrow k+1} - Q\mathbf{y}_{\leftarrow k} - P\Delta\mathbf{u}_{\leftarrow k-1}] \qquad (4.24)$$

Of the optimising sequence of future input increments, only the first is implemented as the optimisation is recomputed at every sampling instant. Hence, the GPC control law is defined by the first block element of $\Delta\mathbf{u}_{\rightarrow k}$, that is,

$$\Delta u_k = \mathbf{E}_1^T \Delta\mathbf{u}_{\rightarrow}, \quad \mathbf{E}_1^T = [I, 0, 0, ..., 0].$$

Substituting this into (4.24):

$$\Delta u_k = \mathbf{E}_1^T (H^T H + \lambda I)^{-1} H^T [\mathbf{r}_{\rightarrow k+1} - Q\mathbf{y}_{\leftarrow k} - P\Delta\mathbf{u}_{\leftarrow k-1}] \qquad (4.25)$$

Summary: The computation (4.25) is recalculated at each sampling instant and therefore the unconstrained GPC control law is:

$$\Delta u_k = P_r \mathbf{r}_{\rightarrow k+1} - N_k \mathbf{y}_{\leftarrow k} - \check{D}_k \Delta\mathbf{u}_{\leftarrow k-1} \qquad (4.26)$$

where

$$
\begin{array}{rcl}
P_r &=& \mathbf{E}_1^T (H^T H + \lambda I)^{-1} H^T \\
N_k &=& \mathbf{E}_1^T (H^T H + \lambda I)^{-1} H^T Q \\
\check{D}_k &=& \mathbf{E}_1^T (H^T H + \lambda I)^{-1} H^T P
\end{array}
\qquad (4.27)
$$

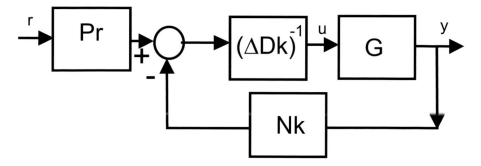

FIGURE 4.3
The effective block diagram structure for GPC with a transfer function model.

4.5.4 Transfer function representation of the control law

For implementation or coding purposes, the formulation given in (4.26) is ideal and no further manipulations are required. However, for the purposes of closed-loop stability analysis, it is easier first to represent (4.26) in transfer function form (as discussed in Appendix Section C.2.4). Given

$$
\begin{aligned}
N_k &= [N_{ko}, N_{k1}, N_{k2}, \ldots, N_{kn}] \\
\check{D}_k &= [D_{ko}, D_{k1}, D_{k2}, \ldots, D_{km}] \\
P_r &= [P_{r1}, P_{r2}, P_{r3}, \ldots, P_{rn_y}]
\end{aligned}
\tag{4.28}
$$

define

$$
\begin{aligned}
N_k(z) &= N_{ko} + N_{k1}z^{-1} + N_{k2}z^{-2} + \cdots + N_{kn}z^{-n} \\
\check{D}_k(z) &= D_{ko} + D_{k1}z^{-1} + D_{k2}z^{-2} + \cdots + D_{km}z^{-m} \\
P_r(z) &= P_{r1}z + P_{r2}z^2 + P_{r3}z^3 + \cdots + P_{rn_y}z^{n_y} \\
D_k(z) &= 1 + z^{-1}\check{D}_k(z)
\end{aligned}
\tag{4.29}
$$

Then, noting the definitions (2.1, 2.49) of $\Delta \mathbf{u}_{\underleftarrow{k-1}}, \mathbf{y}_{\underleftarrow{k}}, \mathbf{r}_{\underrightarrow{k+1}}$, the control law can be implemented in the following fixed closed-loop form (and equivalent block diagram structure is given in Figure 4.3).

$$
D_k(z)\Delta \mathbf{u}(z) = P_r(z)\mathbf{r}(z) - N_k(z)\mathbf{y}(z)
\tag{4.30}
$$

It should be noted that $P_r(z)$ is an anticausal operator; that is, it uses future values of $\mathbf{r}_k$. If these are unknown, simply substitute the best available estimate, for instance the current value.

Remark 4.6 *We note [134] that the default choice for P_r given in GPC is often poor and better alternatives exist when advance knowledge is available. Nevertheless, the most common assumption used is that $\mathbf{r}_{k+i|k} = \mathbf{r}_{k+1|k}, \forall i > 0$ and in this case one can reduce P_r to its steady-state gain, that is $P_r(z) = P_r(1)$. Moreover, due to the implied guarantee of offset-free tracking, one can also show that $P_r(1) = N_k(1)$.*

Summary: In the absence of constraints, GPC reduces to a known and fixed linear feedback law with feedforward.

$$D_k(z)\Delta\mathbf{u}(z) = P_r(z)\mathbf{r}(z) - N_k(z)\mathbf{y}(z) \tag{4.31}$$

where the control parameters are given in (4.27,4.29). Often, where there is no advance knowledge of the target, this is simplified further to:

$$D_k(z)\Delta\mathbf{u}(z) = P_r(1)\mathbf{r}(z) - N_k(z)\mathbf{y}(z) \tag{4.32}$$

STUDENT PROBLEM

By deriving the GPC control law from first principles, prove that an expression for the closed-loop pole polynomial in the SISO case with system $a(z)y(z) = b(z)u(z)$ and control law (4.31) is given as follows:

$$p_c(z) = a(z)\Delta(z)D_k(z) + b(z)N_k(z) \tag{4.33}$$

where $p_c(z)$ is the closed-loop pole poylnomial.
How would this derivation change for the MIMO case?

4.5.5   Numerical examples for GPC with transfer function models and MATLAB code

This section will demonstrate some MATLAB code for generating the GPC control law with an MFD model. Moreover, it will demonstrate the corresponding closed-loop simulations. At this point, no attempt is made to discuss effective tuning and moreover, so as not to complicate discussion, the assumption is made that $\mathbf{r}_{k+i|k} = \mathbf{r}_{k+1|k}, \forall i > 0$ and constraints are not included.

The code to generate the first example is named **gpcsisoexample1.m** and can be used as an exemplar for understanding data entry. This calls m-files **mpc_predmat.m, mpc_simulate_noconstraints.m, mpc_law.m, convmat.m**. A typical call statement is:

[y,u,Du,r,Nk,Dk,Pr] = mpc_simulate_noconstraints(B,A,nu,ny,Wu,Wy,ref,dist, noise);

ILLUSTRATION: GPC for a SISO model

Consider a simple transfer function model.

$$G(z) = \frac{0.1z^{-1} + 0.03z^{-2}}{1 - 1.2z^{-1} + 0.32z^{-2}} \quad (4.34)$$

Choose horizons of $n_y = 15, n_u = 3$ and a weight $\lambda = 1$. The corresponding closed-loop simulations and control law parameters are shown here.

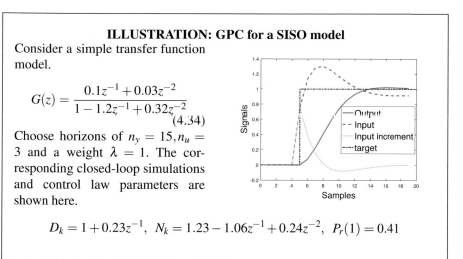

$$D_k = 1 + 0.23z^{-1}, \quad N_k = 1.23 - 1.06z^{-1} + 0.24z^{-2}, \quad P_r(1) = 0.41$$

ILLUSTRATION: GPC for a MIMO model

Consider the 3 by 3 MFD model given in **gpcmimoexample1.m**. Choose horizons of $n_y = 9, n_u = 4$ and a weight $\lambda = 1$. Set different target values for the outputs in each loop. The corresponding closed-loop simulations are shown here and demonstrate good control of the interaction and tracking.

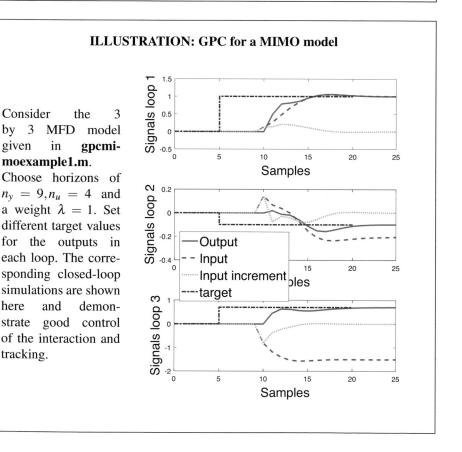

STUDENT PROBLEM

For several SISO transfer function examples of your own choosing, generate the appropriate GPC control law and:

1. Form the closed-loop step responses.

2. Form the closed-loop pole polynomial and find the poles.

Hence investigate the links between closed-loop poles/behaviour and the parameters n_y, n_u, λ. You can perform these tasks using simple edits to the file **gpcsisoexample1.m**.

STUDENT PROBLEM

For several MIMO transfer function examples of your own choosing, generate the appropriate GPC control law and:

1. Form the closed-loop step responses.

2. Form the closed-loop pole polynomial.

and hence investigate the links between closed-loop poles/behaviour and the parameters n_y, n_u, λ. You can also investigate the efficacy and ease of tuning using non-scalar weights, so $W_y \neq I, W_u \neq I$. You can perform these tasks using simple edits to the file **gpcmimoexample1.m**.

4.5.6   Closed-loop transfer functions

In order to do a sensitivity or closed-loop pole analysis, one may want to find the closed-loop transferences. These are of course easy to derive given the control law (4.31). However, for completeness, this section outlines the transference from **r** to **y**. The argument $(.)(z)$ is omitted to improve readability.

Let the MFD model $D\mathbf{y} = N\mathbf{u}$ be written in two forms

$$\hat{N}\hat{D}^{-1} = D^{-1}N \;\Rightarrow\; \begin{cases} \mathbf{y} & = & \hat{N}\hat{D}^{-1}\mathbf{u} \\ \mathbf{y} & = & D^{-1}N\mathbf{u} \end{cases} \tag{4.35}$$

The equations in the loop are:

$$\underbrace{\mathbf{y} = \hat{N}\hat{D}^{-1}\mathbf{u};}_{\text{Model}} \quad \underbrace{D_k \Delta \mathbf{u} = P_r \mathbf{r} - N_k \mathbf{y};}_{\text{Controller}} \tag{4.36}$$

Therefore

$$[D_k\Delta]\mathbf{u} = P_r\mathbf{r} - N_k\hat{N}\hat{D}^{-1}\mathbf{u} \;\Rightarrow\; [D_k\Delta\hat{D} + N_k\hat{N}]\hat{D}^{-1}\mathbf{u} = P_r\mathbf{r} \qquad (4.37)$$

Hence

$$\begin{aligned}
\mathbf{u} &= \hat{D}[D_k(z)\Delta\hat{D} + N_k(z)\hat{N}]^{-1}P_r(z)\mathbf{r} \\
\mathbf{y} &= \hat{N}[D_k(z)\Delta\hat{D} + N_k(z)\hat{N}]^{-1}P_r(z)\mathbf{r}
\end{aligned} \qquad (4.38)$$

Clearly the closed-loop poles depend on $\det(D_k(z)\Delta\hat{D} + N_k(z)\hat{N})$.

Remark 4.7 *In general matrix multiplication is not commutative, i.e., $AB \neq BA$. Hence in the manipulations of this section, one must be careful to preserve the order of the operations.*

Summary: Given (4.31, 4.35), the closed-loop poles for GPC in the nominal case can be derived from

$$P_c(z) = D_k(z)\Delta\hat{D} + N_k(z)\hat{N} \qquad (4.39)$$

4.5.7 GPC based on MFD models with a T-filter (GPCT)

This section looks at the impact of a T-filter on GPC using MFD models. Students of a single course would be better to focus only on the core concepts and omit a detailed study of the algebra.

4.5.7.1 Why use a T-filter and conceptual thinking?

The T-filter is a conceptual or intuitive idea based on classical control ideas of low-pass filtering. In simple terms, it was noted that a standard GPC law seemed to be very sensitive to measurement (high frequency) noise and thus a logical approach is to filter the output measurements. However, filtering introduces lag and it is desired that this lag is not be embedded into the corresponding responses. Hence, a conceptual approach was defined.

1. Filter the output data before using it for predictions. This way the impact of any high frequency noise on the predictions can be reduced, but with minimal detriment to the lower frequency dynamics which are important.
2. Predict using the filtered data.
3. Anti-filter to restore predicted data back to the correct domain.
4. Substitute the *new predictions* into the performance index for optimisation.

The reader is reminded that the procedure above is described more slowly around Section 2.5.4.1.

4.5.7.2   Algebraic procedures with a T-filter

The reader may recall from Section 2.5.4 that the inclusion of a T-filter modified the prediction equations, so that one would use predictions (2.62) in place of (2.49). In effect this covers the items 1-3 from the list above as the predictions are in the original domain but are based on past *filtered* data. Specifically, the part of the predictions depending on past measured data changes as follows:

$$(Q\underset{\leftarrow k}{\mathbf{y}} + P\Delta\underset{\leftarrow k-1}{\mathbf{u}}) \;\;\rightarrow\;\; (\tilde{Q}\underset{\leftarrow k}{\tilde{\mathbf{y}}} + \tilde{P}\Delta\underset{\leftarrow k-1}{\tilde{\mathbf{u}}}) \tag{4.40}$$

Therefore, to form the implied GPC control law, there remains only the need to redo the algebra of Section 4.5.3 using the modified predictions (2.62). Therefore one can simply substitute this change into control law (4.25) and derive the modified control law accordingly. By analogy with (4.26, 4.27), the corresponding control law takes the form:

$$\begin{cases} \Delta\mathbf{u}_k = P_r\underset{\rightarrow k+1}{\mathbf{r}} - \check{D}_k\Delta\underset{\leftarrow k-1}{\tilde{\mathbf{u}}} - \tilde{N}_k\underset{\leftarrow k}{\tilde{\mathbf{y}}} \\ \check{D}_k = [I,0,0,...][H^T H + \lambda I]^{-1}H^T\tilde{P} \\ \tilde{N}_k = [I,0,0,...][H^T H + \lambda I]^{-1}H^T\tilde{Q} \end{cases} \tag{4.41}$$

Substituting $\tilde{\mathbf{u}} = \mathbf{u}/T$, $\tilde{\mathbf{y}} = \mathbf{y}/T$ and rearranging into a more conventional form in terms of z-transforms (as in Section 4.5.4) and omitting the argument $.(z)$, gives:

$$[1 + \frac{\check{D}_k}{T}]\Delta\mathbf{u}_k = P_r\mathbf{r}_k - \frac{\tilde{N}_k}{T}\mathbf{y}_k \tag{4.42}$$

or more conveniently

$$\frac{\tilde{D}_k\Delta}{T}\mathbf{u}_k = P_r\mathbf{r}_k - \frac{\tilde{N}_k}{T}\mathbf{y}_k; \quad \tilde{D}_k = T + \check{D}_k \tag{4.43}$$

Note that the corresponding control parameters for GPC (D_k and N_k) and GPCT ($\tilde{D}_k$ and $\tilde{N}_k$ will be different, as $T \neq 1$ implies $\tilde{P} \neq P$, $\tilde{Q} \neq Q$ — see Equation (2.62).

Lemma 4.1 *In the absence of model uncertainty and disturbances, control laws (4.31, 4.43) must give identical tracking performance even though the control laws have different coefficients and structures.*

Proof: This follows from a simple logical argument. Both control laws are derived from minimising the same performance index with the same d.o.f. and thus, in the nominal case, must give identical choices for the control increments $\Delta\mathbf{u}_k$. In the case of uncertainty, the predictions will now be slightly different and thus the associated control laws will also vary.   □

One can form the closed-loop poles arising by combining model (4.35) with controller (4.43) by analogy to (4.38). Hence:

$$\begin{aligned} \mathbf{u}_k &= \hat{D}T[\tilde{D}_k\Delta\hat{D} + \tilde{N}_k\hat{N}]^{-1}P_r\mathbf{r}_k \\ \mathbf{y}_k &= \hat{N}T[\tilde{D}_k\Delta\hat{D} + \tilde{N}_k\hat{N}]^{-1}P_r(z)\mathbf{r}_k \end{aligned} \tag{4.44}$$

where one notes that the only differences are the insertion of T and the use of $\tilde{D}_k$ and $\tilde{N}_k$ in place of D_k and N_k. The closed-loop poles are given by

$$\tilde{P}_c = \tilde{D}_k(z)\Delta\hat{D} + \tilde{N}_k(z)\hat{N} \tag{4.45}$$

Theorem 4.1 *There is a relationship between $\tilde{P}_c$ and P_c of (4.39); more specifically one can show that in the nominal case $\tilde{P}_c = P_c T$.*

Proof: It was noted in Lemma 4.1 that, in the nominal case, performance must be identical and hence the closed-loop transfer functions with and without T must be identical, that is:

$$\hat{N}T\underbrace{[\tilde{D}_k\Lambda\hat{D}+\tilde{N}_k\hat{N}]^{-1}P_r}_{\tilde{P}_c} = \hat{N}\underbrace{[D_k\Lambda\hat{D}+N_k\hat{N}]^{-1}P_r}_{P_c}$$

Hence the roots of $\tilde{P}_c$ include the roots of T.   □

Remark 4.8 *Performance of (4.38) and (4.44) differs for the uncertain case, that is when $\hat{N}$, $\hat{D}$ do not correspond to those used for forming the prediction equations (2.49) and (2.62) or when there are disturbances. This is because in such a case GPC and GPCT will be using different predictions and hence the optimisations give differing answers.*

ILLUSTRATION: Guidance for T-filter selection

There is not much guidance in the literature [175, 196] on how to design a T-filter beyond the general recommendation that the poles of $T(z)$ should be matched to the dominant poles of the process. Hence, with typical sampling times of about a tenth of the system settling time, one would expect choices similar to T_1, T_2 used in the following illustrations.

ILLUSTRATION: SISO GPC with a T-filter

Consider the SISO model given in **gpcsisoexample1_tfilt.m**

$$G(z) = \frac{0.1z^{-1}+0.03z^{-2}}{1-1.2z^{-1}+0.32z^{-2}}$$

and use a range of T-filters as follows:

$$T_1 = 1 - 0.8z^{-1};$$

$$T_2 = 1 - 0.9z^{-1}; \quad T_3 = T_1^2$$

Compute the closed-loop step responses with a disturbance at the 25th sample and a noisy output from the 40th sample.

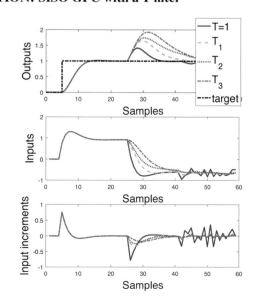

It is clear that as the level of filtering is increased:

- The noise rejection (high frequency uncertainty) improves in that the input is much less active which saves on actuation and wear and tear.

- The disturbance rejection (low frequency uncertainty) is worse; the output is slower to return to the correct target.

- During the set point change (no uncertainty) the responses are identical.

ILLUSTRATION: MIMO GPC with a T-filter

Consider the MIMO model given in **gpcmimoexample1_tfilt.m** and use a T-filter as follows:

$$T_1 = 1 - 0.8z^{-1}$$

Compute the closed-loop step responses with an output disturbance at the 25th sample and a noisy output from the 40th sample. As with the SISO example, it is clear that the T-filter has reduced input responsiveness to measurement noise, but given slower response to output disturbances.

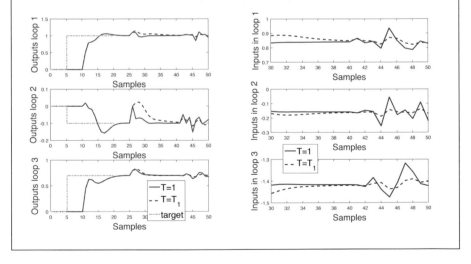

STUDENT PROBLEMS

Using several examples of your choice, investigate the effect of adding a T-filter to GPC on:

   1. the closed-loop step responses with no uncertainty.

2. the closed-loop step responses when output disturbances, noise and parameter uncertainty are introduced.

3. the closed-loop pole polynomial.

You can use the files **gpcsisoexample1_tfilt.m** and **gpcmimoexample1_tfilt.m** as templates to form much of this data.

Summary: The z-transform representation of the control laws for GPC, GPCT are summarised in the table below.

	Control laws	**Closed-loop pole polynomial**
GPC	$D_k \Delta \mathbf{u} = P_r \mathbf{r} - N_k \mathbf{y}$	$D_k(z)\Delta\hat{D} + N_k(z)\hat{N}$
GPCT	$\dfrac{\tilde{D}_k}{T}\Delta \mathbf{u} = P_r \mathbf{r} - \dfrac{\tilde{N}_k}{T}\mathbf{y}$	$\tilde{D}_k(z)\Delta\hat{D} + \tilde{N}_k(z)\hat{N}$

Adding a T-filter does not change nominal performance but has a significant impact on sensitivity. Generally a T-filter is recommended where there is significant measurement noise.

4.5.8   Sensitivity of GPC

The previous section has demonstrated the impact of a T-filter through numerical simulations and argument. For completeness, it is worth noting that sensitivity functions can be computed analytically and thus compared using Bode diagrams or otherwise.

In this section the sensitivity functions will be derived for the MIMO case assuming a GPC controller structure and transfer function (MFD) models. In the SISO case there would be some simplification as commutativity of the matrix functions is no longer an issue.

To simplify the notation somewhat the argument $.(z)$ is omitted and for this section let the MPC control law and plant be given by

$$D\Delta \mathbf{u} = P\mathbf{r} - N\mathbf{y}; \quad A\mathbf{y} = B\mathbf{u} \tag{4.46}$$

We will need the two alternative model and controller forms:

$$\underbrace{A^{-1}B = \tilde{B}\tilde{A}^{-1}}_{model}; \quad \underbrace{D^{-1}N = \hat{N}\hat{D}^{-1}}_{controller} \tag{4.47}$$

4.5.8.1   Complementary sensitivity

Complementary sensitivity is the transference from the set point to the output. This can be computed by solving the model and controller equations of (4.46) as simultaneous equations. An example procedure for deriving the complementary sensitivity is given next.

$$
\begin{aligned}
A\mathbf{y} &= BD^{-1}\Delta^{-1}[P\mathbf{r} - N\mathbf{y}] \\
(A\Delta + BD^{-1}N)\mathbf{y} &= BD^{-1}P\mathbf{r} \\
(B^{-1}A\Delta + D^{-1}N)\mathbf{y} &= D^{-1}P\mathbf{r} \\
(DB^{-1}A\Delta + N)\mathbf{y} &= P\mathbf{r} \\
(D\tilde{A}\Delta\tilde{B}^{-1} + N)\mathbf{y} &= P\mathbf{r} \\
(D\tilde{A}\Delta + N\tilde{B})\tilde{B}^{-1}\mathbf{y} &= P\mathbf{r} \\
\mathbf{y} &= \tilde{B}(\underbrace{[D\tilde{A}\Delta + N\tilde{B}]}_{P_c})^{-1}P\mathbf{r}
\end{aligned}
\tag{4.48}
$$

Hence the complementary sensitivity is defined as

$$
S_c = \tilde{B}P_c^{-1}P
\tag{4.49}
$$

where P_c defines the closed-loop poles:

$$
P_c = [D\tilde{A}\Delta + N\tilde{B}]
\tag{4.50}
$$

If needed, the transference from set point to the input is given as $\mathbf{u} = \tilde{A}P_c^{-1}P\mathbf{r}$.

4.5.8.2   Sensitivity to multiplicative uncertainty

This section rapidly runs through the definitions of sensitivity to multiplicative uncertainty. In this case one is interested in robust stability, that is, for what range of uncertainty the closed-loop system is guaranteed to remain stable. Hence it is based on an analysis of the closed-loop poles. Closed-loop poles can be derived from

$$
[I + GK] = 0; \qquad G = A^{-1}B, \qquad K = D^{-1}\Delta^{-1}N
\tag{4.51}
$$

Multiplicative uncertainty can be modelled as:

$$
G \rightarrow (1 + \mu)G
\tag{4.52}
$$

for μ a scalar (possibly frequency dependent). Substituting this into expression (4.51) the closed-loop poles for the uncertain case are derived from

$$
0 = [1 + GK + \mu GK] = [1 + GK]^{-1}(1 + \mu GK[1 + GK]^{-1})
\tag{4.53}
$$

As it is known that $[1 + GK]$ has stable roots by design (this is the nominal case), the system can be destabilised if and only if $|\mu GK[1 + GK]^{-1}| \geq 1$. Hence the sensitivity to multiplicative uncertainty is defined as

$$
\begin{aligned}
S_G &= [1 + GK]^{-1}GK \\
&= [G^{-1} + K]^{-1}K \\
&= [\tilde{A}\Delta\tilde{B}^{-1} + D^{-1}N]^{-1}D^{-1}N \\
&= \tilde{B}[D\tilde{A}\Delta + N\tilde{B}]^{-1}N \\
&= \tilde{B}P_c^{-1}N
\end{aligned}
\tag{4.54}
$$

Sensitivity to multiplicative uncertainty: MIMO case

$$S_G = \tilde{B} P_c^{-1} N \qquad (4.55)$$

4.5.8.3 Disturbance and noise rejection

This section rapidly runs through the definitions of sensitivity to output disturbances and to measurement noise. An output disturbance (ζ) and noise (n) enter the system model as follows:

$$Ay = Bu + \underbrace{C\frac{\zeta}{\Delta} + An}_{f} \qquad (4.56)$$

For convenience these can be grouped as the term f. Redoing the algebra of Eqn.(4.48) gives:

$$
\begin{aligned}
A\mathbf{y} &= BD^{-1}\Delta^{-1}[P\mathbf{r} - N\mathbf{y}] + \mathbf{f} \\
(A\Delta + BD^{-1}N)\mathbf{y} &= BD^{-1}P\mathbf{r} + \Delta\mathbf{f} \\
(B^{-1}A\Delta + D^{-1}N)\mathbf{y} &= D^{-1}P\mathbf{r} + B^{-1}\Delta\mathbf{f} \\
(DB^{-1}A\Delta + N)\mathbf{y} &= P\mathbf{r} + DB^{-1}\Delta\mathbf{f} \\
(D\tilde{A}\Delta\tilde{B}^{-1} + N)\mathbf{y} &= P\mathbf{r} + DB^{-1}\Delta\mathbf{f} \\
(D\tilde{A}\Delta + N\tilde{B})\tilde{B}^{-1}\mathbf{y} &= P\mathbf{r} + DB^{-1}\Delta\mathbf{f} \\
\mathbf{y} &= \tilde{B}P_c^{-1}[P\mathbf{r} + DB^{-1}\Delta\mathbf{f}]
\end{aligned}
\qquad (4.57)
$$

Hence the sensitivities S_{yd}, S_{yn} of the output to disturbances/noise, respectively, are:

$$S_{yd} = \tilde{B}P_c^{-1}DB^{-1}C; \quad S_{yn} = \tilde{B}P_c^{-1}DB^{-1}A\Delta \qquad (4.58)$$

Using a similar procedure the sensitivities S_{ud}, S_{un} of the input to disturbances/noise can be defined as:

$$S_{ud} = \tilde{A}P_c^{-1}NA^{-1}\frac{C}{\Delta}; \quad S_{un} = \tilde{A}P_c^{-1}NA^{-1}A \qquad (4.59)$$

Given the definitions above, it is easy to do a sensitivity *analysis*.

Summary: The sensitivity functions for a GPC control law are:
Sensitivity to multiplicative uncertainty

$$S_G = \tilde{B}P_c^{-1}N \qquad (4.60)$$

Sensitivity to disturbances

$$\underbrace{S_{yd} = \tilde{B}P_c^{-1}DB^{-1}C}_{\text{Output sensitivity}}; \quad \underbrace{S_{ud} = \tilde{A}P_c^{-1}NA^{-1}\frac{C}{\Delta}}_{\text{Input sensitivity}} \qquad (4.61)$$

Sensitivity to noise

$$S_{yn} = \tilde{B}P_c^{-1}D\Delta B^{-1}A; \qquad S_{un} = \tilde{A}P_c^{-1}N \qquad (4.62)$$

Output sensitivity   Input sensitivity

STUDENT PROBLEM

Write some code to display and investigate the sensitivity of a GPC control law. How are these affected by choices such as the horizons and control weighting?

4.5.8.4 Impact of a T-filter on sensitivity

The previous subsections have been done deliberately with the notation N, D for the compensator. Consequently one can note the changes of Section 4.5.7 in order to compare the sensitivity with and without a T-filter.

$$D_k \;\rightarrow\; \frac{\tilde{D}_k}{T}; \quad N_k \;\rightarrow\; \frac{\tilde{N}_k}{T} \qquad (4.63)$$

The equivalent of (4.50) is:

$$\tilde{P}_c = \frac{\tilde{D}\tilde{A}\Delta + \tilde{N}\tilde{B}}{T} = P_c \qquad (4.64)$$

Hence substituting (4.63, 4.64) into (4.55, 4.58, 4.59) gives:

$$S_G = \tilde{B}P_c^{-1}\frac{\tilde{N}}{T} \qquad (4.65)$$

$$S_{yd} = \frac{\tilde{B}P_c^{-1}\tilde{D}\Delta B^{-1}C}{\Delta T}; \quad S_{yn} = \frac{\tilde{B}P_c^{-1}\tilde{D}\Delta B^{-1}A}{T} \qquad (4.66)$$

$$S_{ud} = \frac{\tilde{A}P_c^{-1}\tilde{N}A^{-1}C}{\Delta T}; \quad S_{un} = \frac{\tilde{A}P_c^{-1}\tilde{N}}{T} \qquad (4.67)$$

Summary: A key observation is that all these sensitivity functions have $T(z)$ in the denominator and thus, assuming $1/T$ is a low pass filter, one might expect these to be less sensitive at high frequencies. However, one cannot give a solid guarantee of these because of the parallel change of $N_k \rightarrow \tilde{N}_k, D_k \rightarrow \tilde{D}_k$ which is difficult to unpick due to the highly nonlinear relationships between $P, \tilde{P}$ and $Q, \tilde{Q}$ (see Eqn.(2.62)).

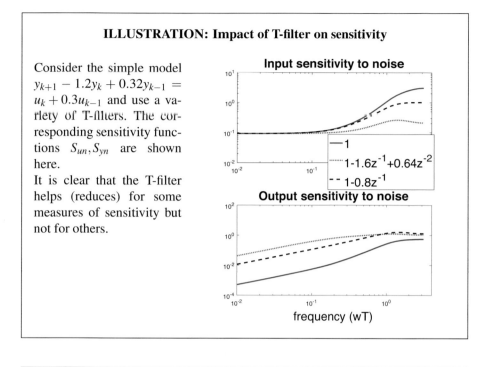

ILLUSTRATION: Impact of T-filter on sensitivity

Consider the simple model $y_{k+1} - 1.2y_k + 0.32y_{k-1} = u_k + 0.3u_{k-1}$ and use a variety of T-filters. The corresponding sensitivity functions S_{un}, S_{yn} are shown here.

It is clear that the T-filter helps (reduces) for some measures of sensitivity but not for others.

STUDENT PROBLEMS

1. Explain the conceptual steps behind the adoption of a T-filter and therefore where the benefits are expected to be.

2. Compute the sensitivity functions for a number for SISO examples of your choice and overlay these functions with and without T-filters to observe the trade offs that need to be managed.

4.5.9 Analogies between PFC and GPC

It is straightforward to show that a PFC law based on prediction equations (2.49) will give a control law with an identical block diagram structure to GPC (see Figure 4.3). For instance, using predictions (2.49), the PFC computation (3.3) reduces to

$$e_{n_y}^T [H\Delta \mathbf{u}_k + P\Delta \underset{\leftarrow k-1}{\mathbf{u}} + Q\underset{\leftarrow k}{\mathbf{y}}] = \lambda^{n_y} \underset{\leftarrow k}{\mathbf{y}} + (1 - \lambda^{n_y})\mathbf{r}_{k+1} \qquad (4.68)$$

This can easily be rearranged into the same form as (4.31) with

$$D_k = [1 + \frac{z^{-1}\mathbf{e}_{n_y}^T P}{\mathbf{e}_{n_y}^T H}]; \quad N_k = \frac{\mathbf{e}_{n_y}^T Q - \lambda^{n_y}}{\mathbf{e}_{n_y}^T H}; \quad P_r = \frac{1 - \lambda^{n_y}}{\mathbf{e}_{n_y}^T H} \qquad (4.69)$$

Summary: The structure of a PFC control law is the same as other common predictive control laws. That is, the structure depends solely on the model used to form the predictions.

STUDENT PROBLEM

Using several SISO examples of your choice (such as those in Table 3.1), investigate the ease of tuning of GPC as compared to PFC for obtaining a desired performance. You can use files such as **gpcsisoexample1.m, pfchighorderexample1.m** as a start point.

4.6 GPC formulation for finite impulse response models and Dynamic Matrix Control

The GPC law with an impulse response model can be stated by inspection using the results of the previous section in this chapter as a finite impulse response (FIR) model is the same as an MFD model with particular choices of D, N. Readers may also like to know that the popular DMC [28] algorithm deployed in industry is, in effect, equivalent to GPC but using a step response model as opposed to an MFD model.

ILLUSTRATION: Differences between GPC and DMC

If one ignores nuances which are not core concepts:

1. GPC and DMC use an equivalent performance index.

2. Historically DMC made use of a step response model whereas GPC uses a transfer function model. The former may be easier to identify in many industrial scenarios.

As noted in Section 4.5.7 on the T-filter, changing the model does not change the nominal performance and thus, could be considered a matter of preference linked to the ease of the modelling procedure. However, as seen in the previous section, the choice of prediction model has a significant impact on sensitivity to uncertainty.

DMC has different sensitivity properties to GPC in general.

The prediction results using an FIR model are given in Section 2.8. To form a GPC control law, we simply use these predictions in place of (2.49) and hence the corresponding GPC algorithm is given next.

Step 1. Take the prediction equation of (2.100)

$$\underset{\rightarrow k+1}{\mathbf{y}} = C_H \Delta \underset{\rightarrow k}{\mathbf{u}} + M \Delta \underset{\leftarrow k-1}{\mathbf{u}} + L\hat{y}_k \qquad (4.70)$$

and note that $C_H = H$ (where H is given in for instance Eqn.(2.49) and Section 4.5.1).

Step 2. Substitute this prediction into the cost (4.15). Hence

$$
\begin{aligned}
J = \ & [\underset{\rightarrow k+1}{\mathbf{r}} - L\hat{y}_k - H\Delta \underset{\rightarrow k}{\mathbf{u}} - M\Delta \underset{\leftarrow k-1}{\mathbf{u}}]^T [\underset{\rightarrow k+1}{\mathbf{r}} - L\hat{y}_k - H\Delta \underset{\rightarrow k}{\mathbf{u}} - M\Delta \underset{\leftarrow k-1}{\mathbf{u}}] \\
& + \lambda \Delta \underset{\rightarrow k}{\mathbf{u}}^T \Delta \underset{\rightarrow k}{\mathbf{u}}
\end{aligned}
$$
$$(4.71)$$

Step 3. Minimise J w.r.t. $\Delta \underset{\rightarrow k}{\mathbf{u}}$ to give

$$\Delta \underset{\rightarrow k}{\mathbf{u}} = [H^T H + \lambda I]^{-1} H^T [\underset{\rightarrow k+1}{\mathbf{r}} - L\hat{y}_k - M\Delta \underset{\leftarrow k-1}{\mathbf{u}}] \qquad (4.72)$$

Step 4. Extract the first element $\Delta \mathbf{u}_k$

$$
\begin{aligned}
\Delta \mathbf{u}_k &= \mathbf{E}_1^T [H^T H + \lambda I]^{-1} H^T [\underset{\rightarrow k+1}{\mathbf{r}} - L\hat{y}_k - M\Delta \underset{\leftarrow k-1}{\mathbf{u}}] \\
&= P_r \underset{\rightarrow k+1}{\mathbf{r}} - \check{D}_k \Delta \underset{\leftarrow k-1}{\mathbf{u}} - N_k \hat{y}_k
\end{aligned}
$$
$$(4.73)$$

where $P_r = \mathbf{E}_1^T [H^T H + \lambda I]^{-1} H^T$, $\check{D}_k = P_r M$, $N_k = P_r L$. One notes that this control law has the same form as (4.26) with the only difference being the orders of $N_k, \check{D}_k$. That is less emphasis is placed on measured outputs (N_k has just one term) and more emphasis is placed on past inputs ($\check{D}_k$ has the same number of terms as the FIR model).

Summary: GPC for FIR models takes the same form as GPC for MFD models except that the compensators have different orders as the predictions are based primarily on past input information as opposed to past output measurements,

The control law of (4.73) is more commonly known as DMC [28]. The only significant difference with GPC [25] is the prediction model used.

STUDENT PROBLEMS

For a number of examples of your choice, compare the closed-loop behaviour that arises from using control laws (4.26, 4.73).

NOTE: If the reader would like to avoid writing separate code for DMC, the basic MFD code from the earlier sections will be *almost equivalent* to DMC if $N(z)$ is replaced by a truncated impulse response and $D(z) = I$.

HINT: The only difference between control laws (4.26, 4.73) should be in the closed-loop sensitivity to uncertainty (see previous section).

4.7   Formulation of GPC with an independent prediction model

The chapter on prediction emphasised that the use of an independent model (IM) formulation (Section 2.7) is a *choice*, although the pros and cons of such a choice are not well understood in general. An independent model can take any form, for example transfer function, state space, step response and so on. The main point is how unbiased predictions are formed (Section 2.7.2).

The formulation of GPC as given in Section 4.5.3 emphasised that this reduces to minimising performance index a J, wrt future control increments $\Delta \underset{\rightarrow k}{\mathbf{u}}$. A key part of this optimisation is the substitution into J of predictions for the future outputs and these predictions can be derived using whichever model format the user desires.

This section will concisely illustrate the use of an independent prediction model of a transfer function type. A briefer subsection will clarify the use of an independent state space model after Section 4.8.

The reader is reminded that using an independent model it is implicit that one has a block diagram structure somewhat like Figure 2.2, so the model is simulated in parallel with the process using the same inputs. All the model states/outputs are assumed to be known precisely as this is a *computer simulation*.

4.7.1   GPC algorithm with independent transfer function model

This is analogous to the earlier derivations so simply stated.

Step 1. Take the prediction equations of (2.76, 2.77)

$$\underset{\rightarrow}{\mathbf{y}}_p(k+1) = H\underset{\rightarrow k}{\Delta \mathbf{u}} + P\underset{\leftarrow k-1}{\Delta \mathbf{u}} + Q\underset{\leftarrow}{\mathbf{y}}_m(k) + L\mathbf{d}_k; \quad \mathbf{d}_k = \mathbf{y}_p(k) - \mathbf{y}_m(k)$$

$$(4.74)$$

where H is the same as used for any GPC approach.

Step 2. Substitute this prediction into the cost (4.15). Hence

$$J = \| \underset{\rightarrow k+1}{\mathbf{r}} - H\Delta\underset{\rightarrow k}{\mathbf{u}} - P\Delta\underset{\leftarrow k-1}{\mathbf{u}} - Q\underset{\leftarrow}{\mathbf{y}}_m(k) - L\mathbf{d}_k] \|_2^2 + \lambda \|\Delta\underset{\rightarrow k}{\mathbf{u}}\|_2^2 \quad (4.75)$$

Step 3. Minimise J w.r.t. $\Delta\underset{\rightarrow k}{\mathbf{u}}$ to give

$$\Delta\underset{\rightarrow k}{\mathbf{u}} = [H^T H + \lambda I]^{-1} H^T [\underset{\rightarrow k+1}{\mathbf{r}} - P\Delta\underset{\leftarrow k-1}{\mathbf{u}} - Q\underset{\leftarrow}{\mathbf{y}}_m(k) - L\mathbf{d}_k] \quad (4.76)$$

Step 4. Extract the first block element $\Delta\mathbf{u}_k$

$$\begin{aligned} \Delta\mathbf{u}_k &= \mathbf{E}_1^T [H^T H + \lambda I]^{-1} H^T [\underset{\rightarrow k+1}{\mathbf{r}} - P\Delta\underset{\leftarrow k-1}{\mathbf{u}} - Q\underset{\leftarrow}{\mathbf{y}}_m(k) - L\mathbf{d}_k] \\ &= P_r\underset{\rightarrow k+1}{\mathbf{r}} - \check{D}_k\Delta\underset{\leftarrow k-1}{\mathbf{u}} - N_k\underset{\leftarrow}{\mathbf{y}}_m(k) - M_k\mathbf{d}(k) \end{aligned} \quad (4.77)$$

where $P_r = \mathbf{E}_1^T [H^T H + \lambda I]^{-1} H^T$, $\check{D}_k = P_r P$, $N_k = P_r Q$, $M_k = P_r L$.

Step 5. Expanding out the offset term $\mathbf{d}_k$, the control law reduces to:

$$\Delta\mathbf{u}_k = P_r\underset{\rightarrow}{\mathbf{r}} - \check{D}_k\Delta\underset{\leftarrow}{\mathbf{u}} - N_k\underset{\leftarrow}{\mathbf{y}}_m(k) + M_k\mathbf{y}_m(k) - M_k\mathbf{y}_p(k) \quad (4.78)$$

Summary: The GPC control law (4.78) based on an independent transfer function model prediction as in Figure 2.2, unsurprisingly is based on two output signals: (i) the output $\mathbf{y}_m$ from the independent model and (ii) the process measurement $\mathbf{y}_p$.

ILLUSTRATION: Differences between GPC based on an independent model prediction and a CARIMA model prediction

Consider the SISO model with the parameters

$$a(z) = 1 - 1.2z^{-1} + 0.32z^{-2}$$

$$b(z) = z^{-1} - 0.3z^{-2}$$

and choose horizons of $n_y = 15$, $n_u = 3$ and a weight $\lambda = 1$. The corresponding closed-loop step responses for the alternative GPC formulations of (4.26, 4.78) for a change in target at the 5th sample and a change in disturbance at the 15th and measurement noise from the 25th sample are shown here.

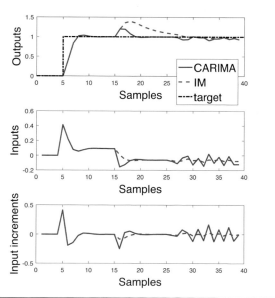

It is clear that the performance is equivalent when there is no uncertainty, but differs in the presence of disturbances and measurement noise.
The corresponding code can be found in **gpcsisoexample1_im.m**.

ILLUSTRATION: Differences between GPC based on an independent model prediction and a CARIMA model prediction

Consider the MIMO model used in **gpcmimoexample1_im.m**.
Choose horizons of $n_y = 9, n_u = 4$ and a weight $\lambda = 1$. The corresponding closed-loop step responses for the two GPC formulations of (4.26,4.78) for a change in target at the 10th sample and a change in disturbance at the 25th and measurement noise from the 35th sample are shown here.

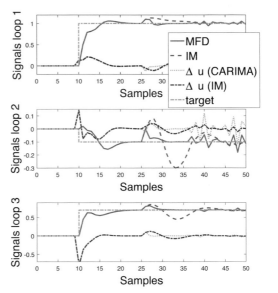

It is clear that the performance is equivalent when there is no uncertainty, but differs in the presence of disturbances and measurement noise.

STUDENT PROBLEMS

For a number of examples of your choice, compare the closed-loop behaviour that arises from using control laws (4.26,4.73,4.78) and demonstrate through simulation that in the absence of uncertainty they are all equivalent, but in the presence of uncertainty, different behaviour results.

You can use the matlab files such as **gpcmimoexample1_im.m** and files from earlier sections as templates to enter your own data.

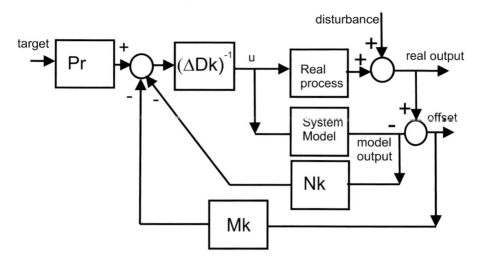

FIGURE 4.4
The effective block diagram structure for independent model GPC with a transfer
function model.

4.7.2 Closed-loop poles in the IM case with an MFD model

This section is largely for completeness as a transfer function formulation is useful
for those who wish to do some sensitivity analysis. Using the same arguments as with
the section on the T-filter, it is easy to argue that nominal performance should not be
affected by the choice of prediction model.

The corresponding block diagram for control law (4.78) takes the form in Figure
4.4. Using equivalent techniques to Equations (4.28, 4.29), express (4.78) in equivalent z-transform format. Hence:

$$D_k(z)\Delta \mathbf{u}(z) = P_r \underset{\rightarrow}{\mathbf{r}}(z) - \hat{N}_k(z)\mathbf{y}_m(z) - M_k(z)\mathbf{y}_p(z); \quad \hat{N}_k(z) = N_k(z) - M_k(z) \quad (4.79)$$

It is implicit in this control law that the IM model update seen in Figure 2.2 is actually
part of the control law calculation, so one must include the following IM equations
in the compensator

$$D(z)\mathbf{y}_m = N(z)\mathbf{u}(z); \quad \mathbf{u}_k = \mathbf{u}_{k-1} + \Delta \mathbf{u}_k \quad (4.80)$$

Summary: For MFD models and an IM prediction, the GPC compensator
takes the form of the following three update equations:

$$
\begin{aligned}
D_k(z)\Delta \mathbf{u}(z) &= P_r \underset{\rightarrow}{\mathbf{r}}(z) - \hat{N}_k(z)\underset{\leftarrow}{\mathbf{y}_m}(z) - M_k \mathbf{y}_p(z) \\
D(z)\mathbf{y}_m &= N(z)\mathbf{u}(z) \\
\mathbf{u}_k &= \mathbf{u}_{k-1} + \Delta \mathbf{u}_k
\end{aligned}
\qquad (4.81)
$$

Equivalently:

$$D_k(z)\Delta\mathbf{u}(z) = P_r\mathbf{r}(z) - \hat{N}_k(z)[D^{-1}N]\mathbf{u}(z) - M_k(z)\mathbf{y}_p(z) \qquad (4.82)$$

One can compute the *nominal* closed-loop poles by combining the compensator Equations (4.81) with the process $N(z)\mathbf{u}(z) = D(z)\mathbf{y}_p(z)$ (nominal case). Collect common terms in (4.82) and hence:

$$D_i(z)\mathbf{u} = P_r\mathbf{r} - M_k\mathbf{y}_p; \quad D_i(z) = [D_k(z)\Delta + \hat{N}_k(z)D^{-1}(z)N(z)] \qquad (4.83)$$

Now compensator (4.83) is in a similar form to the control laws for GPC and GPCT (e.g., (4.31)) and thus we can use analogies to derive the closed-loop poles. The closed-loop poles for model (4.35) in conjunction with compensator (4.83) can be computed from:

$$\begin{aligned} P_c &= D_i\hat{D} + M_k\hat{N} &= [D_k\Delta + \hat{N}_kD^{-1}N]\hat{D} + M_k\hat{N} \\ &= D_k\Delta\hat{D} + \hat{N}_k\hat{N} + M_k\hat{N} \end{aligned} \qquad (4.84)$$

This can be simplified to:

$$P_c = D_k\Delta\hat{D} + \underbrace{[\hat{N}_k + M_k]}_{N_{im}}\hat{N} \qquad (4.85)$$

which is unsurprisingly identical in form to (4.39). [Readers might like to note that the formulation of $\hat{N}_k(z)$ from (4.78) in essence implies it comprises contributions from N_k and M_k and thus N_{im} effectively removes the M_k term from the computation of the implied closed-loop poles so that (4.85) reduces to the same form as (4.33); that is $N_{im}(z) \equiv N_k(z)$.]

STUDENT PROBLEMS

1. Derive the result illustrated in Eqn.(4.79).

2. For a number of examples of your choice, compute the closed-loop poles using the control laws of (4.26) and (4.81) and demonstrate that, for the nominal case, the poles are the same.

3. Investigate the impact of parameter uncertainty.

Summary: The nominal closed-loop poles for GPC compensation based on an IM model are the same as with a more conventional use of the MFD model and the implied compensator can be reduced to a similar form, as given in (4.81).

4.8 GPC with a state space model

This section gives two alternative means of setting up a state space based predictive control law based on either state augmentation and performance index (4.12) or no state augmentation and performance index (4.13). It is assumed that having gained some expertise, readers will be able to modify what is included in the performance index, and thus modify what is presented here, to meet their own aims.

4.8.1 Simple state augmentation

In the transfer function case use is made of an incremental, or CARIMA, model and this automatically is written in terms of input increments which can be substituted into a cost function of the type (4.12). In the state space case it is more usual to use state augmentation of the model, in essence making $\mathbf{u}_k$ an additional internal state and $\Delta\mathbf{u}_k$ the input d.o.f. This was discussed in Section 2.4.6 where one made a simple change. The prediction model is:

$$\underbrace{\begin{bmatrix} \mathbf{x}_{k+1} \\ \mathbf{u}_k \end{bmatrix}}_{\mathbf{z}_{k+1}} = \underbrace{\begin{bmatrix} A & B \\ 0 & I \end{bmatrix}}_{A_a} \begin{bmatrix} \mathbf{x}_k \\ \mathbf{u}_{k-1} \end{bmatrix} + \underbrace{\begin{bmatrix} B \\ I \end{bmatrix}}_{B_a} \Delta\mathbf{u}_k \qquad (4.86)$$

$$\mathbf{y}_m(k+1) = \underbrace{\begin{bmatrix} C & 0 \end{bmatrix}}_{C_a} \begin{bmatrix} \mathbf{x}_k \\ \mathbf{u}_{k-1} \end{bmatrix} + \mathbf{d}_k; \quad \mathbf{d}_k = \mathbf{y}_p(k) - C_a \mathbf{z}_k$$

To ensure no bias in steady-state predictions, this model should satisfy prediction consistency conditions, that is

$$\left.\begin{array}{c} \mathbf{x}_k = \mathbf{x}_{ss} \\ \mathbf{u}_k = \mathbf{u}_{ss} \end{array}\right\} \forall k \geq 0 \quad \Rightarrow \quad \left\{\begin{array}{c} \mathbf{y} = \mathbf{r} \\ \Delta\mathbf{u} = 0 \end{array}\right\} \qquad (4.87)$$

As discussed in Chapter 2, the disturbance estimate $\mathbf{d}_k = \mathbf{y}_p(k) - \mathbf{y}_m(k)$ is used to ensure this consistency and thus could enable appropriate computations for $\mathbf{x}_{ss}, \mathbf{u}_{ss}$, although in this case, those values are not used explicitly as the approach is more analogous to those used for independent model GPC of Section 4.7.

4.8.1.1   Computing the predictive control law with an augmented state space model

The GPC algorithm can now be implemented in a straightforward fashion using similar steps to those for the MFD model (Section 4.5).

> **Step 1.** Find prediction equations for the augmented model (4.86) using (2.14) and substitute these into J of Eqn.(4.12) to give
>
> $$J = \| \underset{\rightarrow k+1}{\mathbf{r}} - H\underset{\rightarrow k}{\Delta \mathbf{u}} - P\mathbf{z}_k - L\mathbf{d}_k \|_2^2 + \lambda \| \underset{\rightarrow k}{\Delta \mathbf{u}} \|_2^2 \qquad (4.88)$$
>
> where L has the usual definition (see Table C.2).
>
> **Step 2.** Perform the optimisation of minimising J w.r.t. $\underset{\rightarrow k}{\Delta \mathbf{u}}$, hence
>
> $$\frac{dJ}{d\underset{\rightarrow k}{\Delta \mathbf{u}}} = 0 \;\Rightarrow\; (H^T H + \lambda I)\underset{\rightarrow k}{\Delta \mathbf{u}} = [H^T \underset{\rightarrow k+1}{\mathbf{r}} - H^T P\mathbf{z}_k - H^T L\mathbf{d}_k] \quad (4.89)$$
>
> **Step 3.** Solve for the first block element of $\underset{\rightarrow}{\Delta \mathbf{u}}$
>
> $$\begin{aligned} \Delta \mathbf{u}_k &= E_1^T (H^T H + \lambda I)^{-1} H^T [\underset{\rightarrow k+1}{\mathbf{r}} - [P,L] \begin{bmatrix} \mathbf{z}_k \\ \mathbf{d}_k \end{bmatrix}] \\ &= P_r \underset{\rightarrow k+1}{\mathbf{r}} - \hat{K} \begin{bmatrix} \mathbf{z}_k \\ \mathbf{d}_k \end{bmatrix} \end{aligned} \qquad (4.90)$$
>
> where $P_r = E_1^T (H^T H + \lambda I)^{-1} H^T$, $\hat{K} = P_r[P,L]$, $E_1^T = [I,0,0,\cdots]$.

Hence predictive control reduces to a state feedback with a feedforward term; the state feedback includes a term which acts on the disturbance estimate $\mathbf{d}_k$ used to ensure offset-free tracking.

Remark 4.9 *It is easy to see that P_r from step 3 above is identical to that of (4.27).*

> **Summary:** Predictive control based on a state space model can take the form of a state feedback (4.90) of the augmented state plus some feedforward. The augmented state includes the disturbance estimate and thus this approach has some analogies to independent model GPC.

ILLUSTRATION: SISO state space GPC using an augmented model

Consider the SISO model with the following parameters

$$A = \begin{bmatrix} 1.4 & -0.105 & -0.108 \\ 2 & 0 & 0 \\ 0 & 1 & 0 \end{bmatrix} ; \; B = \begin{bmatrix} 0.002 \\ 0 \\ 0 \end{bmatrix} ; \; C = [0.05 \; 0.075 \; 0.05]$$

$$(4.91)$$

and choose horizons of $n_y = 15, n_u = 3$ and a weight $\lambda = 10$. The corresponding closed-loop step response for a change in target at the 10th sample and a change in disturbance at the 40th sample are shown here. The corresponding code can be found in **ssgpcmimoexample1.m**.

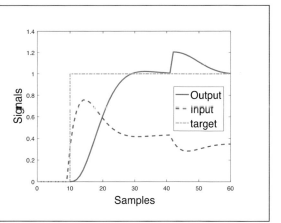

ILLUSTRATION: GPC of a power generation model

Consider the MIMO model with the following parameters

$$A = \begin{bmatrix} 0.91 & 0 & 0.0405 \\ 0.17 & 0.14 & 0.006 \\ 0 & 0 & 0.135 \end{bmatrix}; \quad B = \begin{bmatrix} 0.054 & -0.076 \\ 0.0053 & 0.15 \\ 0.86 & 0 \end{bmatrix}; \quad (4.92)$$

$$C = \begin{bmatrix} 1.8 & 13.2 & 0 \\ 0.82 & 070 \end{bmatrix}$$

and choose horizons of $n_y = 25, n_u = 10$ and a weight $\lambda = 1$. The corresponding closed-loop step responses for asynchronous changes in target and a change in disturbance at the 40th sample are shown here.
The corresponding code can be found in **ssgpcmimoexample2.m**.

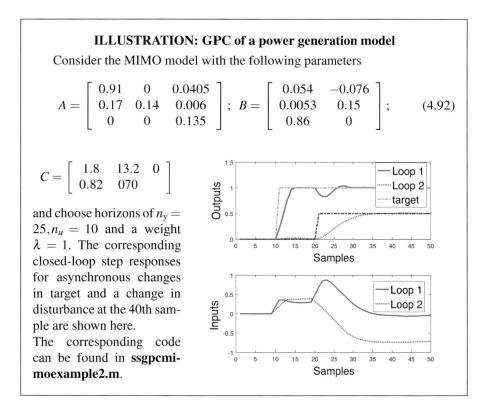

STUDENT PROBLEM

For a number of systems such as those covered in the earlier illustration boxes and the case studies of Appendix A, investigate the efficacy of control law (4.90). You can use the files **ssgpcmimoexample1.m, ssgpcmimoexample2.m** as templates for adding different examples and scenarios.

Also compare the behaviour with that resulting from a transfer function based control law (4.26) on an equivalent model.

4.8.1.2 Closed-loop equations and integral action

The nominal closed-loop can be derived by combining (4.86) and (4.90). Hence

$$\mathbf{z}_{k+1} = A_a\mathbf{z}_k - B_a\hat{K}\begin{bmatrix}\mathbf{z}_k \\ \mathbf{d}_k\end{bmatrix} + B_a P_r \underset{\rightarrow k+1}{\mathbf{r}}; \quad \mathbf{y}_p(k) = C_a\mathbf{z}_k + \mathbf{d}_k \tag{4.93}$$

Hence the closed-loop poles can be derived from the eigenvalues of $[A_a - B_a\hat{K}]$.

The illustrations above validate the assumption about the automatic inclusion of integral action, but it is worth revisiting why this is automatic in this case. offset-free tracking in the steady-state is assured by the consistency of (4.86) in the state estimator (model) and hence also carries over to uncertain plant. Critically, the estimator can be based on the original model (2.4) with state $\mathbf{x}$, as the part of the augmented state; $\mathbf{u}_k$ is known exactly so does not need to be estimated.

Summary:

1. For a state space model, a possible GPC control law takes the form

$$\Delta\mathbf{u}_k = P_r\underset{\rightarrow k+1}{\mathbf{r}} - \hat{K}\begin{bmatrix}\mathbf{z}_k \\ \mathbf{d}_k\end{bmatrix} \tag{4.94}$$

where $\mathbf{z}_k = [\mathbf{x}_k^T, \mathbf{u}_{k-1}^T]^T$ is an augmented state and $\mathbf{d}_k$ is the disturbance estimate.

2. Integral action and offset-free control are included by having an internal model of the disturbance (which is typically estimated as $\mathbf{d}_k = \mathbf{y}_p(k) - \mathbf{y}_m(k)$) within the prediction model of (4.86) and ensuring consistency of (4.87) so that the minimum achievable cost is $J = 0$.

4.8.2 GPC using state space models with deviation variables

Readers will recall that the chapter on prediction presented two alternatives for giving unbiased prediction with a state space model, that is: (i) to augment the model and base predictions on input increments, the state and a disturbance estimate (Section

2.4.6), or (ii) to express predictions in terms of deviations about an estimated steady-state (Section 2.4.5). Here we consider the use of the latter in a GPC like control law. Hence the predictions take the form:

$$\begin{array}{rcl}
\underset{\rightarrow k+1}{\tilde{\mathbf{x}}} & = & P_x \tilde{\mathbf{x}}_k + H_x \underset{\rightarrow k}{\tilde{\mathbf{u}}} \\
\underset{\rightarrow k+1}{\tilde{\mathbf{y}}} & = & P \tilde{\mathbf{x}}_k + H \underset{\rightarrow k}{\tilde{\mathbf{u}}}
\end{array} \quad ; \quad \begin{array}{l} \tilde{\mathbf{x}}_k = \mathbf{x}_k - \mathbf{x}_{ss} \\ \tilde{\mathbf{u}}_k = \mathbf{u}_k - \mathbf{u}_{ss} \end{array} \tag{4.95}$$

It is implicit in this prediction that one has estimates of the steady-state inputs and states which will give no offset in the steady-state. Such estimates [107] were presented in Section 2.4.5, and assume a constant target, that is:

$$\left\{ \begin{array}{l} \mathbf{r}_{k+1} = C\mathbf{x}_{ss} + \mathbf{d}_k \\ \mathbf{x}_{ss} = A\mathbf{x}_{ss} + B\mathbf{u}_{ss} \end{array} \right\} \quad \mathbf{d}_k = \mathbf{y}_p(k) - \mathbf{y}_m(k) \tag{4.96}$$

Readers will also recall from Section 4.4.3 that alternative choices of cost function are possible. In the event that predictions are based on deviation variables, it would seem reasonable to use a cost function such as (4.13) which is also based on deviation variables.

$$\min_{\underset{\rightarrow k}{\mathbf{u}}} \ J = [\underset{\rightarrow k+1}{\mathbf{x}} - L\mathbf{x}_{ss}]^T Q[\underset{\rightarrow k+1}{\mathbf{x}} - L\mathbf{x}_{ss}] + [\underset{\rightarrow k}{\mathbf{u}} - L\mathbf{u}_{ss}]^T R[\underset{\rightarrow k}{\mathbf{u}} - L\mathbf{u}_{ss}] \tag{4.97}$$

where L has the usual flexible definition (see Table C.2). Note, a more compact form of this optimisation is given as:

$$\min_{\underset{\rightarrow k}{\tilde{\mathbf{u}}}} \ J = [\underset{\rightarrow k+1}{\tilde{\mathbf{x}}}]^T Q[\underset{\rightarrow k+1}{\tilde{\mathbf{x}}}] + [\underset{\rightarrow k}{\tilde{\mathbf{u}}}]^T R[\underset{\rightarrow k}{\tilde{\mathbf{u}}}] \tag{4.98}$$

4.8.2.1 GPC algorithm based on deviation variables

A GPC algorithm can now be implemented in a straightforward fashion using similar steps to those in Section 4.8.1.1.

Step 1. Find prediction equations (4.95) for the state space model using deviation variables and substitute these into J of Eqn.(4.98) to give:

$$\min_{\underset{\rightarrow k}{\tilde{\mathbf{u}}}} \ J = [P_x \tilde{\mathbf{x}}_k + H_x \underset{\rightarrow k}{\tilde{\mathbf{u}}}]^T Q[P_x \tilde{\mathbf{x}}_k + H_x \underset{\rightarrow k}{\tilde{\mathbf{u}}}] + [\underset{\rightarrow k}{\tilde{\mathbf{u}}}]^T R[\underset{\rightarrow k}{\tilde{\mathbf{u}}}] \tag{4.99}$$

Step 2. Perform the optimisation of minimising J w.r.t. $\underset{\rightarrow k}{\tilde{\mathbf{u}}}$, hence:

$$\frac{dJ}{d\underset{\rightarrow}{\tilde{\mathbf{u}}}} = 0 \ \Rightarrow \ (H_x^T Q H_x + R)\underset{\rightarrow k}{\tilde{\mathbf{u}}} = [H_x^T Q P_x]\tilde{\mathbf{x}}_k \tag{4.100}$$

Step 3. Solve for the first block element of $\underset{\rightarrow k}{\tilde{\mathbf{u}}}$

$$\begin{array}{rcl}
\underset{\rightarrow k}{\tilde{\mathbf{u}}} & = & (H_x^T Q H_x + R)^{-1} H_x^T Q P_x \tilde{\mathbf{x}}_k \\
\mathbf{u}_k & = & -K(\mathbf{x} - \mathbf{x}_{ss}) + \mathbf{u}_{ss}
\end{array} \tag{4.101}$$

where $K = E_1^T (H_x^T Q H_x + R)^{-1} H_x^T P_x$.

Hence predictive control reduces to a state feedback with a feed forward term which is based on estimates of the expected steady-state values of the input and state to achieve the desired target.

Core differences between GPC control laws based on an augmented model and deviation variables

State feedback (4.101) differs, possibly to a significant extent, from (4.90) for two major reasons.

Predictions: Firstly, because it uses prediction equations based on (2.4) instead of predictions based on (4.86). That is, there is no state augmentation. Moreover the disturbance estimate is incorporated via $\mathbf{x}_{ss}, \mathbf{u}_{ss}$. Nevertheless, the key difference is a core prediction assumption about the future input trajectory:

(i) State feedback law (4.101) is based on the embedded assumption within the predictions that:

$$\mathbf{u}_{k+i|k} = \mathbf{u}_{ss}, \quad i \geq n_u$$

(ii) Conversely, state feedback law (4.90) uses a prediction assumption that:

$$\Delta\mathbf{u}_{k+i|k} = 0, \quad i \geq n_u$$

That is, one fixes the terminal value of the predicted input at $\mathbf{u}_{ss}$, whereas the other implicitly allows any terminal value and thus is more flexible which may be an advantage during constraint handling.

Performance index: it will also be apparent to the reader that the control laws (4.101, 4.90) were based on the optimisation of different performance indices, that is (4.12, 4.13). As different objectives are set, it will be unsurprising if different control laws result.

Remark 4.10 *Optimisation (4.99) emphasises the distance of the input from its expected steady-state and thus does not directly put emphasis on input increments/activity. Moreover, the pressure for the input to go directly to its steady-state puts an implied pressure on this control law to behave in an open-loop fashion if R is large. Conversely, optimisation (4.88) focuses on input changes, and thus the weighting parameter λ allows a trade-off between the speed of input changes and the expected tracking errors. In simplistic terms, the choice of (4.99) seems to be popular due to its strong links with LQR but perversely, this format may be less flexible in general.*

STUDENT PROBLEM

For a number of systems such as those covered in the earlier illustration boxes investigate the efficacy of control law (4.102) and in particular, determine the extent to which R is a useful tuning parameter with low n_u. You can use the file **ssgpcslsoexample2.m** as a template for adding different examples and scenarios.

Summary: The predictive control law using cost function (4.13) and the input constraint that $\mathbf{u}_{k+i|k} = \mathbf{u}_{ss}, i \geq n_u$ can be implemented as

$$\mathbf{u}_k - \mathbf{u}_{ss} = -E_1^T[H_x^T Q H_x + R]^{-1}H_x^T Q P_x(\mathbf{x} - \mathbf{x}_{ss}) \qquad (4.102)$$

This control law gives markedly different behaviour to (4.90) because it uses a different terminal constraint on the inputs and a different performance index.

ILLUSTRATION: Comparing GPC using deviations variables or an augmented model

Consider the SISO model with the following parameters

$$A = \begin{bmatrix} 1.4 & -0.105 & -0.108 \\ 2 & 0 & 0 \\ 0 & 1 & 0 \end{bmatrix}; \; B = \begin{bmatrix} 0.002 \\ 0 \\ 0 \end{bmatrix}; \; C = [0.05\ 0.075\ 0.05]$$

$$(4.103)$$

and choose horizons of $n_y = 15, n_u = 3$ and a weight $\lambda = 10$. The corresponding closed-loop step response for a change in target at the 10th sample and a change in disturbance at the 40th sample are shown here and overlaid for control laws (4.90) [labelled output 1] and (4.102) [labelled output 2]. The corresponding code can be found in **ssgpcsisoexample2.m**.

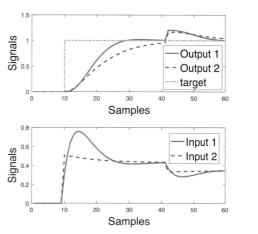

It is clear that different closed-loop behaviours arise from using the different laws, but this should not be surprising as different cost functions have been used which implies different definitions of optimum. The user is encouraged to think carefully about their choice of cost function (definition of optimum) before deciding which to use.

4.8.2.2   Using an observer to estimate steady-state values for the state and input

This section and indeed the chapter on prediction have tacitly assumed that state values, either of the process or an independent model, are available for feedback purposes. In practice, it is common to use an observer to estimate the states from a smaller number of output measurements. Consequently, this section gives a brief reminder of how to ensure unbiased estimates of the steady-state values by using an appropriate disturbance model in the observer. If one can get an unbiased estimate of the disturbance, then the predictions will be unbiased, in the steady-state.

1. Add a disturbance to the state space model as follows:

$$\mathbf{x}_{k+1} = A\mathbf{x}_k + B\mathbf{u}_k; \quad \mathbf{d}_{k+1} = \mathbf{d}_k; \quad \mathbf{y}_m(k) = C\mathbf{x}_k; \quad (4.104)$$

where $E[\mathbf{y}_p(k)] = \mathbf{y}_m(k) + \mathbf{d}_k$ and thus $\mathbf{d}_k$ is an unknown signal representing disturbances and model mismatch.

2. Build an observer to estimate $\mathbf{x}_k, \mathbf{d}_k$. Let the observer take the form

$$\mathbf{z}_{k+1} = A_o\mathbf{z}_k + B_o\mathbf{u}_k - F(C_o\mathbf{z}_k - \mathbf{y}_p(k)) \quad (4.105)$$

where

$$\mathbf{z}_{k+1} = \begin{bmatrix} \mathbf{x}_{k+1} \\ \mathbf{d}_{k+1} \end{bmatrix}; \quad A_o = \begin{bmatrix} A & 0 \\ 0 & I \end{bmatrix}; \quad B_o = \begin{bmatrix} B \\ 0 \end{bmatrix}; \quad C_o = \begin{bmatrix} C & I \end{bmatrix}$$
$$(4.106)$$

Clearly $\mathbf{z}$ contains the estimates of the state $\mathbf{x}$ and disturbance $\mathbf{d}$. F is the observer gain to be designed either by Kalman methods or otherwise.

3. Given that $\mathbf{d}$, $\mathbf{r}$ are known, estimate the required steady-state values of $\mathbf{x}, \mathbf{u}$ to get desired output, that is $\mathbf{y}_{ss} = \mathbf{r}$:

$$\mathbf{y}_{ss} = \mathbf{r} = C\mathbf{x}_{ss} + \mathbf{d}; \quad \mathbf{x}_{ss} = A\mathbf{x}_{ss} + B\mathbf{u}_{ss} \quad (4.107)$$

4. One can compute $\mathbf{x}_{ss}$, $\mathbf{u}_{ss}$ by solving (4.107) as simultaneous equations.

5. The control law is implemented as

$$\mathbf{u}_k = -K(\mathbf{x}_k - \mathbf{x}_{ss}) + \mathbf{u}_{ss} \quad (4.108)$$

Theorem 4.2 *If one drives state estimates $\mathbf{z}$ to $\mathbf{x}_{ss}$, $\mathbf{u}_{ss}$, then the output $\mathbf{y}$ must converge to $\mathbf{r}$, even if the model is uncertain, that is $A \neq A_o, B \neq B_o$.*

Proof: The observer update equation is given from

$$\mathbf{z}_{k+1} = A_o\mathbf{z}_k + B_o\mathbf{u}_k - F(C_o\mathbf{z}_k - \mathbf{y}_p(k)) \qquad (4.109)$$

This uses $\mathbf{y}_p(k)$ from the actual plant. Assume the process output is steady, then the state estimate $\mathbf{z}$ will settle if and only if

$$C_o\mathbf{z}_k = \mathbf{y}_p(k); \quad \mathbf{u}_k = \mathbf{u}_{k-1} = \cdots \qquad (4.110)$$

This in turn implies that the observer output must equal to the process output. However, steady-state also implies (from 4.107) that $\mathbf{x} = \mathbf{x}_{ss}$, $\mathbf{u} = \mathbf{u}_{ss}$ and therefore the values $\mathbf{x}_{ss}$, $\mathbf{u}_{ss}$ are consistent with an observer output equal to the set point $\mathbf{r}$. This in turn must imply that the process output is $\mathbf{r}$. $\qquad\qquad\square$

Remark 4.11 *One can also set up the disturbance model on the states or inputs if that is more appropriate see [107].*

> **Summary:** Consistent steady-state values $\mathbf{x}_{ss}$, $\mathbf{u}_{ss}$ can be computed from (4.107). Use of these in conjunction with observer (4.105) and control law (4.108) gives offset-free tracking in the steady-state.

4.8.3 Independent model GPC using a state space model

Independent model GPC was discussed more fully in Section 4.7 but with a focus on the use of transfer function or MFD models. This section outlines the differences when using a state space model.

Conceptually, nothing changes in the implied block diagram and formulation of the control law, that is the control law has two components:

1. A standard feedback law which uses the states of the independent model.
2. A correction term for the offset between the independent model and the process output.

As it happens, the control law of for example (4.90) is already in the appropriate format as it is implicit that both a *state estimate* and a *disturbance estimate* will be used with a state space model. Whether the state estimate derives from an observer or an independent model is to some extent irrelevant in the formulation of the GPC law, although of course this will impact on sensitivity.

4.9 Chapter summary and general comments on stability and tuning of GPC

In the unconstrained case, the closed-loop poles for *unconstrained* GPC can be computed for both the state space and transfer function prediction models (e.g., 4.39,

4.93). This is because the control law reduces to a fixed linear feedback law. However, there are few generic a priori stability results with GPC; that is to say, for an arbitrary set of n_y, n_u, λ the resulting control law may be destabilising or give poor performance.

The next chapter will discuss various means of avoiding this apparent weakness and in this chapter we have simply stated how the closed-loop poles can be computed. However, two observations are in order.

1. Assuming the same performance index, same degrees of freedom and the same information (say full states can be observed), then each algorithm should give identical closed-loop poles and identical tracking performance in the nominal case. This is obvious because if one minimises the same performance index with the same d.o.f., the optimum control increment must be the same. Assuming full information and no uncertainty, each model will give identical predictions and hence the performance indices will be equivalent.

2. In the presence of uncertainty, either parameter or disturbances or measurement noise, this uncertainty affects the predictions in different ways depending upon the model and therefore the choice of model has an effect on the loop sensitivity.

3. The closed-loop pole polynomial tends to have as many non-zero poles as there are open-loop process poles (including the integrator); this conclusion is obvious from (4.93) which shows that the number of poles is given by the dimension of the closed-loop state matrix $A - BK$. If this is not true, there is a good chance you have a bug in your code. You will note that the order of P_c arising from, for instance Eqn.(4.39) appears to be greater than this; in this case polynomial P_c must have have several zero coefficients.

STUDENT PROBLEMS

For a number of systems of your choice:

1. Compare the behaviour arising from using control laws (4.90, 4.102).

2. Compare the behaviour with GPC control laws (4.26) and (4.81) based on an equivalent MFD model.

3. For both the above, also investigate the sensitivity to output disturbances and measurement noise.

You can use the MATLAB code listed in Section 4.10 to help with this task.

Summary: When coding, you should check for errors by ensuring consistency of behaviour and pole positions in the nominal case, regardless of the model structure adopted.

4.10 Summary of MATLAB code supporting GPC simulation

This code is provided so that students can learn immediately by trial and error, entering their own parameters and thus get a feel for the impact of different choices.

Readers should note that this code is not *professional quality* and is deliberately simple, short and free of comprehensive *error catching* so that it is easy to understand and edit by users for their own scenarios. Please feel free to edit as you please.

4.10.1 MATLAB code to support GPC with a state space model and a performance index based on deviation variables

ssgpc_simulatedev_noconstraints.m	Simulates a GPC control law based on an state space model using deviation variables $x - x_{ss}, u - u_{ss}$ and performance index (4.13), as in Section 4.8.2.1.
ssgpc_predmatdev.m	Forms prediction matrices as given in (4.95).
ssgpc_costfunctiondev.m	Forms the cost function parameters as given in Section 4.8.2.1.
ssgpcsisoexample2.m	Compares the closed-loop responses of GPC based on Sections 4.8.1.1 and 4.8.2.1.

4.10.2   MATLAB code to support GPC with an MFD or CARIMA model

mpc_simulate_noconstraints.m	Simulates a GPC control law based on a MIMO CARIMA model. Includes input variables to allow for target changes, output disturbances and measurement noise.
mpc_simulate_tfilt_noconstraints.m	Simulates a GPC control law based on a MIMO CARIMA model with a T-filter. Includes input variables to allow for target changes, output disturbances and measurement noise.
mpc_predmat.m	Forms prediction matrices as given in (2.49).
mpc_predtfilt.m	Forms prediction matrices as given in (2.62).
mpc_law.m	Forms the parameters of the control law given in (4.26).
caha.m	Forms Hankel and Toeplitz matrices as presented in (2.43).
convmat.m	Does multiplication of matrix polynomials, assuming a specified syntax.
gpcsisoexample1.m	Template file showing how to call mpc_simulate_noconstraints.m and also to allow easy manipulation of all data such as model, tuning parameters, target signals and plotting.
gpcsisoexample1_tfilt.m	As above but including a T-filter.
gpcmimoexample1.m	As above but with a MIMO MFD model.
gpcmimoexample1_tfilt.m	As above but including a T-filter.

4.10.3 MATLAB code to support GPC using an independent model in MFD format.

imgpc_simulate_noconstraints.m	Simulates a GPC control law based on a MIMO with independent MFD model as in (4.78). Includes input variables to allow for target changes, output disturbances and measurement noise.
gpcsisoexample1_im.m	An example simulation of a GPC control law based on a SISO independent transfer function model as in Section 4.7.1.
gpcmimoexample1_im.m	An example simulation of an GPC control law based on an independent MIMO MFD as in Section 4.7.1.
imgpc_tf_predmat.m	Forms prediction matrices as given in (4.74) specifically for use with IM GPC.
imgpc_tf_law.m	Forms the parameters of the control law given in (4.78).

4.10.4 MATLAB code to support GPC with an augmented state space model.

ssgpc_simulateDu.m	Simulates an GPC control law based on an augmented state space model and performance index (4.12), as in Section 4.8.1.1.
ssgpc_predmatDu.m	Forms prediction matrices as given in (2.14) but using the augmented model (4.86).
ssgpc_costfunctionDu.m	Forms the cost function of (4.88).
ssgpc_constraintsDu.m	Subroutine used by ssgpc_simulateDu.m for constraint handling.
ssgpcmimoexample1.m	Demonstrates data entry and simulation for a SISO state space model using the algorithm of Section 4.8.1.1.
ssgpcmimoexample2.m	Demonstrates data entry and simulation for a MIMO state space model using the algorithm of Section 4.8.1.1.

5

Tuning GPC: good and bad choices of the horizons

CONTENTS

5.1　Introduction

The previous chapter did not engage with any discussion on how parameters n_y, n_u, λ should be chosen. This chapter gives a series of examples and demonstrates how GPC might be tuned quickly and effectively. The aim is to give the reader some insight into what effects the parameters n_y, n_u, λ have on the resulting behaviour and thus what constitute wise and unwise choices and moreover, why the simple guidance adopted in industry is often sufficient to give reasonable performance.

The chapter will use a number of case studies with different dynamics (well-damped, under-damped, non-minimum phase and open-loop unstable) to demonstrate how the recommended guidance to be proposed in Section 5.6 does not depend on system dynamics. SISO examples will be used for simplicity as the same insights automatically carry across to the MIMO case.

In particular however, the insight gained in this chapter will give a better understanding as to why there has been a move in the academic literature, which is more interested in guarantees and rigour, towards infinite horizons.

Any discussion of constraint handling is deliberately omitted from this chapter, as a logical premise for effective constraint handling is that the underlying optimisation is well-posed in the unconstrained case. Once students understand how to set up the optimisation safely, constraints will be introduced in Chapter 7.

Short cut: A summary of the observations and useful MATLAB® code are given in sections 5.9, 5.10 for the reader who wants simply to scan the chapter.

5.2　Guidance for the lecturer/reader

This chapter is comprised mainly of numerous illustrations which can be reproduced using the MATLAB code summarised in Section 5.10. The author would recommend that readers and introductory courses in predictive control engage with the majority of the chapter, with the exception of the sections focussing on recommendations for unstable systems and alternative input parameterisations.

Section 5.3 gives some motivating examples followed by Section 5.4 which introduces the concepts of *poor* and *well* posed optimisations. These are followed by a section with a large number of illustrations focussing on alternative strategies for choosing n_y, n_u; the insights from these are used to formulate a general tuning guidance in Section 5.6. Later sections look at special cases such as open-loop unstable processes.

Assessment could be focused on two aspects:

- **Exam type questions** would focus on a discussion of tuning guidance and using *common-sense* examples to demonstrate why this guidance makes good sense. [Numerical demonstrations are not generally possible in an examination scenario.]

- **Assignments** could encourage students to demonstrate, through several case studies, the efficacy of or need for some tuning guidance.

ILLUSTRATION 1: GPC behaviour strongly linked to n_y, n_u

A simple but unstable first-order model is used here:

$$[1 - 1.2z^{-1}]y(z) = [0.1z^{-1} + 0.03z^{-2}]u(z)$$

and the different choices illustrated are:

- Choice 1: $n_y = 3, n_u = 1, \lambda = 1$.

- Choice 2: $n_y = 6, n_u = 1, \lambda = 1$.

The corresponding closed-loop responses show that the behaviour is acceptable with choice 2 but not with choice 1. The corresponding code can be found in **poorbehaviourexample1.m**.

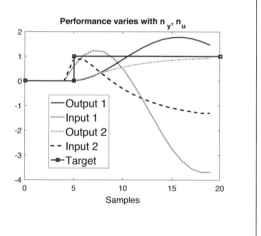

5.3 Poor choices lead to poor behaviour

The first thing to establish is a motivation for gaining a deeper understanding. One concern to the author is that many practitioners take an off-the-shelf MPC algorithm and deploy almost random choices for the parameters n_y, n_u and assume that good behaviour will result *just because* MPC makes sense. However, the few examples shown next will illustrate that in many cases extremely poor behaviour can result from using GPC. More generally the core principle is established that good closed-loop behaviour may happen by accident with unwise choices, but it is far more reliable to make good choices in the first place. This latter point is particularly important when constraints become active and one cannot fall back on *a posteriori* linear analysis to validate the behaviour.

ILLUSTRATION 2: GPC behaviour strongly linked to n_y, n_u

An under-damped model is used here:

$$[1 - 0.9z^{-1} + 0.6z^{-2}]y(z) = [z^{-1} + 0.4z^{-2}]u(z)$$

and the different choices illustrated are:

- Choice 1: $n_y = 6, n_u = 1, \lambda = 1$.

- Choice 2: $n_y = 8, n_u = 2, \lambda = 1$.

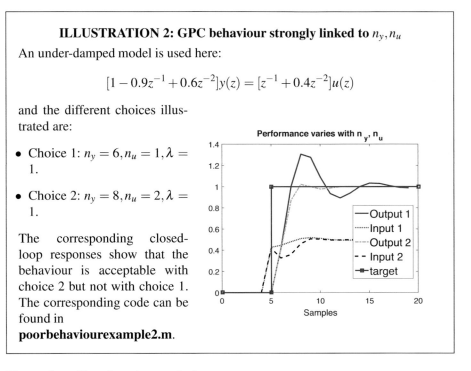

The corresponding closed-loop responses show that the behaviour is acceptable with choice 2 but not with choice 1. The corresponding code can be found in **poorbehaviourexample2.m**.

The reader will perhaps be wondering:

- Why does GPC work well for some choices and not well for others?

- Is there some underpinning explanation which would enable systematic design?

- Early authors tried to find mathematical relationships between stability and the horizons n_y, n_u, but that approach is largely considered invalid nowadays and so it is not pursued here.

It is noted that, intuitively, if GPC is poor in the unconstrained case, it will not work well in the constrained case. Moreover, if particular SISO attributes cause problems, the same attributes and more could cause problems in the MIMO case and thus for simplicity of illustration this chapter uses mainly SISO examples.

ILLUSTRATION 3: GPC behaviour strongly linked to n_y, n_u

A non-minimum phase model is used here:

$$[1 - 1.7z^{-1} + 0.72z^{-2}]y(z) = [z^{-1} - 2z^{-2}]u(z)$$

and the different choices illustrated are:

- Choice 1: $n_y = 3, n_u = 1, \lambda = 1$.

- Choice 2: $n_y = 6, n_u = 1, \lambda = 1$.

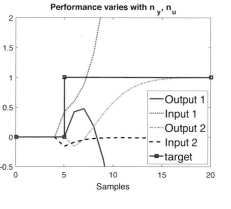

The corresponding closed-loop responses show that the behaviour is acceptable with choice 2 but unstable with choice 1. The corresponding code can be found in **poorbehaviourexample3.m**.

Summary: The obvious point here is that the choices for n_u, n_y, λ matter enormously. Poor choices can lead to poor or even unstable closed-loop behaviour. Hence there is a need for guidance as to what constitutes good choices.

5.4 Concepts of well-posed and ill-posed optimisations

Something that was not adequately considered in the early literature on MPC was whether the *optimisation problem* was well defined in the first place. Instead there was a rather naive assumption that because it used prediction and aimed to minimise squares of predicted future errors, which accords with human behaviour (see Chapter 1), then surely the answers must be sensible.

The weakness of the expectation is a lack of critical reflection on human behaviour:

- Humans do not use arbitrary output horizons.

- Humans do not use arbitrary input horizons.

- Largely through experience humans learn the horizons required for different contexts and are fastidious in keeping to these for a very good reason!

This section will use some simple examples to demonstrate how humans determine the horizons required to achieve adequate control and the consequence of not doing this well. These examples will give insight into the requirements for an automated algorithm, that is, GPC.

5.4.1   Examples

Before moving to recommendations, it is worth demonstrating the potentially detrimental affects of *limited predictions*.

1. The first illustration demonstrates the dangers of having too few degrees of freedom over which to optimise future predicted behaviour. [That is n_u is too small.]

2. The second illustration demonstrates the danger of not predicting far enough ahead. [That is n_y is too small.]

ILLUSTRATION: Navigation example

Imagine that you have to plan a route from **A** to **T** through a maze of houses as in the figure below. However, your decision making is constrained to routes that require at most one turn en route.

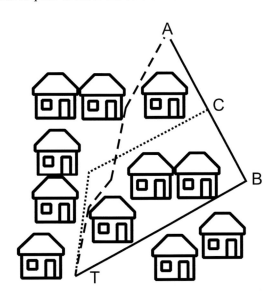

- Your initial choice would be to turn at **B** (solid line path) as this is the only route with just one turn.

- However, thereafter you can re-plan as long as any new plan still has only one future turn.

- When you get to **C** you notice that you can plan an alternative route with just one turn that is shorter. So you change your route and now follow the dotted line.

Of course all the time, there was an even better route (the dashed line), should you be allowed to plan routes with even more turns. Hence, the planning restriction to use just one turn in any planned path has the effect that a far longer path than necessary is taken.

A key point to note here is that the walker may **keep changing their mind** because as they move forward they in effect have access to more degrees of freedom which may facilitate better solutions. The downside of this is that the initial trajectory may be very poor as it was based on a faulty assumption that only one turn would be allowed whereas in fact, this restriction applies only to instantaneous plans and not to reality. The *arbitrary* constraint on planning may prevent the walker from obtaining a solution which is close to optimal and indeed may result in very poor plans. To obtain a good or close to optimal prediction, sufficient turns must be allowed in the plans, that is n_u must be large enough.

ILLUSTRATION: Driving in the fog example

Imagine that you have an objective to drive as fast as possible, but at any time you can only see 20m ahead.

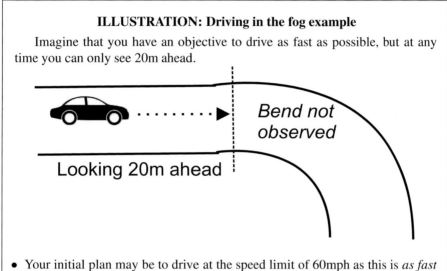

- Your initial plan may be to drive at the speed limit of 60mph as this is *as fast as possible*.

- You first observe the bend when 20m away, which is too late to slow down and take the turn safely.

Of course no sensible human would do this, but it demonstrates the dangers of a mismatch between the objective (to go as fast as possible over the next 20m) and repercussions on the long-term behaviour.

STUDENT PROBLEM

Come up with some examples of your own based on common sense and human behaviour which indicate objectives, d.o.f. and scenarios which are likely to lead to poor decision making.

This section has demonstrated the danger of not predicting far enough into the future. Because the plans do not take account of long-term effects, they may seem *optimal* in the short term whereas in fact they are severely suboptimal in the long term. To avoid this problem, drivers typically look quite far ahead (that is n_y is large) and/or slow down significantly when vision is restricted.

5.4.2   What is an ill-posed objective/optimisation?

An objective/optimisation can be ill-posed for several reasons:

- If the degrees of freedom are parameterised in such a way that what you really want is not a possible outcome from the optimisation (such as in the navigation example). In effect, for GPC, this means n_u is too small.

- If your objective function does not capture the overall behaviour that you want but rather only captures a snippet or snapshot of the overall picture. In effect, for GPC, this means n_y is too small.

   The first of these weaknesses is linked to poorly chosen degrees of freedom (d.o.f.) or choices within your predictions. The second is linked to the design of the performance index which should be chosen to capture the overall picture.

5.4.2.1   Parameterisation of degrees of freedom

A core part of any planning is the parameterisation of the freedom within future predictions. Within GPC the d.o.f. are contained within $\Delta\underset{\rightarrow k}{\mathbf{u}}$ as defined in (4.19). Consider what would happen if the *optimal* or most desired future input involved numerous changes in the input, but we used $n_u = 1$.

ILLUSTRATION: Mismatch between input parameterisation and desired trajectory

Let the desired/optimal trajectory have a number of changes as shown here. Next, the optimisation tries to match this but with $n_u = 1$ and thus *the optimised solution* (**or best**) is actually not very close and does not capture the desired dynamics even closely at all.

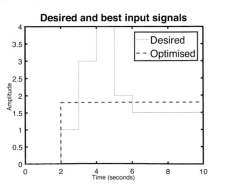

Summary: If the parameterisation of the input does not include sufficient d.o.f., also placed in suitable positions, then any optimisation trying to obtain the unknown *optimal* trajectory must fail. Moreover, the output of the optimisation has questionable value as it is based on a potentially seriously erroneous assumption which could send you in totally the wrong direction (as shown in the navigation illustration).

5.4.2.2 Choices of performance index

The performance index should capture overall behaviour and not just a snapshot as optimising behaviour over a short time interval may imply very poor behaviour outside that interval. Consider the next illustration.

ILLUSTRATION: Mismatch between optimisation and overall behaviour

In this case the optimisation has focused on minimising tracking errors over the first 5 samples (up to the *horizon* marked with the vertical line) and consequently has balanced positive and negative errors appropriately in this domain. However, the overall behaviour beyond 5 samples is clearly very poor and far from target, but this component was not considered in the optimisation.

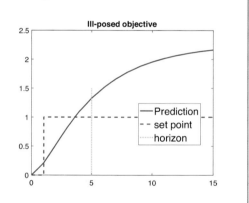

The result is that the overall prediction is quite poor, even though the *snapshot* over the first 5 samples looks okay.

ILLUSTRATION: Using a fire for warmth

Another example could be based on open fires. You are cold and want the fire to give off more heat quickly. So you pile on (carefully to avoid killing existing flames) lots of wood. In the short term, parts of all the new wood catch alight and the fire is lovely. Unfortunately, 5 minutes later, much more of the new wood has caught alight and now you have a raging bonfire - a little too hot and dangerous. Because your decision making only considered the short-term future and not the long term, consequently it was flawed.

STUDENT PROBLEM

Come up with a number of everyday scenarios which indicate the importance of planning for the long term and the risks of only planning over a short horizon.

Summary: The performance index should capture all the behaviour which means n_y should be large, to ensure any ignored components are unimportant (for example close to zero error).

5.4.2.3   Consistency between predictions and eventual behaviour

The final point made in this section is to propose a diagnostic tool rather than a design tool, that is a means of checking whether the optimisation the user has chosen is well-posed in the first place. A key measure of a well-posed objective in the context of predictive control is ***recursive consistency***, that is the decision making at every sample should reinforce the decisions made at previous samples as being good ones.

ILLUSTRATION: Inconsistent decision making

Consider the optimised input trajectories at two successive samples.

$$\underset{\rightarrow k+1|k}{\Delta \mathbf{u}} = \begin{bmatrix} 2 \\ 1 \\ 0.5 \\ -0.5 \\ -0.5 \\ 0 \end{bmatrix} ; \quad \underset{\rightarrow k+1|k+1}{\Delta \mathbf{u}} = \begin{bmatrix} 1 \\ 0.4 \\ 0 \\ -1 \\ -1 \\ 0.4 \end{bmatrix} \quad (5.1)$$

It is clear that $\underset{\rightarrow k+1|k}{\Delta \mathbf{u}} \not\approx \underset{\rightarrow k+1|k+1}{\Delta \mathbf{u}}$ and thus there has been a radical change of mind over what the optimal strategy should be. In other words, the supposed *optimum* at sample k was far from optimal and thus may not even have been useful!

An easy test of consistency is to overlay the *optimised predictions* with the closed-loop behaviour that ensues. If these are close together then the decision making is consistent whereas if they are far apart there must have been some major decision changes which implies inconsistency (as in the navigation illustration). An obvious corollary of this is that the optimised input trajectory at any given sample must be sufficiently flexible to closely capture the desired and actual closed-loop input trajectory.

STUDENT PROBLEM

Come up with some examples of your own which demonstrate how inconsistency between predictions and eventual behaviour can result in poor behaviour.

ILLUSTRATION: Inconsistent decision making in GPC

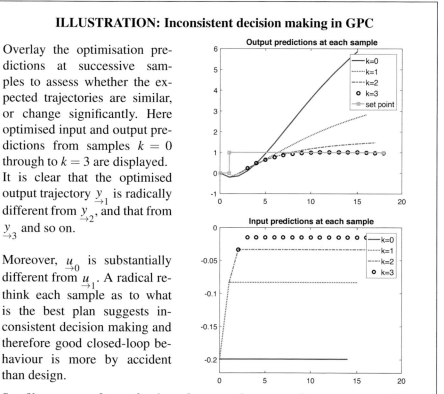

Overlay the optimisation predictions at successive samples to assess whether the expected trajectories are similar, or change significantly. Here optimised input and output predictions from samples $k = 0$ through to $k = 3$ are displayed. It is clear that the optimised output trajectory $\underset{\rightarrow 1}{y}$ is radically different from $\underset{\rightarrow 2}{y}$, and that from $\underset{\rightarrow 3}{y}$ and so on.

Moreover, $\underset{\rightarrow 0}{u}$ is substantially different from $\underset{\rightarrow 1}{u}$. A radical rethink each sample as to what is the best plan suggests inconsistent decision making and therefore good closed-loop behaviour is more by accident than design.

See file **varynyex2ex.m** for these figures and **varynyex1ex.m, varynyex3ex.m** for other figures not presented here on different systems.

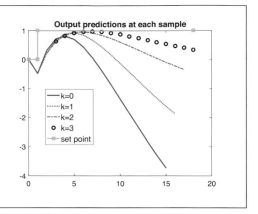

With a different choice of tuning parameters on the same example, one can again see worrying inconsistency in the output predictions.

Summary: Consistent decision making from one sample to another is a useful measure of whether the MPC optimisation is giving useful results. Evidence of inconsistent decision making would suggest the MPC optimisation is poorly posed and needs changing.

5.4.2.4   Summary of how to avoid ill-posed optimisations

The following sections will use a number of illustrations to emphasise the points made in this subsection and hence move towards systematic guidance for GPC tuning. This section has shown that ill-posed optimisations can arise from simple design errors. Emphasis has been given here on three main aspects which enable users to avoid this, that is to ensure well-posed decision making.

Overall Summary:

1. Insufficient or poorly placed d.o.f. in the input parameterisation means the *desired* trajectory cannot be reached, even approximately. In general small n_u should be used with caution.

2. Insufficient attention given to overall behaviour means too much focus on a small aspect of the behaviour which could be seriously detrimental to other parts. In general small n_y should be avoided.

3. Evidence of repeated *changes of mind*, that is changes in the optimised input trajectory, suggest potential chaos in the decision making which consequently cannot be optimal.

One would not expect GPC to give good control if either the basic set-up of the optimisation objective and/or the parameterisation of the d.o.f. is poor.

5.5 Illustrative simulations to show impact of different parameter choices on GPC behaviour

This section uses simulation examples to give insight into the efficacy of the GPC optimisation and control law. This insight is used to derive good practice guidelines on how to choose n_y, n_u to ensure effective and reliable control.

It should be noted that this section assumes there is no advance information on the target, so in essence the feedforward P_r is a constant and only the future target value $\mathbf{r}_{k+1|k}$ is known (this is a common assumption in practice). Hence, discussions on how *tuning guidance* might change when advance information on the target is available is delayed until later (Appendix B.5) so as not to confuse core concepts. However, viewers should note that there is a very strong link between n_u and behaviour when advance information of target changes is provided, see [134] and Appendix B.5.

5.5.1 Numerical examples for studies on GPC tuning

A range of examples with different attributes are defined here as the case studies used to investigate tuning of GPC. G_1 is a simple over damped system, G_2 is a non-minimum phase system, G_3 is an open-loop unstable system and G_4 is an under damped system.

$$G_1 = \frac{z^{-1}+0.3z^{-2}}{1-1.2z^{-1}+0.32z^{-2}} \quad G_2 = \frac{z^{-1}-2z^{-2}}{1-1.7z^{-1}+0.72z^{-2}}$$

$$G_3 = \frac{z^{-1}+0.4z^{-2}}{1-1.7z^{-1}+0.6z^{-2}} \quad G_4 = \frac{z^{-1}+0.4z^{-2}}{1-0.9z^{-1}+0.6z^{-2}}$$

$$(5.2)$$

The following subsections will use these examples to illustrate the impact of small and large choices of n_y, n_u and also whether the insights gained are impacted by changes in the control weighting λ.

5.5.2 The impact of low output horizons on GPC performance

First, this section considers the impact of choosing relatively small values of n_y on the GPC optimisation and control law.

ILLUSTRATION: GPC with $n_y = 5$ on G_1

The figure shows the optimised predictions at $k = 0$ with various choices of n_u but $n_y = 5$ throughout. The closed-loop line plot corresponds to the behaviour that results with the use of the control from $n_u = 1, n_y = 5$. It is clear that in all cases, the predictions do a good job over the prediction horizon (first 5 samples), but the overall *predicted* behaviour is poor, and notably beyond the prediction horizon of 5.

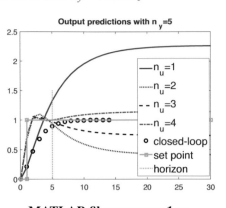

MATLAB file: varynyex1.m

Moreover, as noted in Section 5.4.2.3, there is a significant mismatch between the optimised prediction for $n_u = 1$ at sample $k = 0$ and the corresponding final closed-loop behaviour with $n_u = 1$ which suggests inconsistent decision making.

CONCLUSION: With a small value for n_y, good closed-loop behaviour is more by chance than design! The following three illustrations on different examples give similar conclusions.

ILLUSTRATION: GPC with $n_y = 6$ on G_2

The figure shows the optimised predictions at $k = 0$ with various choices of n_u but $n_y = 6$ throughout. The closed-loop line plot corresponds to the control law with $n_u = 1$.

It is clear that in all cases, the predictions do a good job over the prediction horizon (first 6 samples), but the overall predicted behaviour is poor. Moreover, there is a significant mismatch between the prediction

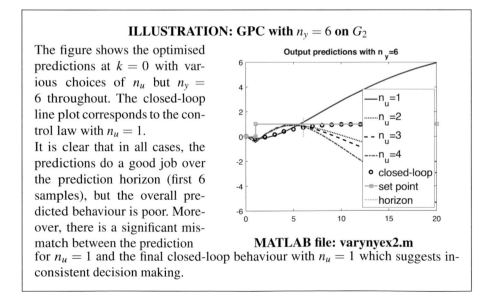

MATLAB file: varynyex2.m

for $n_u = 1$ and the final closed-loop behaviour with $n_u = 1$ which suggests inconsistent decision making.

ILLUSTRATION: GPC with $n_y = 6$ on G_3

The figure shows the optimised predictions with various choices of n_u but $n_y = 6$ throughout. The closed-loop line plot corresponds to $n_u = 1$. It is clear that in all cases, the predictions do a good job over the prediction horizon (first 6 samples), but the overall predicted behaviour is poor. In this case the significant mismatch between optimised predictions, **which are divergent**, and the final closed-loop behaviour is very apparent and emphasises the possible arbitrariness (not systematic) of the implied decision making.

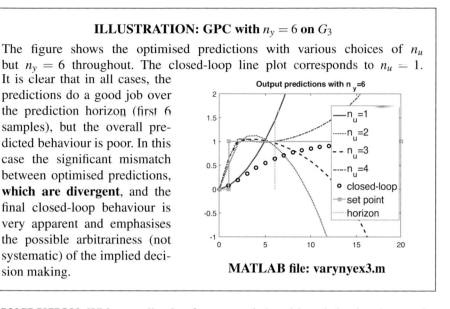

MATLAB file: varynyex3.m

CONCLUSION: With a small value for n_y, good closed-loop behaviour is more by chance than design because there is strong evidence of inconsistent decision making from one sample to the next. The evidence is that the optimised predictions are not close to the final behaviour which results.

ILLUSTRATION: GPC with $n_y = 4$ on G_4

The figure shows the optimised predictions with various choices of n_u but $n_y = 4$ throughout. The closed-loop line plot corresponds to $n_u = 1$. It is clear that in all cases, the predictions do a good job over the prediction horizon (first 4 samples), but the overall predicted behaviour is poor. Moreover, there is a significant mismatch with the final closed-loop behaviour which suggests inconsistent decision making.

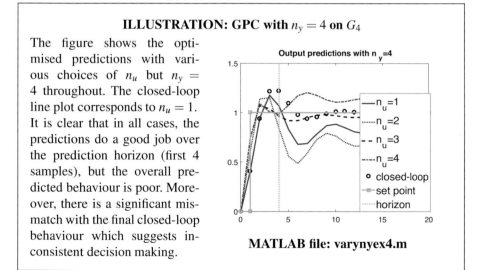

MATLAB file: varynyex4.m

Summary: This section has demonstrated through examples that for many cases, a low output horizon almost always leads to very poor predictions as the optimised predictions may not give a good match to the target over a longer horizon. This is specifically a problem for systems with non-minimum phase characteristics, unstable open-loop behaviour or oscillatory open-loop poles.

The use of the associated performance index can be misleading as there is no penalty on longer term tracking errors and thus it may not imply optimality in any reasonable sense. One possible consequence is the control law **keeps changing its mind** over what is a good (supposedly optimal) input trajectory and that is not a sound basis for reliable behaviour.

STUDENT PROBLEMS

1. This section has deployed horizons $n_y = 4, 5, 6$; readers may like to investigate the impact of using horizons lower than this and decide whether the corresponding optimisations become even more poorly posed and unreliable. You can edit the MATLAB files **varynyex1-4.m** from the illustrations as an easy starting point.

2. Show that with the non-minimum phase example G_2, a choice of $n_y = 1, 2$ will inevitably lead to closed-loop instability. Why is this?

3. You can also try examples of your own choosing and decide whether you reach similar conclusions.

5.5.3   The impact of high output horizons on GPC performance

The implicit argument in Section 5.4.2 is that the performance index should consider the overall or whole behaviour and not just a snapshot. The obvious way to do this is to choose n_y to be large. Of course the questions remain: **What is large n_y? Can n_y be too large?** This section gives a number of illustrations before drawing some conclusions.

5.5.3.1   GPC illustrations with n_y large and $n_u = 1$

It will quickly be clear from the first and second of the following illustrations that good closed-loop behaviour is more likely to be obtained systematically with high n_y! However, the same observation does not follow for the third illustration using G_3 and the reader needs to ask: why not?

ILLUSTRATION: GPC with various n_y and $n_u = 1$ on G_1

The figure shows the optimised predictions with various choices of n_y but $n_u = 1$ throughout. The closed-loop line plot corresponds to $n_y - 20$. The predictions here are much more representative of the desired behaviour than when n_y was small and indeed, have asymptotic values closer to the target for higher n_y. There is no longer significant mismatch between the final closed-loop behaviour and the optimised predictions which suggests more consistent decision making by the control law.

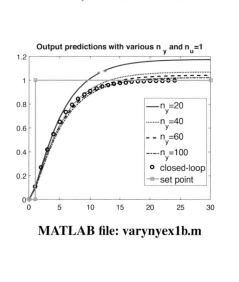

MATLAB file: varynyex1b.m

ILLUSTRATION: GPC with various n_y and $n_u = 1$ on G_2

The figure shows the optimised predictions with various choices of n_y but $n_u = 1$ throughout. The closed-loop line plot corresponds to $n_y = 40$. For $n_y \geq 40$, but not $n_y = 20$, the optimised predictions are now a good match to desirable long-term behaviour and indeed the open-loop predictions are also a good match to the resulting closed-loop behaviour which suggests consistent decision making.

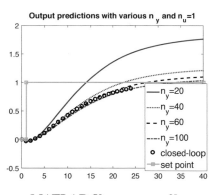

MATLAB file: varynyex2b.m

In this case good closed-loop behaviour is more likely to be obtained systematically with $n_y \geq 40$ as then prediction consistency is clear.

In some cases the use of high n_y alone is not enough to ensure a well-posed GPC optimisation and indeed can be counter productive as will be seen in the following illustration.

ILLUSTRATION: GPC with various n_y and $n_u = 1$ on G_3

The figure shows the optimised predictions with various choices of n_y but $n_u = 1$ throughout. The closed-loop line plot corresponds to $n_y = 40$. This picture suggests a disaster. None of the predictions are anywhere close to the behaviour that we would really like and the closed-loop behaviour seems to be almost not moving! Clearly the optimisation is failing badly (very poorly posed) in this case and indeed closed-loop instability is highly likely.

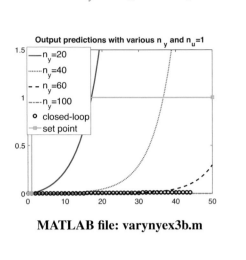

MATLAB file: varynyex3b.m

ILLUSTRATION: GPC with various n_y and $n_u = 1$ on G_4

The figure shows the optimised predictions with various choices of n_y and $n_u = 1$ throughout. The closed-loop line plot corresponds to $n_y = 40$. It is clear that in all cases, the predictions are closed matched to the final closed-loop behaviour which suggests consistent decision making. However, the performance is still poor.

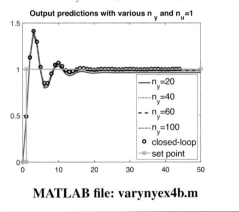

MATLAB file: varynyex4b.m

CONCLUSION: Stable closed-loop behaviour and consistent decision making is more likely with large n_y, but large n_y alone need not imply good behaviour.

STUDENT PROBLEMS

1. The illustrations in this section have deployed horizons $n_u = 1$ simply to save space. Do you get the same conclusions with higher n_u? You can edit the MATLAB files

varynyex1-4b.m from the illustrations above as an easy starting point.

2. Why do problems (significant mismatch or poor performance) still persist with examples G_2, G_3, G_4 and how would you rectify this?

Summary: The use of large n_y has improved consistency between optimised predictions and the eventual closed-loop behaviour in all cases except G_3. Consistency in decision making by the control law is a *requirement* for any simple stability and performance arguments and therefore is assumed hereafter.

Readers should note that open-loop unstable systems such as G_3 are a special case which needs special treatment; this will be covered in the following chapter and Section 5.8.

5.5.3.2   When would a designer use large n_y and $n_u = 1$?

This section gives some analysis which supports insight into when the choice of large n_y with $n_u = 1$ is sensible.

Lemma 5.1 *For open-loop stable processes, long output horizons reduce the implied steady-state error in the predictions compared to very short output horizons.*

Proof: Consider a typical performance index with $n_u = 1$:

$$J = \left\{ \sum_{i=1}^{n_y} \mathbf{e}_{k+i}^T W_y \mathbf{e}_{k+i} \right\} + \Delta \mathbf{u}_k^T W_{\Delta u} \Delta \mathbf{u}_k \tag{5.3}$$

The use of $n_u = 1$ implies the predicted behaviour is in effect open-loop as the future input is maintained at a fixed value. Therefore the predictions will reach close to steady-state after the system settling time n (assumed much less than n_y). Consequently, the performance index J will be dominated by steady-state errors if $n_y \gg n$ which in turn implies that the optimised prediction will ensure that $\lim_{i \to \infty} \mathbf{e}_{k+i|k} \approx 0$.
□

Theorem 5.1 *For a stable process, the use of very large n_y implies the optimal control move will be close to the desired steady-state and thus the closed-loop behaviour will be very close to open-loop dynamics.*

Proof: If the predictions at any sample are such that the asymptotic predicted error is close to zero, and it is assumed that within the predictions the future input does not change, then re-optimisation will give approximately the same value of future input, that is the one which implies $\lim_{i \to \infty} \mathbf{u}_{k+i|k} \approx \mathbf{u}_{ss}$. If the input is constant, open-loop behaviour will result.
□

Corollary 5.1 *The following observations about using large n_y with $n_u = 1$ are obvious corollaries of the analysis above.*

1. Where open-loop dynamics are good, then this tuning will be effective as the closed-loop behaviour will in effect be open-loop dynamics with integral action. Moreover, the optimisation is well-posed in that the decision making is consistent from one sample to the next. This is seen in examples G_1 and G_2.

2. If open-loop dynamics are bad, then this choice of tuning will be ineffective as open-loop behaviour is embedded: (i) this can be seen with example G_4 where the the oscillatory modes are embedded into the closed-loop behaviour and (ii) if the open-loop dynamics are unstable, this can be disastrous and indeed the optimisation is now very poorly posed and may give nonsense results as seen in example G_3 above!

Remark 5.1 *For an open-loop unstable process, a single control move results in divergent predictions. Consequently, it is not possible to form a meaningful prediction close to the target and thus any optimisation is not well-posed. If the open-loop dynamics are unstable, the use of $n_u = 1$ can be disastrous for almost any choice of n_y and should be avoided! GPC stabilising this case would be more by accident than design and should not be trusted.*

STUDENT PROBLEM

Use MATLAB simulations to demonstrate that with large n_y and $n_u = 1$, almost irrespective of the weights on the inputs, the closed-loop input is almost constant (that is close to a step function) and thus open-loop behaviour results.

How big a weight ($W_{\Delta u}$ in eqn. (5.3) or λ in a conventional GPC performance index) is needed before this observation begins to fail?

Summary: The use of long output horizons seems to make good sense and avoids any possibility of ignoring important dynamics within the performance index. With $n_u = 1$, optimised predictions for stable open-loop systems match the resulting closed-loop behaviour quite closely as any ignored dynamics in the predictions are at steady-state.

However, the danger now is that the steady-state errors can swamp the transient errors. Although J must consider the steady-state, we may want more emphasis on the transient errors as these are the only thing affected by the current control choice. Moreover, for some systems, open-loop dynamics are unsatisfactory.

What is big enough n_y?

The input prediction is assumed constant after n_u steps and thus the output (open-loop stable cases) will settle after $n + n_u$ samples where n is the system settling time. Thus any predictions beyond the $(n + n_u)^{th}$ sample will essentially be constant. In consequence, typical guidance is $n_y > n + n_u$, but not usually substantially greater (infinite horizons and the associated issues are discussed in the next chapter) to avoid the steady-state errors dominating and excessive computation. Hence, the output horizon must include enough of the steady-state errors so that the long-term predictions match what is really wanted. The obvious test off-line is to check the optimised predictions do indeed go close to the desired steady-state for the chosen n_y.

5.5.3.3   Remarks on the efficacy of DMC

It was demonstrated through numerous examples that where the open-loop dynamics are satisfactory, then a choice of large n_y with $n_u = 1$ will give effective closed-loop control as it retains the open-loop dynamics but adds in the integral action to ensure offset-free tracking. It is unsurprising therefore that such choices are not uncommon with many industrial applications of DMC on chemical processes often described as *first-order with a delay*. A further advantage of using $n_u = 1$ is that the corresponding optimisation is simpler with few d.o.f. and this could be invaluable for managing any constraint handling.

5.5.4   Effect on GPC performance of changing n_u and prediction consistency

It has been noted that the selection n_y is large is not a choice but essential to ensure consistent decision making. Hence, this will be assumed hereafter. However, it was also noted that using $n_u = 1$ was effective, if and only if the open-loop dynamics were acceptable. Consequently, this section considers the impact of including higher values of n_u and more specifically, whether a systematic choice can be made.

Consistency check and choice of n_u

A good indicator of whether GPC is likely to work well is to compare open-loop predictions with the closed-loop behaviour that results (nominal case). If the two are similar, then the optimisation has clearly delivered a sensible or consistent answer. If the two are different, then re-optimisation at subsequent samples is causing a change of mind and this undermines the validity of any of the earlier results. This section indicates, through illustration, that as long as n_u is large enough such consistency can be delivered.

Illustration insights: The following four illustrations show the optimised predictions $\underset{\rightarrow k}{\mathbf{u}}, \underset{\rightarrow k+1}{\mathbf{y}}$ with various choices of n_u but $n_y = 20$ throughout. The closed-loop output line plot corresponds to $n_u = 4$. In most cases the **conclusion** is that good closed-loop behaviour is more likely with high n_u and high n_y, especially as the decision making is now consistent from one sample to the next.

ILLUSTRATION: GPC with $n_y = 20$ and various n_u on G_1

It is clear that in all cases except where $n_u = 1$ (where there is clear inconsistency between the closed-loop input and the initial optimal), the predictions do a good job of matching the desired behaviour and also being consistent with the final closed-loop behaviour which suggests consistent decision making.

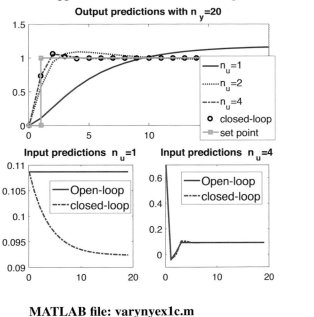

The consistent decision making is emphasised in the lower plot which shows the closed-loop input and the initial optimised prediction are almost the same with $n_u = 4$.

MATLAB file: varynyex1c.m

ILLUSTRATION: GPC with $n_y = 20$ and various n_u on G_2

It is clear that in all cases except where $n_u = 1$ (where there is clear inconsistency between the closed-loop input and the optimal), the predictions do a reasonable job of matching the desired behaviour and also being consistent with the final closed-loop behaviour which suggests consistent decision making. The consistent decision making is emphasised in the lower plot which shows the closed-loop input and the initial optimised prediction are almost the same with $n_u = 4$.

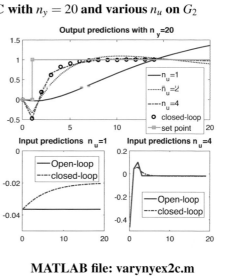

MATLAB file: varynyex2c.m

ILLUSTRATION: GPC with $n_y = 20$ and various n_u on G_3

This case is more interesting as the predictions, except for $n_u = 1$ (where there is clear inconsistency between the closed-loop input and the optimal) do a reasonable job of matching the desired behaviour within the prediction horizon of n_y. However, the predictions are divergent beyond n_y! Nevertheless, despite this inconsistency in the far horizon, GPC gives good closed-loop behaviour and again consistent decision making is emphasised in the lower plot which shows the closed-loop input and the initial optimised prediction are almost the same with $n_u = 4$. The key point here is that the update of the decision making each sample is enough to remove the divergent components.

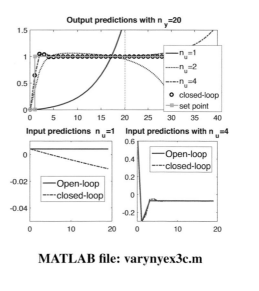

MATLAB file: varynyex3c.m

CONCLUSION: Extending n_u above 1 allows GPC to choose input/output trajectory pairs which are *almost convergent asymptotically* but certainly good in transients and close to ideal trajectories; the feedback corrections each sample save the day.

ILLUSTRATION: GPC with $n_y = 20$ and various n_u on G_4

In this case both $n_u = 1$ and $n_u = 2$ give similar and poor predictions and poor consistency between initial optimal prediction for $\mathbf{u}_{\to k}$ and the final closed-loop behaviour. On the other hand $n_u = 4$ gives satisfactory predictions and good consistency with the final closed-loop behaviour which is emphasised in the lower plot which shows the closed-loop input and the initial optimised prediction are almost the same with $n_u = 4$, but not with $n_u = 2$.

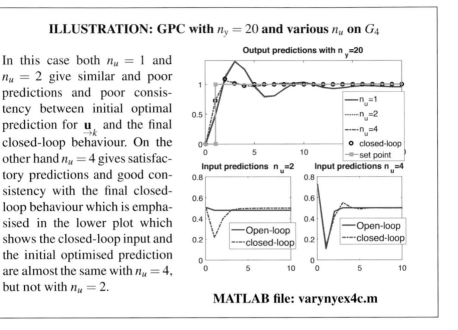

MATLAB file: varynyex4c.m

We have already established that the use of low output horizons n_y is invalid in general, irrespective of the choice of n_u. Long output horizons in combination with larger input horizons allow better predicted performance and in most cases shown, indicate good performance both within the horizon window and beyond which is one indicator of a well-posed optimisation (an exception is open-loop unstable processes). For the examples shown, the differences in the first input value were minimal for n_u equal to around 3 or greater and indeed the closed-loop behaviour of the inputs demonstrated that the initial optimised input trajectory was very close to the updated trajectories made by optimising at later samples. The suggestion therefore is that there is rarely a need for n_u to be very large, although of course some trial and error is needed to establish a sufficient value.

> **Summary:** In general terms, prediction consistency with closed-loop behaviour seems to improve as n_u increases and thus larger n_u are to be preferred where this is an option.

STUDENT PROBLEM

Use MATLAB simulations on various systems of your choice to

demonstrate that with large n_y and large n_u, almost irrespective of the weights on the inputs, the closed-loop and optimised open-loop behaviour behaviour are similar and also, likely to be good. Use trial and error to establish how large n_u needs to be. How does this value vary with the choice of weight λ?

5.5.5 The impact of the input weight λ on the ideal horizons and prediction consistency

So far, we have left the input weights at unity so as not to introduce too many concepts simultaneously. This section uses more illustrations to demonstrate how the choice of weights has a potentially significant impact on the choice of the horizons.

The assumption taken here is that our goal is **PREDICTION CONSISTENCY**, that is, we want the optimised predictions to be close to the closed-loop behaviour that results, because this ensures our decision making is consistent from one sample to the next and that is a requirement for the optimisations to be well-posed rather than chaotic. It is automatic that the prediction horizon n_y is well beyond the system settling time to ensure that any *ignored dynamics* are close to the asymptotic target.

The following **illustrations** show the optimised input predictions with $n_y = 20$, various choices of n_u and two choices of λ.

ILLUSTRATION: GPC with $n_y = 20$ and various λ on G_1

In the case with $\lambda = 1$, it is noted that the optimised input prediction with $n_u = 6$ and $n_u = 8$ are about the same. This suggests that there is minimal benefit increasing n_u beyond 6 as the extra control increments that result are negligible.

In the case with $\lambda = 100$, it is noted that the optimised input predictions keep changing as n_u increases and indeed even $n_u = 8$ is too low to capture the *optimum* trajectory so that a much higher n_u would be required to be confident of consistent decision making.

Input predictions with $n_y = 20$ $\lambda = 1$

$n_u = 1$
$n_u = 2$
$n_u = 4$
$n_u = 6$
$n_u = 8$

Input predictions with $n_y = 20$ $\lambda = 100$

MATLAB file: varynyex1d.m

CONCLUSION: The required n_u to capture an input prediction close to the desired closed-loop optimal will vary with the choice of λ.

ILLUSTRATION: GPC with $n_y = 20$ and various λ on G_2

In the case with $\lambda = 0.2$, it is noted that the optimised input prediction with $n_u = 4$ is actually not too far from what is achieved with higher n_u which suggests that there is likely to be minimal benefit increasing n_u beyond 4.

In the case with $\lambda = 100$, it is noted that the optimised input predictions keep changing as n_u increases but seem to be reaching an asymptotic pattern around $n_u = 8$.

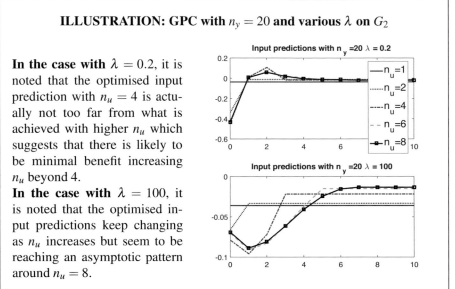

MATLAB file: varynyex2d.m

ILLUSTRATION: GPC with $n_y = 20$ and various λ on G_4

In the case with $\lambda = 1$, it is noted that the optimised input prediction with $n_u = 6$ and $n_u = 8$ are about the same. This suggests that there is minimal benefit increasing n_u beyond 6.

In the case with $\lambda = 100$, it is noted that the optimised input predictions keep changing as n_u increases and indeed even $n_u = 8$ is probably too low to capture the *optimum* trajectory.

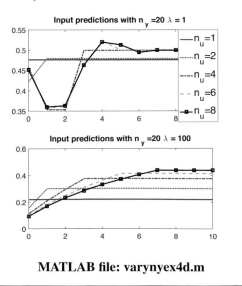

MATLAB file: varynyex4d.m

The reader will note that the required choice of n_u to ensure a *well-posed* optimisation or equivalently a well-defined GPC control law, is rarely immediately obvious and varies with the choice of λ. Some form of trial and error is required to find a suitable value.

> **Summary:** The required value for n_u varies with the process dynamics and also the choice of weights in the performance index. In general, trial and error and off-line investigations are required to give confidence that a chosen value gives a well-posed objective (consistent decision making).
>
> The input horizon must include enough terms so that the control increments can: (i) get the system to correct steady-state; (ii) counter any undesirable dynamics in the open-loop and (iii) have sufficient flexibility remaining to optimise transient performance, that is, to capture the *ideal* closed-loop input trajectory.
>
> In practise, for many cases a smaller n_u may give a similar first choice of $\mathbf{u}_k$ to a larger n_u, but is this accident rather than design? Will you come unstuck when constraints are added?

STUDENT PROBLEM

This section has not considered the interaction between the weighting λ and the choice of horizon n_y. Using simulation examples of your own, demonstrate that the efficacy of λ for achieving a compromise between speed of response (tracking accuracy) and input rate activity may vary significantly with the choice of n_y for some choices of n_u.

5.5.6 Summary insights on choices of horizons and weights for GPC

This section has given a number of insights which are now drawn together before we progress to the development of more systematic guidance. Critically it is observed that the horizons n_y and n_u are not tuning parameters or indeed *options*; if not chosen correctly and systematically, then the associated optimisation will be ill-posed in that optimising J need not imply sensible or good overall trajectories.

- If n_y is small, the emphasis is all on transient behaviour but with no consideration of the values and shape of the long-term predictions. Consequently, the associated optimisation is likely to be ill-posed and good closed-loop behaviour is more luck that design.

- To ensure all the *ignored* parts of the prediction are constant and included in some sense, the constraint $n_y - n_u \gg n$ is used where n is the open-loop settling time. This will usually lead to a well-posed optimisation in some sense, but may not give consistent decision making if n_u is poorly chosen.

- To ensure the optimisation gives consistent answers from one sample to the next, it is necessary that each optimisation considers the entire output trajectory (large n_y) and also has a large enough n_u to capture the core input dynamics.

- The required n_u will vary with the weights as, for example, high λ implies heavy penalties on input rates and thus the input could be slow to reach steady-state and thus can only be described with a large number of samples.

- The choices of large n_y and $n_u = 1$ are appropriate where open-loop dynamics are satisfactory. Such a choice has the advantage of computational simplicity and transparency.

STUDENT PROBLEM

Look at all the student problems in this section and replicate them with MIMO examples. Comment on your findings. Also, did you discover any additional *patterns* which would effect the selection of suitable parameters n_y, n_u, λ?

Remark 5.2 *The reader is reminded that necessarily all these insights will carry across to the MIMO case although the number of figures and responses to check will increase and thus even more care is required.*

Summary:

1. Optimising J tells you the best with restricted d.o.f. over a restricted horizon and this could be seriously flawed if n_y, n_u are poorly chosen.

2. For open-loop stable processes, one can optimise expected closed-loop performance by choosing both n_y and n_u to be as large as possible, although what is large enough may require some off-line testing.

3. However, often these choices are undesirable as they are computationally intractable especially once constraint handling is introduced (see later chapter).

4. Open-loop unstable systems often seem to be controlled effectively notwithstanding the poor (divergent) long-term predictions. Nevertheless, readers are encouraged to recognise this fundamental flaw as potentially dangerous; better alternatives exist!

5.6   Systematic guidance for "tuning" GPC

The observations of this chapter were not fully understood at first and it was considered that the issues of both tuning and stability were major problems with GPC and variants. Although often in practice good control and good robustness margins were achieved, rigorous proofs were scarce and it was still only a posteriori stability checks that could be performed; that is, find the control law and then compute the implied closed-loop poles.

Nevertheless, it should be clear to the reader by now that effective choices of n_y, n_u, λ are important and inter-dependent. They are not tuning parameters but rather parameters that need to be chosen correctly to ensure GPC delivers the desired closed-loop performance. Indeed, one can make a strong argument for the statement that:

The optimisation of J is only meaningful if the optimised trajectories at each sample are close to the closed-loop responses that result (for a fixed target).

If the optimised predictions differ significantly from the eventual closed-loop behaviour, then the predictions are meaningless and thus the optimisation is meaningless as they do not represent what is actually going to happen. Obviously, there must also be a close match between predictions and the behaviour we actually want, otherwise our optimisation cannot lead us to the behaviour we want! Examples have shown that there is a close match between predictions and closed-loop behaviour, if and only if some conditions are met.

- The output horizon should be well beyond the system open-loop settling time so that ignored tracking errors are zero.

- The input horizon must be long enough to capture the entire desired closed-loop input trajectory, otherwise there is a mismatch between what the optimisation can offer and what you want.

- The required horizons may change with the control weighting, as the weighting has an impact on the desired speed of response.

This section uses the above observations to derive *tuning guidance* to help ensure that the parameters are always chosen well.

Summary: It is possible to use common sense criteria in the list above as a basis for a systematic tuning method for GPC.

5.6.1   Proposed offline tuning method

This section summarises guidance based on *common-sense* observations in the chapter so far; a more rigorous approach is in the following chapter.

1. Choose n_u to be large enough to capture the expected optimal closed-loop input trajectory.

2. Choose n_y to be significantly greater than n_u, for example $n_y - n_u \gg n$, where n is the system settling time. [For an open-loop unstable system, use the slowest pole to estimate this.]

3. Look at corresponding closed-loop responses and determine the *actual* settling time n_s and number of significant input moves n_{us}. Then choose $n_y = n_s, n_u = n_{us}$.

It is noted that some trial and error may be required for the first 2 steps unless one defaults to just using very large values, especially given these values have an unclear relationship with the chosen λ.

5.6.2   Illustrations of efficacy of tuning guidance

This section will show how easily the tuning guidance can be implemented and critically, that the resulting horizons for implementation online need not be large. This latter point means that the implied online computation need not be onerous.

ILLUSTRATION: Using GPC tuning guidance on G_1

In the case with $\lambda = 1$, it is noted that the optimised closed-loop behaviour with $n_y = 40, n_u = 20$ suggests that the output has largely settled $n_s = 10$ samples after the transient begins and the input has mostly stopped changing after $n_{us} = 5$ samples.

Closed-loop responses with $n_y = 10, n_u = 5$ are similar enough to be considered identical in reality.

One could easily argue that smaller n_y, n_u would also be effective here. **TRY IT!**

Closed-loop responses for G_1 (λ=1)

— y (n_y=40, n_u=20)
o y (n_y=10, n_u=5)
---- u (n_y=40, n_u=20)
□ u (n_y=10, n_u=5)
set point

Samples

MATLAB file: varynyex1g.m

In the case with $\lambda = 100$, it is noted that the optimised closed-loop behaviour with $n_y = 40, n_u = 20$ suggests that the output has largely settled after $n_s = 15$ samples and the input has mostly stopped changing after $n_{us} = 7$ samples. Closed-loop responses with $n_y = 15, n_u = 7$ are similar enough to be considered identical in reality.

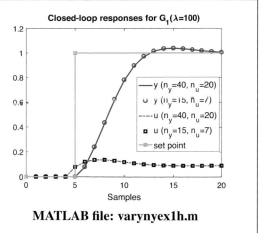

MATLAB file: varynyex1h.m

Remark 5.3 *This chapter focusses on the unconstrained case. One could argue that the presence of constraints could have a significant impact on the required n_y, n_u and thus undermine the choices made. This is true, but a counter argument is that the variety of scenarios that could occur with constraints make offline analysis difficult in general and thus typical solutions in the literature such as parametric approaches [8] are non-trivial and would be pursued only if there was an evident need.*

ILLUSTRATION: Using GPC tuning guidance on G_2

In the case with $\lambda = 1$, it is noted that the optimised closed-loop behaviour with $n_y = 40, n_u = 10$ suggests that the output has largely settled by $n_s = 10$ and the input has mostly stopped changing by $n_{us} = 5$.

Closed-loop responses with $n_y = 10, n_u = 5$ are similar enough to be considered identical in reality.

One could easily argue that smaller n_y, n_u would also be effective here. **TRY IT!**

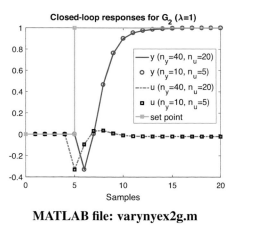

MATLAB file: varynyex2g.m

ILLUSTRATION: Using GPC tuning guidance on G_3

In the case with $\lambda = 1$, it is noted that the optimised closed-loop behaviour with $n_y = 40, n_u = 20$ suggests that the output has largely settled by $n_s = 8$ and the input has mostly stopped changing by $n_{us} = 6$.

Closed-loop responses with $n_y = 8, n_u = 6$ are similar enough to be considered identical in reality.

One could easily argue that smaller n_y, n_u would also be effective here. **TRY IT!**

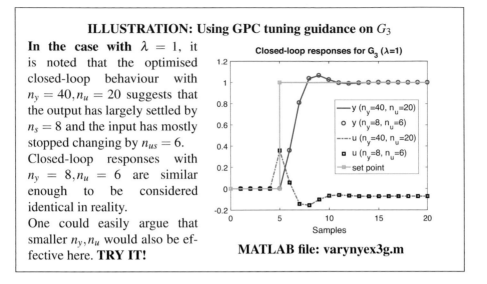

Closed-loop responses for G_3 (λ=1)

— y (n_y=40, n_u=20)
○ y (n_y=8, n_u=6)
---- u (n_y=40, n_u=20)
□ u (n_y=8, n_u=6)
■ set point

Samples

MATLAB file: varynyex3g.m

STUDENT PROBLEM

For a large number of problems of your own choosing with varying dynamics, investigate the efficacy of the tuning guidance provided in this section. Also investigate how much smaller you can make the horizons, below the advisable magnitude, before significant performance degradation/prediction mismatch begins to occur. Files such as **varynyex1g.m** provide a useful template for such investigations.

ILLUSTRATION: Using GPC tuning guidance on G_4

In the case with $\lambda = 1$, it is noted that the optimised closed-loop behaviour with $n_y = 40, n_u = 20$ suggests that the output has largely settled by $n_s = 8$ and the input has mostly stopped changing by $n_{us} = 6$.

Closed-loop responses with $n_y = 8, n_u = 6$ are similar enough to be considered identical in reality.

Closed-loop responses for G_3 (λ=1)

— y (n_y=40, n_u=20)
○ y (n_y=8, n_u=6)
---- u (n_y=40, n_u=20)
□ u (n_y=8, n_u=6)
■ set point

Samples

MATLAB file: varynyex4g.m

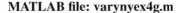

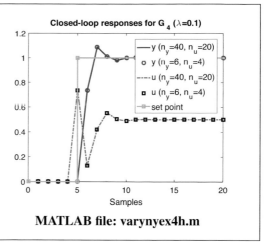

In the case with $\lambda = 0.1$, it is noted that the optimised closed-loop behaviour with $n_y = 40, n_u = 20$ suggests that the output has largely settled by $n_s = 6$ and the input has mostly stopped changing by $n_{us} = 4$. Closed-loop responses with $n_y = 6, n_u = 4$ are similar enough to be considered identical in reality.

MATLAB file: varynyex4h.m

The reader will have noticed that in all these cases it is implied that $n_u \gg 1$. This is a useful insight, as it tells us that if we want anything other than to replicate open-loop dynamics ($n_u = 1, n_y = \text{large}$), then a choice of small n_u will typically give rise to an ill-posed optimisation in the sense that there will be substantial changes in the perceived optimum strategy from one sample to the next.

Summary: One can ensure that GPC tuning is systematic and gives rise to a well-posed optimisation simply by choosing the horizons to be large. Some simple offline analysis to determine the associated closed-loop settling times can be used to determine reasonable minima for the horizons and thus to minimise the associated online computation.

5.7 MIMO examples

This chapter contains only a brief summary of applications to MIMO examples, as most of the conclusions are the same as for the SISO case. The focus is on what changes and what else the designer must bear in mind.

Due to the use of a performance index, GPC gives an optimal management of interaction, at least by way of performance specified in J. Hence if the designer wishes to trade off performance in one loop with another, the systematic route is via the weightings in J. That is, if one increases the relative weight on one loop w.r.t. to another, then the relative errors will reduce. However, there is no simple analytic link to the actual weights, at least not when J is quadratic. So if the variance in one loop was too high, one could increase the weight on that loop but this would be a trial and error process and the repercussions on errors in the other loops would not be known a priori.

An example is given next based on a simplified boiler model (**powerstation_model.m**) for electricity generation. This is a stiff system whose state space matrices are given as:

$$A = \begin{bmatrix} -0.0042 & 0 & 0.0050 \\ 0.0192 & -0.1000 & 0 \\ 0 & 0 & -0.1000 \end{bmatrix}; \quad B = \begin{bmatrix} 0 & -0.0042 \\ 0 & 0.0191 \\ 0.1000 & 0 \end{bmatrix}$$

$$C = \begin{bmatrix} 1.6912 & 13.2160 & 0 \\ 0.8427 & 0 & 0 \end{bmatrix}$$

(5.4)

The examples given here will demonstrate that:

1. As indicated in Section 5.5.6, the use of large enough n_y, n_u will result in effective tuning and, as this system is multivariable, it will also handle the interaction.

2. The choice of weights is very significant as these allow relatively different emphasis on different outputs and input activity. However, in general, it is difficult to find a systematic link between the choice of weights and the closed-loop behaviour that results.

Closed-loop responses for a step demand in power output (the first output) are given in the following illustration for tuning parameters $n_y = 15$, $n_u = 3$ and different output weights.

- It is clear that changing the emphasis on the second output (the pressure) has changed the performance substantially.

- However, it is difficult from this to say what the weighting should be. It is evident however, that this tuning process is somewhat ad hoc.

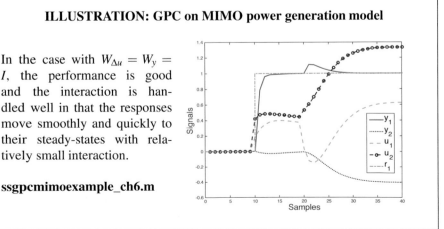

ILLUSTRATION: GPC on MIMO power generation model

In the case with $W_{\Delta u} = W_y = I$, the performance is good and the interaction is handled well in that the responses move smoothly and quickly to their steady-states with relatively small interaction.

ssgpcmimoexample_ch6.m

In the case with $W_{\Delta u} = I, W_y = diag([1,4])$, the behaviour is now **unacceptable** as there is significant oscillation during transients. It is surprising here to note that the change in weights to bring about this marked change in behaviour is relatively small.

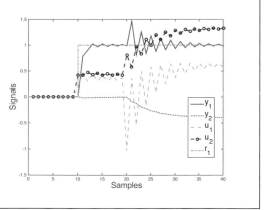

STUDENT PROBLEMS

1. Investigate by trial and error different weighting choices for system (5.4) and hence propose sensible pairings. Attempt to form a link with performance criteria such as the respective rise times and interactions in each loop.
2. Perform similar investigations on MIMO problems of your choice.

Summary: MIMO design is linked directly to the cost function J. The predicted performance is always optimised w.r.t. J and thus concepts such as overshoot or relative bandwidth are not considered directly. You change the design by changing the weights in J.

The cost function J assesses performance by way of a weighted time domain 2-norm measure. The means of selecting weights to achieve the best performance by way of criteria such as bandwidth, damping or interaction is unclear and iteration of weights to achieve this can be a time consuming part of an MPC design.

5.8   Dealing with open-loop unstable systems

As noted earlier in this chapter, GPC makes use of open-loop predictions which in the case of open-loop unstable systems will necessarily be divergent in the long term. An optimisation which seeks to form a smooth and convergent predicted output from a linear combination of a small number of divergent outputs is clearly not well thought through! This section will illustrate the dangers briefly and also give some insights into the accepted approaches in the literature. Detailed algebra is omitted as being far beyond a first and even a second course in MPC and readers are referred to the references for more discussion [100, 138, 161].

5.8.1   Illustration of the effects of increasing the output horizon with open-loop unstable systems

Take the unstable example and perform some simulations with different n_y.

$$G_5 = \frac{z^{-1} + 0.4z^{-2}}{1 - 1.5z^{-1}}$$

The reasons for the odd behaviour noted in the following illustrations is nothing to do with the concepts being deployed, rather it is a simple numerical issue. An investigation of the elements of the prediction matrices H, P, Q in Eqn.(2.49) will show that for open-loop unstable models, as n_y increases the higher rows have larger and larger numbers as the system open-loop predictions are divergent. At some point, the computer's ability to handle the small numbers and the large numbers in the same matrix exceeds the precision available and then accuracy is lost. Very quickly the numerical algebra implied in the computation of the GPC law will become arbitrary/chaotic due to these rounding errors and thus the control law also becomes arbitrary.

ILLUSTRATION: Ill-conditioning with open-loop unstable plant

Plot the closed-loop responses for G_5 with $n_u = 3$, various n_y and $\lambda = 1$. The figure shows that both the optimised predictions and the closed-loop output behaviour begin to deviate from the expected patterns once $n_y > 40$.

Despite the high n_y the predictions *settle* at the incorrect steady-state whereas this was not the case with lower n_y.

The closed-loop behaviour begins to diverge for $n_y = 54$ and the control law cannot be computed at all for higher n_y!

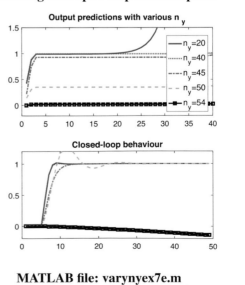

MATLAB file: varynyex7e.m

So, the expected improvements in behaviour as n_y increase occur up to a point, but if n_y is too large, these improvements are lost and chaos seems to ensue.

ILLUSTRATION: Dependence of control parameters on n_y

Plot the (1,1) parameter of control law numerator N_k as n_y varies. Initially as n_y increases the parameters settle but then as n_y gets larger still, ill-conditioning causes the parameter values to become chaotic. For $n_y \geq 54$ MATLAB fails altogether and returns NaN.

Thus ill-conditioning due to divergent predictions has caused the GPC algorithm to fail for high n_y. One can only safely use values of n_y where the range of magnitudes in the parameters of the predictions is well within the precision limits of the processor.

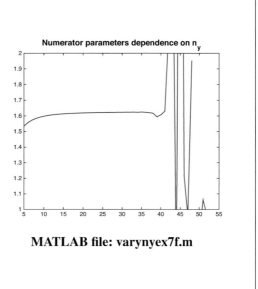

MATLAB file: varynyex7f.m

5.8.2 Further comments on open-loop unstable systems

The previous section indicates that there is a sort of *goldilocks* zone, that is n_y should be big enough to capture overall behaviour but small enough so that ill-conditioning is avoided. Of course, in both cases one will have divergent predictions in the long term, although the updates at each sample will generally correct for this. Thus, in practice one can normally obtain effective tuning as long as some simple guidance is followed.

- For $n_u = 1$ the system predictions cannot be *stabilised*, even approximately, and thus this choice is likely to be poor for any n_y. The use of $n_u = 1$ with high n_y will result in very small initial control moves due to the relatively high values of some of the output in the predictions, thus results in a slow initial response.

- For higher $n_u = 2$ the system predictions can be *psuedo-stabilised* as the input has sufficient d.o.f. to approximately *cancel* the unstable mode. In consequence the associated predictions are often effective enough.

Summary: Use of high n_y in a *standard* MPC design may cause severe performance degradation for an open-loop unstable plant. This seems counter-intuitive given the insights of Section 5.5.6.

For unstable processes it may be difficult to obtain good performance with $n_u = 1$. A typical guideline is that $n_u \geq p+1$, where p is the number of open-loop unstable poles.

STUDENT PROBLEMS

1. Using examples of your choice, demonstrate the ranges of prediction horizon for which GPC gives a reasonable solution. The file **varynyex7f.m** can be used as a template.

2. MATLAB stores a large number of significant figures. How would the insights of this section be affected by processors which have much lower precision?

3. The results of this section have been illustrated with transfer function models [129]. Demonstrate that similar conclusions will apply to GPC based on state space models.

5.8.3   Recommendations for open-loop unstable processes

Early suggestions in the literature suggested *cancellation of unstable modes* in the predictions [52, 116]. This means that the future input is re-parameterised into two parts: (i) a state dependent component which ensures that the dependence of the predictions on the current state is stabilised and (ii) d.o.f. which are multiplied with the unstable poles and thus give predictions which must be convergent. The algebra is somewhat cumbersome and although interesting, these approaches have largely been superseded in the more recent literature and thus are not covered here [52, 116, 161].

These days, most authors would recommend the use of *pre-stabilisation* or *cascade*. That is, first stabilise the system with some feedback, and then apply MPC as the outer loop of the cascade. As it happens, this approach is almost equivalent to the dual-mode approaches of the following chapter and indeed the prediction methods in Section 2.9 and so are not discussed further here.

5.9 Chapter summary: guidelines for tuning GPC

In order to ensure that the optimisation is well-posed, that is the minimisation bears a direct relationship to the expected closed-loop behaviour, one should choose:

1. The output horizon n_y should be larger than n_u plus the system settling time n. For a rigorous proof one should choose $n_y = \infty$ but in practice $n_y > n_u + n$ will make negligible difference.

2. Ideally the input horizon should be as large as the expected transient input behaviour. In practice a value of $n_u \geq 3$ often seems to give performance close to the '*global optimal*' achievable with larger n_u, but this could be embedding inconsistent decision making by the control law, so low values of n_u should be used with caution.

3. To achieve closed-loop behaviour close to open-loop behaviour, $n_u = 1$ will often be sufficient as long as n_y is large.

4. The weighting matrices can be used to shift the emphasis onto different loops but the efficacy of any change in the weights depends on the corresponding gradients in the cost function; sometimes an apparently large change in weight may have negligible effect on the optimum and vice versa. In practice the choice of the weights to achieve a desired compromise across inputs/outputs and different loops is unlikely to be achieved systematically.

5. With unstable processes one should proceed with caution and ideally check that there is no ill-conditioning in the matrix algebra. Alternatively use a better conditioned approach to be presented in Chapter 6.

THE ELEPHANT IN THE ROOM:
Readers who can still recall the first chapter will perhaps be thinking that the previous two chapters have made negligible reference to constraints and yet surely a major reason for the uptake of MPC in industry is its ability to handle constraints. Nevertheless, this discussion is deliberately deferred to the later Chapter 7, as a logical premise for effective constraint handling is that the underlying optimisation is well-posed in the unconstrained case and combining too many issues can confuse insight for the reader.

5.10   Useful MATLAB code

The utility files are the same as used in Chapter 4 and listed in section 4.10. For convenience this chapter has focussed on SISO transfer function representations which are simulated using a command of the form:

[y,u,Du,r] = mpc_simulate_noconstraints(b,a,nu,ny,Wu,Wy,ref,dist,noise)

Detailed instructions on the inputs and outputs for this file are evident from the illustration script files in the table below.

poorbehaviourexample1,2,3.m	Simulates a GPC control law and overlays responses with two choices of tuning; users can change simulation parameters with a direct edit to these files.
varynyex1*,2*,3*,4*,7*.m	Computes simulation data and forms the figures in the illustrations based on G_1, G_2, G_3, G_4.
powerstation_model.m	State space model for a MIMO example.
ssgpcmimoexample_ch6	Illustration of closed-loop behaviour with a MIMO state space model.

6

Dual-mode MPC (OMPC and SOMPC) and
stability guarantees

CONTENTS

6.1 Introduction

The previous few chapters have focussed on finite horizon MPC approaches such as GPC and PFC. In both cases, while it is often possible to gain good closed-loop behaviour, the reader may have been somewhat uncomfortable with some of the assumptions made and in particular the theoretical weaknesses associated to these. Of particular importance it is noted that a well-posed optimisation should meet a number of criteria such as the prediction class including the *most desired one* and the horizons being long enough to capture the whole picture; however, in practice these criteria are only met partially by GPC.

A researcher is likely to ask the question: **how do I define a well-posed optimisation which necessarily will give a good outcome?** If I can do this, then I am far more likely to be able to obtain some rigorous assurances of stability and performance and not be subject to the risks of poor choices illustrated in Section 5.4.

This chapter will do precisely that and build the foundation of MPC on the common sense requirements set out in the previous chapter and demonstrate that so doing gives an algorithm on a much more secure theoretical footing.

6.2 Guidance for the lecturer

This chapter could be considered on the advanced side for students doing a single course in predictive control, but in that case it would still be useful for students to engage with the principal concepts discussed in Sections 6.3, 6.4, 6.8.4 even if the detailed algebra is not tackled there. For those who have space in the curriculum, I would still advise a focus more on illustration of concepts rather than assessing detailed algebra as the former is what students are likely to find useful in future careers; the flavour of the questions provided in Appendix A reflects this. Consequently, this chapter is organised by presenting concepts before algebra so that a natural stopping point is obvious for those who do not need to be able to derive or modify the detailed algebra.

As in previous chapters, numerous MATLAB® files are provided so that students and staff can engage in assignment activities where dual-mode approaches are compared and contrasted with finite horizon approaches.

A discussion of constraint handling is deferred to Chapters 7 and 8 so as not to confuse the reader with different interacting issues.

Typical learning outcomes for an examination assessment: It would be reasonable to focus on effective communication of core concepts in an examination assessment. The author would be less fussy about precision in the algebra given it is a big ask for students to engage and learn this alongside other aspects of MPC in a single module.

Typical learning outcomes for an assignment/coursework: Following the same thinking as with GPC, the author would provide students with working code and thus ask them to demonstrate core understanding through illustrations on case studies where they can also contrast the efficacy/complexity with GPC and/or PFC.

6.3 Foundation of a well-posed MPC algorithm

The previous chapter highlighted the issues with selecting parameters n_y, n_u, λ to ensure good performance and stability of GPC although ironically those insights became more readily understood in the community only after the development and dissemination of dual-mode approaches. Dual-mode approaches are a more generic solution to this frustrating issue which became common knowledge in the early 1990s (e.g., [26, 75, 103, 116, 132, 135, 171]) (though actually in the literature from the late 1970s, e.g., [68, 84]). This section will set out the underpinning thinking of dual-mode approaches by first building on what we have learnt from playing with finite horizon algorithms. The following sections will then take this conceptual thinking and develop some MPC algorithms.

The previous chapter set out common sense arguments for defining a well-posed optimisation based on the predictions of future system behaviour.

1. The cost function should contain large output horizons which in effect capture all future behaviour so that any ignored behaviour is at the correct steady-state.

2. The d.o.f. in the input trajectory should be sufficient to capture the ideal closed-loop behaviour so that the *optimal* behaviour is a possible solution of the optimisation.

3. The d.o.f. in the optimisation should be parameterised in such a way that they include the optimum trajectory from the previous sample to ensure that one is not *forced into* a change of mind.

These criteria are still somewhat loose and thus this chapter will tighten them up so that more rigour can be applied to the corresponding analysis. More specifically it is noted that the easiest way to meet the horizon requirement is to choose $n_y = n_u = \infty$. This removes all prediction mismatch and ensures that the prediction class, in principle, can capture any desired trajectory and also enables strong stability results. This then only leaves the 3rd point above which is a rather subtle one.

The following subsections demonstrate how these simple criteria enable a guarantee of stability for the nominal case.

6.3.1   Definition of *the tail* and recursive consistency

The tail is a shorthand notation used to capture the requirement for consistent decision making from one sample to the next. In simple terms, *the tail* is the part of the optimised trajectory from the previous sample that has yet to occur.

Let the optimal predictions at sampling instant k be given by the pair:

$$\underset{\rightarrow k|k}{\Delta \mathbf{u}} = \begin{bmatrix} \Delta \mathbf{u}_k \\ \Delta \mathbf{u}_{k+1|k} \\ \Delta \mathbf{u}_{k+2|k} \\ \vdots \end{bmatrix} ; \quad \underset{\rightarrow k+1}{\mathbf{y}} = \begin{bmatrix} \mathbf{y}_{k+1|k} \\ \mathbf{y}_{k+2|k} \\ \mathbf{y}_{k+3|k} \\ \vdots \end{bmatrix} \qquad (6.1)$$

At the next sampling instant, that is $k+1$, the first components of this pair have already occurred and hence can no longer be called predictions. The part that still constitutes a prediction is called the tail, i.e.,

$$
\underset{\rightarrow k+1,\text{tail}}{\Delta\mathbf{u}} = \begin{bmatrix} \Delta\mathbf{u}_{k+1|k} \\ \Delta\mathbf{u}_{k+2|k} \\ \Delta\mathbf{u}_{k+3|k} \\ \vdots \end{bmatrix} ; \quad \underset{\rightarrow k+2,\text{tail}}{\mathbf{y}} = \begin{bmatrix} \mathbf{y}_{k+2|k} \\ \mathbf{y}_{k+3|k} \\ \mathbf{y}_{k+4|k} \\ \vdots \end{bmatrix} \tag{6.2}
$$

It is important to note that the predictions given in the tail at $k+1$ were those computed at the previous sampling instant k.

Consistent decision making requires that the class of possible predictions at $k+1$ must be parameterised such that, for the nominal case, one can choose for example (should you want to):

$$
\underset{\rightarrow k+1|k+1}{\Delta\mathbf{u}} = \underset{\rightarrow k+1,\text{tail}}{\Delta\mathbf{u}}; \quad \underset{\rightarrow k+2|k+1}{\mathbf{y}} = \underset{\rightarrow k+2,\text{tail}}{\mathbf{y}} \tag{6.3}
$$

In more detail this implies, for example, that one could choose

$$
\underset{\rightarrow k+1|k+1}{\Delta\mathbf{u}} = \begin{bmatrix} \Delta\mathbf{u}_{k+1|k+1} \\ \Delta\mathbf{u}_{k+2|k+1} \\ \Delta\mathbf{u}_{k+3|k+1} \\ \vdots \end{bmatrix} = \begin{bmatrix} \Delta\mathbf{u}_{k+1|k} \\ \Delta\mathbf{u}_{k+2|k} \\ \Delta\mathbf{u}_{k+3|k} \\ \vdots \end{bmatrix} \tag{6.4}
$$

ILLUSTRATION: GPC and the tail

GPC automatically includes the tail in its prediction class. This is self-evident from a comparison of the tail with the d.o.f. at the current sample.

$$
\underset{\rightarrow k|tail}{\Delta\mathbf{u}} = \begin{bmatrix} \Delta\mathbf{u}_{k|k-1} \\ \Delta\mathbf{u}_{k+1|k-1} \\ \vdots \\ \Delta\mathbf{u}_{k+n_u-2|k-1} \\ 0 \end{bmatrix} ; \quad \underset{\rightarrow k}{\Delta\mathbf{u}} = \begin{bmatrix} \Delta\mathbf{u}_{k|k} \\ \Delta\mathbf{u}_{k+1|k} \\ \vdots \\ \Delta\mathbf{u}_{k+n_u-2|k} \\ \Delta\mathbf{u}_{k+n_u-|k} \end{bmatrix} \tag{6.5}
$$

STUDENT PROBLEM

Come up with some everyday illustrations where failure to incorporate *the tail* in planning procedures has led to disaster.

Summary: The tail is those parts of the predictions made at the previous sample which have still to take place. For a well-posed MPC law, the tail must be included in the parameterisation of the predictions, as failure to do so means one is unable to continue with a previously designed optimum strategy, that is, one is forced into a *change of mind*. GPC automatically includes the tail and thus in principle can make consistent decision making from one sample to the next.

6.3.2 Infinite horizons and the tail imply closed-loop stability

Consider the infinite horizon cost (for simplicity the weights here are scalar but clearly arguments transfer easily to matrix weights) and optimisation

$$J_k = \min_{\Delta\mathbf{u}_{k+i}, \ i=0,1,\dots,\infty} \quad J = \sum_{i=1}^{\infty} \|\mathbf{r}_{k+i} - \mathbf{y}_{k+i}\|_2^2 + \lambda \|\Delta\mathbf{u}_{k+i-1}\|_2^2 \qquad (6.6)$$

After optimisation, a whole input trajectory is defined; let this trajectory be given as

$$\Delta\underset{\rightarrow k|k}{\mathbf{u}}^T = [\Delta\mathbf{u}_{k|k}^T, \Delta\mathbf{u}_{k+1|k}^T, \dots] \qquad (6.7)$$

Of this trajectory, the first element $\Delta\mathbf{u}_{k|k}$ is implemented. Now consider the optimisation to be performed at the next sampling instant $k+1$. Clearly, when dealing with the nominal case and selecting the input increments for all future time, the optimal values cannot differ from those computed previously (this is a well-known observation in optimal control theory) as the underlying objective is the same. Hence

$$\begin{aligned} \Delta\underset{\rightarrow k+1|k+1}{\mathbf{u}}^T &= \left[\, \Delta\mathbf{u}_{k+1|k+1}^T, \quad \Delta\mathbf{u}_{k+2|k+1}^T, \quad \dots \,\right] \\ &= \left[\, \Delta\mathbf{u}_{k+1|k}^T, \quad \Delta\mathbf{u}_{k+2|k}^T, \quad \dots \,\right] \end{aligned} \qquad (6.8)$$

That is the new optimum must coincide with the tail: $\Delta\underset{\rightarrow k+1|k+1}{\mathbf{u}} = \Delta\underset{\rightarrow k+1,tail}{\mathbf{u}}$.

Now consider the implications on the cost function J using the notation that J_k is defined as the minimum of J at the kth sampling instant. From consideration of (6.6) it is clear that:

$$J_{k+1} = J_k - \|\mathbf{r}_{k+1} - \mathbf{y}_{k+1}\|_2^2 - \lambda \|\Delta\mathbf{u}_k\|_2^2 \qquad (6.9)$$

That is J_k and J_{k+1} share common terms apart from the values associated to sample k which do not appear in J_{k+1}. Hence it is clear from (6.9) that

$$J_{k+1} \le J_k \qquad (6.10)$$

That is, the optimal predicted cost J_k can never increase. Moreover it is monotonically decreasing. This can be demonstrated by a simple contradiction.

If $J_k = J_{k+1}$, then $\|\mathbf{r}_{k+1} - \mathbf{y}_{k+1}\|_2^2 + \lambda \|\Delta\mathbf{u}_k\|_2^2 = 0$ which implies the output is at set point and no input increment was required. This can only happen repeatedly if the plant is in steady-state at the desired set point. If $J_k \ne J_{k+1}$, then $J_{k+1} < J_k$; that is there is a decrease in cost. In summary, if one uses $n_y = n_u = \infty$, then the cost function J becomes a Lyapunov function which in turn implies convergence/stability because a monotonically reducing J_k must imply reducing values of both tracking errors and input increments.

What is most significant here is that the stability proof does not entail computation of implied closed-loop poles (e.g., 4.39). That is, one can state in advance of computing the control law that it will be stabilising. This is called an *a priori* guarantee and clearly differs from the *a posteriori* checks required by GPC in general.

Remark 6.1 *It is implicit in the above arguments that J_k is bounded, that is within the optimised predictions:*

$$\lim_{k \to \infty} (\mathbf{r}_k - \mathbf{y}_k) = \Delta\mathbf{u}_k = 0 \qquad (6.11)$$

This is an automatic outcome of using infinite horizons as the failure of condition (6.11) would imply an unbounded cost which is a contradiction of this being a minimum.

Remark 6.2 *The reader is reminded that the arguments in this section apply to the nominal case only. It is assumed that if a law is stabilising for the nominal case, then the gain and phase margins are likely to be large enough to imply some degree of robustness. A detailed discussion of how to handle uncertainty in a systematic fashion within MPC is far beyond the remit of this book.*

Summary: Using J_k as a potential Lyapunov function has now become an accepted method for establishing a priori stability of MPC control laws.

1. Use of infinite horizons guarantees that J_k is Lyapunov and thus guarantees stability of the associated MPC law.

2. Implicit in the proof that J_k is Lyapunov is the incorporation of *the tail* into the class of possible predictions.

6.3.3   Only the output horizon needs to be infinite

The above conclusions require only infinite output horizons because even if n_u is finite, the same arguments will follow. Consider the following optimisation

$$J_k = \min_{\Delta\mathbf{u}_{k+i-1},\ i=1,\dots,n_u} \quad J = \sum_{i=1}^{\infty} \|\mathbf{r}_{k+i} - \mathbf{y}_{k+i}\|_2^2 + \sum_{i=1}^{n_u} \lambda\|\Delta\mathbf{u}_{k+i-1}\|_2^2 \qquad (6.12)$$

Now enforce condition (6.8) which implies that $\Delta\mathbf{u}_{k+n_u|k+1} = 0$ and

$$J_{k+1} = J_k - \|\mathbf{r}_{k+1} - \mathbf{y}_{k+1}\|_2^2 - \lambda\|\Delta\mathbf{u}_k\|_2^2 \qquad (6.13)$$

that is J_k is still Lyapunov. More importantly, at sampling instant $k+1$, one has freedom to select $\Delta\underset{\rightarrow k+1|k+1}{\mathbf{u}}$ such that (6.8) is not satisfied if this makes J_{k+1} smaller still.

Remark 6.3 *It is worth noting that if n_u is small, although one can obtain a guarantee of stability there may still be significant prediction mismatch (as highlighted in Section 5.3) due to repeated changes of mind and this could result in poor performance.*

Remark 6.4 *The proofs given are tacitly assuming that condition (6.11) is possible. If not, the stability proof breaks down. Such an assumption may not be automatic where n_u is small:*

>   *1. The combination of tight constraints alongside large target changes or large disturbances can make (6.11) infeasible (see Chapters 7 and 8).*

>   *2. In the case of open-loop unstable plant, n_u must be large enough to ensure convergent predictions [116].*

However, these issues do not belong here but rather in a discussion on feasibility (e.g., [119, 130]) in Chapter 8.

>   **Summary:** Stability can be guaranteed by the appropriate use of infinite horizons and inclusion of the tail.
>   1. Alongside the mild requirement that (6.11) is feasible, infinite output horizons imply that the cost function is Lyapunov in the nominal case, which implies closed-loop stability.
>   2. This result relies on the d.o.f. being such that condition (6.8) can always be satisfied (that is, inclusion of the tail).

6.3.4   Stability proofs with constraints

The significance of the above a priori result is the constraint handling case. Stability analysis in the presence of constraints is non-simple in general as linear analysis does not apply; the implied control law is nonlinear in the presence of constraints. While one may be able to determine a tailored analysis for a specific system and constraint, it is more difficult to cater for a multitude of constraints and different system types, especially where an a priori result is desired.

   However, a significant advantage of the Lyapunov stability proof above is that it also applies to the nonlinear case and in particular, carries over to the constrained case in a very simple manner. The only requirement is that the constrained optimisation is always feasible; that is, there always exists a future input trajectory $\Delta\underset{\rightarrow k}{\mathbf{u}}$ such that constraints are predicted to be satisfied over the entire future, as well as satisfaction of (6.11). This requirement can be a demanding one in some cases but more often it is satisfied and thus the nominal stability assurance is valid.

Summary: The Lyapunov stability proof also applies during constraint handling, subject to feasibility. More details in Chapter 8.

6.3.5 Are infinite horizons impractical?

So far this section has not discussed the practicality, or not, of optimising Eqn.(6.6). At first sight this is intractable as it requires optimisation over an infinite number of d.o.f. and moreover, requires the computation and minimisation of an infinite number of output tracking errors. There is a need to represent this problem in a finite dimensional form before it becomes computationally tractable.

The most important thing for the reader to know is that this is possible, that is, the infinite dimensional optimisation:

$$J_k = \min_{\Delta \mathbf{u}_{k+i}, \ i=0,1,\ldots,\infty} \quad J = \sum_{i=1}^{\infty} \|\mathbf{r}_{k+i} - \mathbf{y}_{k+i}\|_2^2 + \sum_{i=0}^{\infty} \lambda \|\Delta \mathbf{u}_{k+i-1}\|_2^2 \qquad (6.14)$$

can be solved efficiently. Details will follow in later sections so here it suffices to give a little insight into how.

OPTIMAL CONTROL

The solution of the infinite horizon optimal control [15] problem has been well known from the 1960s and straightforward to solve in the unconstrained linear case using the appropriate Ricatti equations. So in the unconstrained case with linear models, the use of infinite horizons is straightforward both algebraically and numerically.

Nevertheless, as will be seen in the next chapter, practical MPC algorithms also deal with constraints so that the actual optimisation is slightly more challenging and takes the form:

$$\min_{\Delta \mathbf{u}_{\rightarrow k}} \quad J \quad \text{s.t.} \quad M\Delta \mathbf{u}_{\rightarrow k} - \mathbf{v}_k \leq 0 \qquad (6.15)$$

where $M\Delta \mathbf{u}_{\rightarrow k} - \mathbf{v}_k \leq 0$ are linear inequalities representing the constraints. This is a quadratic programming (QP) problem which clearly cannot be solved with an infinite dimensional vector of d.o.f.

Hence a key aim of this chapter is set up a framework which enables us to convert the apparently infinite dimensional optimisation of (6.14) into a finite dimensional one. In order to do this, it is easier and clearer for the reader if constraints are omitted in the first instance, hence the deferral of these to the following chapter.

Summary: In the presence of constraints a simplistic implementation of MPC based on large (infinite) horizons may appear to be intractable. However, optimisations (6.14, 6.15) can be converted into an equivalent finite dimensional form and thus are tractable.

Remark 6.5 *It is interesting that historical developments in MPC started from simple controllers such as minimum variance control in the 1970s; and gradually increased complexity, for instance GPC in the 1980s; and then in the 1990s has returned to a form of optimal (infinite horizon) control which was popular in the 1960s, though in a more tractable formulation. This increase in the complexity of the control strategy also reflects the increase in computing power available and hence problems such as (6.15) look increasingly tractable, whereas in the 1970s they were not.*

6.4 Dual-mode MPC – an overview

Having set out the desire to use infinite horizons, the next step is to consider how these can be managed practically. The accepted technique in the MPC community is the use of so-called dual-mode prediction structures and hence this section gives an overview of the principal components in a dual-mode strategy.

Readers should bear in mind that the key driving principle in this section is how we re-parameterise infinite dimensional prediction vectors in terms of a finite number of variables. How we do this *is a choice*, so this book will focus on the popular and successful choices in the literature although readers might like to consider what alternative parameterisations could be chosen.

6.4.1 What is dual-mode control in the context of MPC?

In the context of MPC, dual mode is a control strategy which is based on predictions with two distinct modes. One mode is used when the system is far away from steady-state (target operating point). The second mode is used when close to the desired operating point. Hence there is an implied switching between one mode of operation and another as the process converges to the desired state.

In MPC, the notation dual mode does not imply a switching between modes in real time, but rather is a description of how the predictions are set up. Hence, consider a prediction over n_y steps; then one could take the first n_c steps to be within mode 1 and the remaining steps to be in mode 2.

$$\underset{\rightarrow k}{\mathbf{x}} = [\underbrace{\mathbf{x}_{k+1|k}, \mathbf{x}_{k+2|k}, \ldots, \mathbf{x}_{k+n_c|k}}_{\text{Mode 1}}, \underbrace{\mathbf{x}_{k+n_c+1|k}, \mathbf{x}_{k+n_c+2|k}, \ldots}_{\text{Mode 2}}] \qquad (6.16)$$

One must emphasise that this switching/structure is in the predictions only. The

closed-loop control law is based on the input that arises from any associated optimisation and thus in effect is a conventional single-mode control law.

ILLUSTRATION: dual-mode prediction structure

It is useful to have a picture in one's head of what a dual-mode prediction looks
like. The dynamics in mode 1 may be different from mode 2 as mode 2 has fixed
dynamics whereas mode 1 is *totally free* [see Eqn. (6.18) for an example].

In this illustration you can
see the change in *prediction shape* as the prediction
goes from mode 1 to mode
2. The dotted line is looking forward to constraint
handling in the next chapter. In general mode 2 is
only feasible within a specified set and cannot be used
safely outside that set.

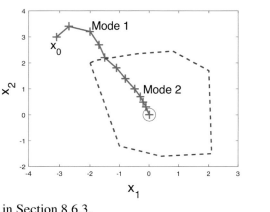

A further illustration is given in Section 8.6.3.

6.4.2   The structure/parameterisation of dual-mode predictions

The most usual parameterisation, at least in MPC, is that, within the predictions, the
first n_c control moves $\Delta\mathbf{u}_{k+i}$, $i = 0, \ldots, n_c - 1$, are free and that the remaining moves
$\Delta\mathbf{u}_{k+i}$, $i \geq n_c$ are given by a fixed feedback law, such as:

$$\mathbf{u}_{k+i-1|k} - \mathbf{u}_{ss} = -K[\mathbf{x}_{k+i-1|k} - \mathbf{x}_{ss}], \quad i > n_c \tag{6.17}$$

This assumption is the sleight of hand which reduces the problem to a finite dimen-
sional one as the far future inputs are no longer d.o.f. being dependent on $\mathbf{x}_{k+n_c+1|k}$
via (6.17) and the system model.

Remark 6.6 *Readers may realise that constraint (6.17) as well as reducing the number of d.o.f. also means that the optimisation (6.14) may be unable to locate the
global optimum due to the choice of K which embeds a given closed-loop dynamic
into the predictions. This discussion is deferred for a few sections as simple remedies
exist.*

An example of dual-mode predictions with a state space model are as follows:

$$\left\{ \begin{array}{c} \mathbf{x}_{k+i} - \mathbf{x}_{ss} = A[\mathbf{x}_{k+i-1} - \mathbf{x}_{ss}] + B[\mathbf{u}_{k+i-1} - \mathbf{u}_{ss}] \\ \mathbf{u}_{k+i-1} \text{ are d.o.f.} \end{array} \right\} \quad i = 1,2,...,n_c \quad \text{Model}$$

$$\left\{ \begin{array}{c} \mathbf{x}_{k+i} - \mathbf{x}_{ss} = [A - BK][\mathbf{x}_{k+i-1} - \mathbf{x}_{ss}] \\ \mathbf{u}_{k+i-1} - \mathbf{u}_{ss} = -K[\mathbf{x}_{k+i-1} - \mathbf{x}_{ss}] \end{array} \right\} \qquad i > n_c \qquad \text{Mode2}$$

$$(6.18)$$

Efficient realisations of this and modification for other model forms and offset-free prediction are omitted as straightforward (see Chapter 2); some elaboration will be given later in this chapter.

> **Summary:**
>
> 1. Dual-mode control refers to the predictions being separated into near transients (mode 1 behaviour) and asymptotic predictions (mode 2 behaviour).
>
> 2. It is normal for mode 2 behaviour to be given by a known control law. In effect this is how the d.o.f. are reduced to a finite number as the asymptotic values of the future input are now dependent on the prediction $\mathbf{x}_{k+n_c|k}$, the model and the chosen feedback K, as is clear in Equation (6.18).
>
> 3. The feedback K is often referred to as the terminal feedback, that is the feedback implicit in the asymptotic phase of mode 2 of the predictions.

6.4.3   Overview of MPC dual-mode algorithms: Suboptimal MPC (SOMPC)

Dual-mode MPC describes a philosophy and hence many different algorithms can be developed. Consequently, this section illustrates a reasonably generic variant based on a state space model. This variant is denoted Suboptimal MPC (SOMPC) as it may not capture the global optimal; the variant which captures the global optimal (OMPC) is more restricted which could be a disadvantage at times.

Algorithm 6.1 SOMPC for the unconstrained case

1. Define a stabilising control law, say $\mathbf{u}_k - \mathbf{u}_{ss} = -K[\mathbf{x}_k - \mathbf{x}_{ss}]$.

2. Define n_c *and compute the prediction equations as discussed in (6.18) and the vector of d.o.f.*

$$\underset{\rightarrow k}{\mathbf{u}} = [\mathbf{u}_k^T, \ldots, \mathbf{u}_{k+n_c-1}^T]^T \qquad (6.19)$$

3. *Define an on-line performance measure such as (4.13)*

$$J = \sum_{i=1}^{\infty} \|\mathbf{x}_{k+i} - \mathbf{x}_{ss}\|_2^2 + \lambda \|\mathbf{u}_{k+i-1} - \mathbf{u}_{ss}\|_2^2 \qquad (6.20)$$

4. *At each sampling instant, minimise the infinite horizon cost as follows:*

$$\min_{\underset{\rightarrow k}{\mathbf{u}}} \quad J \quad \text{s.t.} \quad \mathbf{u}_{k+i} - \mathbf{u}_{ss} = -K[\mathbf{x}_{k+i} - \mathbf{x}_{ss}], \quad i \geq n_c \qquad (6.21)$$

We will leave until a later section the discussion of the algebra which facilitates the optimisation of (6.21) so as not to obscure concepts at this point.

> **Summary:** SOMPC is a generic dual-mode approach because no attempt has been made to define the terminal feedback K beyond the requirement for it to be stabilising. Different choices of K imply MPC algorithms with different properties. However, readers will note that *knowledge* of an appropriate stabilising feedback is a pre-requisite to apply a standard dual-mode MPC approach.

6.4.4   Is SOMPC guaranteed stabilising?

It was stated in Section 6.3.2 that the inclusion of components such as an infinite horizons in the performance index and the tail in the parameterisation of the predictions, is sufficient to guarantee nominal stability. Consequently if we can show these two components are embedded within SOMPC, then the algorithm is guaranteed stabilising for the nominal case.

Firstly it is clear that SOMPC deploys infinite horizons and moreover the associated predictions are convergent to zero error because of the definition of the stabilising feedback $\mathbf{u}_k - \mathbf{u}_{ss} = -K[\mathbf{x}_k - \mathbf{x}_{ss}]$ or in deviation variables $\tilde{\mathbf{u}}_k = -K\tilde{\mathbf{x}}_k$. Hence J is bounded and can serve as a candidate Lyapunov function as required.

Therefore it remains only to show that the parameterisation of the input predictions in (6.18) implies the inclusion of the tail.

Theorem 6.1 *The prediction class deployed in SOMPC includes the tail.*

Proof: Given the terminal feedback is K, the closed-loop dynamic in the far future is given by $\tilde{\mathbf{x}}_{k+i+1} = \Phi\tilde{\mathbf{x}}_{k+i}, \quad \Phi = A - BK$. Compare the input predictions at two consequent sampling instants:

$$
\underset{\rightarrow k+1,tail}{\tilde{\mathbf{u}}} =
\begin{bmatrix}
\tilde{\mathbf{u}}_{k+1|k} \\
\vdots \\
\tilde{\mathbf{u}}_{k+n_c-1|k} \\
-K\tilde{\mathbf{x}}_{k+n_c|k} \\
-K\Phi\tilde{\mathbf{x}}_{k+n_c|k} \\
-K\Phi^2\tilde{\mathbf{x}}_{k+n_c|k} \\
\vdots
\end{bmatrix}
\quad ; \quad
\begin{bmatrix}
\tilde{\mathbf{u}}_{k+1|k+1} \\
\vdots \\
\tilde{\mathbf{u}}_{k+n_c-1|k+1} \\
\tilde{\mathbf{u}}_{k+n_c|k+1} \\
-K\tilde{\mathbf{x}}_{k+n_c+1|k+1} \\
-K\Phi\tilde{\mathbf{x}}_{k+n_c+1|k+1} \\
\vdots
\end{bmatrix}
= \underset{\rightarrow k+1}{\tilde{\mathbf{u}}}
\qquad (6.22)
$$

In order for these two vectors to be the same, that is, for the tail to be an option, one needs to be able to choose

$$
\begin{aligned}
\tilde{\mathbf{u}}_{k+i|k+1} &= \tilde{\mathbf{u}}_{k+i|k}, & i = 1,...,n_c - 1 \\
\tilde{\mathbf{u}}_{k+n_c|k+1} &= -K\tilde{\mathbf{x}}_{k+n_c|k} \\
\tilde{\mathbf{x}}_{k+n_c+1|k+1} &= \Phi\tilde{\mathbf{x}}_{k+n_c|k}
\end{aligned}
\qquad (6.23)
$$

Clearly the first of these is straightforward, as $\tilde{\mathbf{u}}_{k+i|k+1}$, $i = 1,...,n_c - 1$ constitute d.o.f. Second, one can force $\tilde{\mathbf{u}}_{k+n_c|k+1} = -K\tilde{\mathbf{x}}_{k+n_c|k}$, as this also is a d.o.f. Finally, the satisfaction of the first two implies that $\tilde{\mathbf{x}}_{k+n_c+1|k+1} = \Phi\tilde{\mathbf{x}}_{k+n_c|k}$. □

> **Summary:** The SOMPC algorithm is guaranteed stabilising because it meets the basic requirements of deploying infinite horizons, convergent predictions and incorporates the tail in the class of predictions.

6.4.5   Why is SOMPC suboptimal?

The main motivation for using dual-mode predictions is that they give a handle on the predictions over an infinite horizon and in effect allow the horizons to be infinite; it was noted in Section 6.3 that the use of infinite horizons is a pre-requisite to capture the global optimum and thus to allow MPC to obtain truly optimum behaviour.

The part of the prediction in mode 2 is governed by a known control law and thus can be evaluated in its entirety through the dependence on the initial state $\mathbf{x}_{k+n_c|k}$; however, this is also a potential weakness. The predictions of mode 2 are based on *pre-selected* dynamics linked to the choice of K and these may not be optimal or even close to optimal if K is poorly chosen. Hence one is embedding into the predictions a dynamic that may not be desirable and this would be a contradiction of the requirements for a well-posed optimisation.

Summary: Dual-mode predictions allow a reduction in the number of d.o.f. to capture the predictions over an infinite horizon, but this is achieved by embedding a closed-loop dynamic linked to the state feedback K and thus with a poor choice of K, these predictions could be far from the desired behaviour and thus would embed the sort of inconsistency which is a main weakness of GPC with small n_u.

6.4.6 Improving optimality of SOMPC, the OMPC algorithm

The weakness of SOMPC is the choice of terminal feedback K in (6.18). Unsurprisingly therefore one can improve SOMPC significantly by making a good choice for K. This section sets out what is now taken to be the gold standard in the literature of a well-posed MPC algorithm [170].

Consider the objective which we really want to optimise, for example:

$$\min_{\tilde{\mathbf{u}}_{k+i}, \ i=0,1,\ldots,\infty} J = \sum_{i=1}^{\infty} \tilde{\mathbf{x}}_{k+i}^{T} Q \tilde{\mathbf{x}}_{k+i} + \tilde{\mathbf{u}}_{k+i-1}^{T} R \tilde{\mathbf{u}}_{k+i-1} \tag{6.24}$$

It is known from optimal control theory [15] that this problem has a simple and fixed analytical answer, that is:

$$\begin{aligned} \tilde{\mathbf{u}}_k &= -K\tilde{\mathbf{x}}_k; \\ K &= (R + B^T S B)^{-1} B^T S A \\ A^T S A - S - (A^T S B)(R + B^T S B)^{-1}(B^T S A) + Q &= 0 \end{aligned} \tag{6.25}$$

Finding the optimal state feedback

It is not the purpose of this book to discuss optimal control. However, the solution of (6.25) is well known and thus provided by MATLAB for convenience using a simple built-in file:

[K]= dlqr(A,B,Q,R);

In consequence, one can easily determine the optimum state feedback K should this be desired. Of course one should remember that *optimal* is defined relative to the cost function J and thus is still, in some sense, *arbitrary*. However, what is more critical here is that the optimisation of (6.24) is well-posed and thus gives answers from one sample to the next which reinforce each other; this is a prerequisite for expectations that MPC will behave in a sensible and consistent manner.

Summary: If the state feedback is chosen as the one arising from a standard optimum control problem with the same performance index, then the associated MPC optimisation is well-posed in that the ideal closed-loop trajectories are now embedded as a possible solution.

ANOTHER ELEPHANT IN THE ROOM: Readers will rightly be wondering that if we can find the feedback which optimises performance index (6.24) using simple analytic solutions available via optimal control theory (e.g., Eqn. (6.25)), then why bother with MPC and all the associated prediction algebra at all which surely by comparison is *suboptimal* by definition and numerically more demanding?

The answer is that optimal control gives the optimal unconstrained solution. Although we have yet to introduce constraints in this book, a quick review of Chapter 1 demonstrates that constraints are a critical factor in industrial applications and optimal control theory does not cater for constraints, whereas the MPC frameworks we are setting up will enable simple incorporation of constraints as demonstrated in the next chapter.

STUDENT PROBLEM

How would you communicate, in an exam scenario, the core concepts and steps in setting up a dual-mode MPC algorithm as well as demonstrate that this leads to a guarantee of closed-loop stability in the nominal case?

6.5 Algebraic derivations for dual-mode MPC

The previous sections focussed primarily on a conceptual introduction to SOMPC and OMPC. This section will derive the corresponding algebra and thus can be omitted by those happy to take the concepts and go straight to implementation using the provided MATLAB files.

This section will be split into four parts. The first shows how to compute a quadratic cost function over an infinite horizon with single mode linear predictions. The second shows how to construct a performance index with dual-mode predictions and these parts are then united to define the dual-mode MPC algorithm.

The key ingredient in dual-mode strategies is the separation: mode 1 is totally free, whereas mode 2 is pre-determined. This separation limits the number of d.o.f. to those in mode 1 hence giving tractable optimisations. Second, by pre-determining the behaviour of mode 2 predictions, one opens them up to linear analysis. In particular, it is noted that the predictions in mode 2 are deterministic given the predicted value of the state $\mathbf{x}_{k+n_c}$ which has an affine dependence on the d.o.f. $\underset{\rightarrow k}{\mathbf{u}}$ and hence is easy to handle.

Remark 6.7 *For convenience, to reduce excessive notation, this section will express everything in terms of* $\mathbf{x}, \mathbf{u}$ *rather than* $\tilde{\mathbf{x}}, \tilde{\mathbf{u}}$ *and assuming that* $\mathbf{u}_{ss} = \mathbf{x}_{ss} = 0$.

6.5.1 The cost function for linear predictions over infinite horizons

Assume that predictions are deterministic, then one can evaluate the corresponding infinite horizon cost function using a Lyapunov type of equation. For instance, assume

$$\mathbf{x}_{k+i} = \Phi \mathbf{x}_{k+i-1} = \Phi^i \mathbf{x}_k; \quad \mathbf{u}_{k+i} = -K \mathbf{x}_{k+i} = -K \Phi^i \mathbf{x}_k; \quad \Phi = A - BK \quad (6.26)$$

and that

$$J_k = \sum_{i=0}^{\infty} \mathbf{x}_{k+i+1}^T Q \mathbf{x}_{k+i+1} + \mathbf{u}_{k+i}^T R \mathbf{u}_{k+i} \quad (6.27)$$

Substitute into (6.27) from (6.26):

$$
\begin{aligned}
J &= \sum_{i=0}^{\infty} \mathbf{x}_k^T (\Phi^{i+1})^T Q \Phi^{i+1} \mathbf{x}_k + \mathbf{x}_k^T K^T (\Phi^i)^T R \Phi^i K \mathbf{x}_k \\
&= \sum_{i=0}^{\infty} \mathbf{x}_k^T \underbrace{[(\Phi^{i+1})^T Q \Phi^{i+1} + K^T (\Phi^i)^T R \Phi^i K]}_{P} \mathbf{x}_k \\
&= \mathbf{x}_k^T P \mathbf{x}_k
\end{aligned}
\quad (6.28)
$$

where

$$P = \sum_{i=0}^{\infty} (\Phi^{i+1})^T Q \Phi^{i+1} + (\Phi^i)^T K^T R K \Phi^i$$

It can be shown by simple substitution that one can solve for P using a Lyapunov equation:

$$\Phi^T P \Phi = P - \Phi^T Q \Phi - K^T R K \quad (6.29)$$

Summary: For linear predictions (6.26), the quadratic cost function (6.27) takes the form below where P is determined from a Lyapunov Equation (6.29).

$$J = \mathbf{x}_k^T P \mathbf{x}_k \quad (6.30)$$

The MATLAB file **dlyap.m** can be used to solve equations of the type (6.29).

6.5.2 Forming the cost function for dual-mode predictions

The state evolves according to model (6.18). Hence it is convenient to separate cost (6.27) into two parts for modes 1 and 2.

$$J = J_1 + J_2; \qquad \begin{aligned} J_1 &= \sum_{i=0}^{n_c-1} \mathbf{x}_{k+i+1}^T Q \mathbf{x}_{k+i+1} + \mathbf{u}_{k+i}^T R \mathbf{u}_{k+i} \\ J_2 &= \sum_{i=0}^{\infty} \mathbf{x}_{k+n_c+i+1}^T Q \mathbf{x}_{k+n_c+i+1} + \mathbf{u}_{k+n_c+i}^T R \mathbf{u}_{k+n_c+i} \end{aligned} \qquad (6.31)$$

The cost J_1 can be constructed using the usual predictions of Section 2.4. For instance

$$\mathbf{x}_{\underset{\rightarrow}{k+1}} = P_x \mathbf{x}_k + H_x \mathbf{u}_{\underset{\rightarrow}{k}} \qquad (6.32)$$

where $\mathbf{x}_{\underset{\rightarrow}{k+1}}, \mathbf{u}_{\underset{\rightarrow}{k}}$ are defined for a horizon n_c. Hence

$$J_1 = [P_x \mathbf{x}_k + H_x \mathbf{u}_{\underset{\rightarrow}{k}}]^T \text{diag}(Q)[P_x \mathbf{x}_k + H_x \mathbf{u}_{\underset{\rightarrow}{k}}] + \mathbf{u}_{\underset{\rightarrow}{k}} \text{diag}(R) \mathbf{u}_{\underset{\rightarrow}{k}} \qquad (6.33)$$

Using the result of the previous section (Equation (6.28)), it is clear that the cost J_2 depends only on $\mathbf{x}_{k+n_c|k}$ and can be represented as

$$J_2 = \mathbf{x}_{k+n_c|k}^T P \mathbf{x}_{k+n_c|k}; \qquad \Phi^T P \Phi = P - \Phi^T Q \Phi - K^T R K \qquad (6.34)$$

One can use the last block rows of prediction (6.32) to find a prediction for $\mathbf{x}_{k+n_c|k}$; define this as

$$\mathbf{x}_{k+n_c|k} = P_{n_c} \mathbf{x}_k + H_{n_c} \mathbf{u}_{\underset{\rightarrow}{k}} \qquad (6.35)$$

where P_{n_c}, H_{n_c} are the n_c^{th} block rows of P_x, H_x, respectively. Hence

$$J_2 = [P_{n_c} \mathbf{x}_k + H_{n_c} \mathbf{u}_{\underset{\rightarrow}{k}}]^T P [P_{n_c} \mathbf{x}_k + H_{n_c} \mathbf{u}_{\underset{\rightarrow}{k}}] \qquad (6.36)$$

Finally one can combine J_1, J_2 from (6.33, 6.36) to give:

$$\begin{aligned} J = \; & [P_x \mathbf{x}_k + H_x \mathbf{u}_{\underset{\rightarrow}{k}}]^T \text{diag}(Q)[P_x \mathbf{x}_k + H_x \mathbf{u}_{\underset{\rightarrow}{k}}] + \mathbf{u}_{\underset{\rightarrow}{k}} \text{diag}(R) \mathbf{u}_{\underset{\rightarrow}{k}} \\ & + [P_{n_c} \mathbf{x}_k + H_{n_c} \mathbf{u}_{\underset{\rightarrow}{k}}]^T P [P_{n_c} \mathbf{x}_k + H_{n_c} \mathbf{u}_{\underset{\rightarrow}{k}}] \end{aligned} \qquad (6.37)$$

This can be simplified and is summarised next.

Summary: The dual-mode performance index takes a simple quadratic form with n_c block d.o.f.

$$J = \mathbf{u}_{\underset{\rightarrow}{k}}^T S \mathbf{u}_{\underset{\rightarrow}{k}} + \mathbf{u}_{\underset{\rightarrow}{k}}^T T \mathbf{x}_k + w \qquad (6.38)$$

where $S = H_x^T \text{diag}(Q) H_x + \text{diag}(R) + H_{n_c}^T P H_{n_c}$, $T = 2[H_x^T \text{diag} Q P_x + H_{n_c}^T P P_{n_c}]$ and w does not depend on the d.o.f. $\mathbf{u}_{\underset{\rightarrow}{k}}$.

6.5.3 Computing the SOMPC control law

We now have all the components required to define a dual-mode MPC algorithm. The cost is given in (6.38) and the minimisation of this wrt to the d.o.f. $\underset{\rightarrow k}{\mathbf{u}}$ follows analogous algebra to that for MPC in Section 4.5.3; in consequence this is presented concisely.

The relevant optimisation is:

$$\min_{\underset{\rightarrow k}{\mathbf{u}}} \quad \underset{\rightarrow k}{\mathbf{u}}^T S \underset{\rightarrow k}{\mathbf{u}} + \underset{\rightarrow k}{\mathbf{u}}^T T \mathbf{x}_k \tag{6.39}$$

where the reader is reminded that constraints (6.18) are satisfied implicitly in the choices for S, T. The solution is:

$$\mathbf{u}_k = E_1^T \underset{\rightarrow k}{\mathbf{u}} = \underbrace{-0.5 E_1^T S^{-1} T}_{K_{SOMPC}} \mathbf{x}_k \tag{6.40}$$

Unsurprisingly, SOMPC takes the form of a fixed state feedback K_{SOMPC} which thus is analogous in structure to GPC which also gives a fixed linear control law.

Remark 6.8 *Recall that* $\underset{\rightarrow k}{\mathbf{u}}$ *in the optimisation comprises the first n_c control moves only. The choice of n_c therefore controls the number of d.o.f. and the dimension of the optimisation (6.39).*

STUDENT PROBLEM

The derivation here has focussed solely on performance index (6.24). Naturally one could also consider other performance indices such as discussed in Section 4.4. How would you modify the algebra to support the use of other performance indices?

Summary: SOMPC reduces to a simple fixed state feedback (6.40), in the unconstrained case.

6.5.4 Definition of terminal mode control law via optimal control

It was noted earlier that the embedded feedback K in the long-term predictions could be chosen as that arising from the optimal control solution of (6.25). The arguments on well-posed optimisations in for example Section 5.4 proposed the need to ensure consistency between the optimised predictions and the global optimum behaviour that is achievable (here optimum is measured relative to the chosen performance index).

However, due to the need to distinguish the different design options, this section will define two performance indices and their roles.

1. **Measure of performance:** The performance index used by MPC will be taken as (6.27) with weighting matrices Q, R, that is:

$$J = \sum_{i=0}^{\infty} \mathbf{x}_{k+i+1}^T Q \mathbf{x}_{k+i+1} + \mathbf{u}_{k+i}^T R \mathbf{u}_{k+i} \qquad (6.41)$$

2. **Embedded feedback in mode 2:** The performance index used in an optimal control design to determine a terminal feedback K will be given as:

$$J = \sum_{i=0}^{\infty} \mathbf{x}_{k+i+1}^T Q_2 \mathbf{x}_{k+i+1} + \mathbf{u}_{k+i}^T R_2 \mathbf{u}_{k+i} \qquad (6.42)$$

where the reader will note the main difference is the substitution of Q, R by Q_2, R_2.

> **Summary:** It is not necessary to assume that the terminal feedback arises from an optimal control problem (6.42), but such a choice is convenient, enables well-posed MPC optimisations and is widely adopted in the literature.

6.5.5   SOMPC reduces to optimal control in the unconstrained case

Taking the observations/design parameters of (6.41,6.42) together, the following important result falls out.

Theorem 6.2 *In the absence of constraints, and under the assumption that the terminal feedback of (6.18) is chosen using (6.42) with $Q_2 = Q$, $R_2 = R$, then the optimised control law (6.40) must be such that:*

$$K_{SOMPC} = K \qquad (6.43)$$

Proof: Rather than dealing with this too mathematically, instead here a common sense proof is provided. With $Q_2 = Q$, $R_2 = R$, SOMPC minimises the same performance index (6.41) as optimal control, but with fewer d.o.f. due to the embedded terminal mode (mode 2). However, by definition, this embedded mode 2 matches precisely the corresponding solution provided by optimal control and moreover SOMPC is also able to use the d.o.f. in mode 1 to match the mode 1 part of optimal control. In consequence both will give the same solution. □

Corollary 6.1 *In the absence of constraints, and under the assumption that the terminal feedback of (6.18) is chosen NOT to equal that of (6.25) (that is $Q_2 \neq Q$, $R_2 \neq R$), then the optimised control law (6.40) must be different so that:*

$$K_{SOMPC} \neq K \qquad (6.44)$$

Proof: Because the terminal mode (mode 2) embedded into the predictions of SOMPC cannot match the equivalent part of the optimal control solution, then necessarily the two optimisations cannot be equivalent and thus must give different answers. □

> **Summary:** SOMPC is equivalent to optimal control, in the unconstrained case, if and only if $Q_2 = Q$, $R_2 = R$; this special case is denoted as OMPC, in other words optimal MPC [171].

6.5.6 Remarks on stability and performance of SOMPC/OMPC

A remarkable result has been demonstrated in this chapter when considered in the context of the tuning challenges for GPC illustrated in Chapter 5.

SOMPC is always stable in the unconstrained case

With the minor constraint that the underlying K used in (6.18) is stabilising, the control law given in (6.40) is guaranteed stabilising for the nominal case, irrespective of the choice of K and n_c (and implicitly for any Q, Q_2, R, R_2). As stability is assured, the designer can focus their energies on other aspects such as performance and constraint handling.

The reader is then left with some interesting questions.

1. How should I choose n_c?

2. How should I choose the underlying feedback K (equivalently the weights Q_2, R_2 although in fact K can be chosen by any design method)?

3. How should I choose the weights Q, R in the performance index J?

The next section of this chapter will use numerical illustrations to indicate how these decisions might be taken, but alongside a warning that *generic* guidance should always be used with caution as each different scenario has its own requirements which may imply different solution paths are logical. As always in this book the reader is encouraged to understand the likely repercussions of any decisions they take by understanding fully what an algorithm is doing; that way the decisions are more likely to be on solid foundations.

Some brief observations would be helpful at this point.

- If the choice of terminal control law K is poor, then there could still be significant prediction mismatch and when n_c is small, this could lead to poor performance.

[This is analogous to the problems with small n_u with GPC illustrated in for example Sections 5.5.2 and 5.5.4.]

- Ideally one would want to keep n_c as small as possible as this reduces the online computational load. For very small n_c one would want K to be chosen according to (6.25) as this ensures a well-posed optimisation.

- In general, the literature will always take K from (6.25), as to do otherwise seems to be embedding a poorly posed optimisation to no end. Positive reasons for not doing this are largely related to constraint handling [31] and will be discussed in the following chapter.

- As will be seen in the following chapter, there is often a trade-off between the choice of n_c, the choice of K (and implied performance through conflicts between Q, R and Q_2, R_2) and the volume of the associated operating region. As with all trade-offs, these may have not a simple systematic solution.

Summary:

1. The use of infinite horizons and incorporation of the tail allows stability guarantees.

2. The dual-mode/SOMPC paradigm gives an implementation of infinite horizon MPC which is computationally tractable and relatively straightforward to implement.

3. However, SOMPC may not have either a good performance or a large region of applicability. A good compromise between large operating regions (SOMPC) and good performance (OMPC) is a significant design issue which is discussed in Chapter 8. This chapter focuses solely on performance.

6.6 Closed-loop paradigm implementations of OMPC

The previous sections have shown that dual-mode predictions facilitate the design of MPC algorithms with both guaranteed stability and the potential for good performance. This section will consider a related theme which is: *What is a good way to set up a dual-mode MPC algorithm?* The argument made here is that the closed-loop paradigm (CLP), to be introduced next:

1. Gives better numerical conditioning [129, 138] of the optimisation which is essential for open-loop unstable plant (see Section 5.8).

2. Gives useful insight into the structure of dual-mode strategies in that it shows how far one is away from optimum due to constraint handling.

3. Makes robustness analysis more straightforward (e.g., [50, 81, 90, 139]) even for the constrained case.

Consequently, many recent algorithms in the literature [114, 181] tacitly assume the CLP structure due to its beneficial properties.

In the nominal case, the CLP control law is identical to an equivalent open-loop paradigm (OLP) strategy such as in Section 6.5, hence ultimately whether one uses a CLP or an OLP approach will depend a little on personal preference and the particular scenario.

> **Summary:** The closed-loop paradigm is an alternative mechanism for implementing SOMPC/OMPC which can have advantages and improve insight.

6.6.1   Overview of the CLP concept

The basic idea with the CLP is analogous to that in Equation (6.18), that is to embed a stabilising control law (which could be arbitrary but usually is chosen with some aim). However, the core difference here is that the embedding is also done within mode 1 of the predictions. Alternatively one could take the view that the control law is hardwired into the entire prediction computation which implies one has pseudo closed-loop predictions throughout. Thus the predictions are:

$$
\begin{aligned}
\mathbf{x}_{k+i} &= A\mathbf{x}_{k+i-1} + B\mathbf{u}_{k+i-1}; \quad \mathbf{u}_{k+i-1} = -K\mathbf{x}_{k+i-1} + \mathbf{c}_{k+i-1}; \quad i = 1,2,\dots,n_c \\
\mathbf{x}_{k+i} &= A\mathbf{x}_{k+i-1} + B\mathbf{u}_{k+i-1}; \quad \mathbf{u}_{k+i-1} = -K\mathbf{x}_{k+i-1}; \quad\quad\quad\; i > n_c
\end{aligned}
\tag{6.45}
$$

where here the variables $\mathbf{c}_{k+i-1}$ constitute the d.o.f. within the predictions. It will be easier to understand the previous two sections after seeing the input prediction structure so this is illustrated next. The OLP (as given in (6.18)) and CLP prediction (6.45) structures are given together to aid comparison.

$$
\underset{\rightarrow OLP}{\mathbf{u}} =
\begin{bmatrix}
\mathbf{u}_{k|k} \\
\mathbf{u}_{k+1|k} \\
\vdots \\
\mathbf{u}_{k+n_c-1|k} \\
\hline
-K\mathbf{x}_{k+n_c|k} \\
\vdots \\
-K\Phi^{n_y-n_c}\mathbf{x}_{k+n_c|k}
\end{bmatrix}
; \quad
\underset{\rightarrow CLP}{\mathbf{u}} =
\begin{bmatrix}
-K\mathbf{x}_k + \mathbf{c}_k \\
-K\mathbf{x}_{k+1|k} + \mathbf{c}_{k+1} \\
\vdots \\
-K\mathbf{x}_{k+n_c|k} + \mathbf{c}_{k+n_c-1} \\
-K\mathbf{x}_{k+n_c|k} \\
\vdots \\
-K\Phi^{n_y-n_c}\mathbf{x}_{k+n_c|k}
\end{bmatrix}
\tag{6.46}
$$

From here it is clear that the d.o.f. in the OLP predictions are the first n_c control moves; that is $\mathbf{u}_{k+i-1}, i = 0,\dots,n_c-1$, whereas in the CLP predictions the d.o.f. are the perturbations $\mathbf{c}_{k+i-1}, i = 0,\dots,n_c-1$. The mode 2 predictions are the same, that is, based on $\mathbf{x}_{k+n_c|k}$.

Lemma 6.1 *The class of predictions in (6.18) and (6.45) are exactly equivalent but are simply different ways of parameterising the same flexibility in the predictions.*

Proof: It is clear that the mode 2 predictions are identical. The mode 1 predictions differ only in that one uses $\mathbf{u}_{k+i-1}$ as the free variable and the second structures $\mathbf{u}_{k+i-1} = -K\mathbf{x}_{k+i-1} + \mathbf{c}_{k+i-1}$ and thus uses $\mathbf{c}_{k+i-1}$ as the free variable. Clearly for any choice of $\mathbf{u}_{k+i-1}$ there exists an equivalent $\mathbf{c}_{k+i-1}$ and vice versa.   □

The significance of the new parameterisation will be seen later in the formulation of the cost function J using both prediction sets; in general the CLP predictions give a better conditioned optimisation.

ILLUSTRATION: CLP structure

It is useful to have a picture in one's head of how the CLP paradigm might be implemented. As there is an implied underlying control law throughout the prediction structure, this could be viewed as hardwired into the system; that is, one could actually implement this control law.

The OMPC algorithm then supplies the perturbations **c** (marked with dotted lines as not hard wired) to improve performance and/or for constraint handling; the OMPC component utilises state, input and target information.

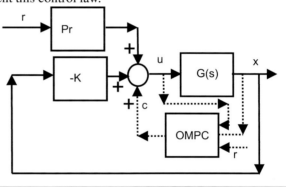

Remark 6.9 *The informed reader will notice a strong link with reference governor strategies [43, 46, 47]. This is unsurprising, as reference governor strategies are often MPC algorithms that simply deploy an alternative parameterisation of the d.o.f. to simplify computation [144].*

Remark 6.10 *If the predictions are set up as being based on a stabilising control law and hence based on a stable closed-loop, then one has a potentially trivial stability proof in the unconstrained case; simply use the control trajectories associated to this underlying law and choose $\mathbf{c}_k = 0$! However, this strategy is unlikely to be reasonable unless the underlying control law itself is well chosen; perhaps another argument for ensuring the embedded feedback is well chosen.*

The beauty of the CLP approach is that the d.o.f. can be set up as perturbations about the most desirable performance [171], that is, the performance that arises from unconstrained control. Hence one gains insight by viewing the magnitude of these perturbations. That is, the larger the perturbations, the further behaviour is away from optimal unconstrained behaviour.

Remark 6.11 *Stability can be demonstrated with an equivalent proof that the terms* $\mathbf{c}_k$ *tend to zero as one then reverts to a known stabilising feedback. However, this is sufficient but not necessary for stability as in cases where K is a suboptimal choice, non-zero values of* $\mathbf{c}_k$ *may be beneficial!*

Predictions with CLP

The reader is reminded that the prediction matrices for (6.45) are provided in Section 2.9.4:

$$\underset{\rightarrow k+1}{\mathbf{x}} = P_{cl}\mathbf{x}_k + H_c \underset{\rightarrow k}{\mathbf{c}}; \quad \underset{\rightarrow k}{\mathbf{u}} = P_{clu}\mathbf{x}_k + H_{cu}\underset{\rightarrow k}{\mathbf{c}} \tag{6.47}$$

The state after n_c steps will be denoted as

$$\mathbf{x}_{k+n_c|k} = P_{cl2}\mathbf{x}_k + H_{c2}\underset{\rightarrow k}{\mathbf{c}} \tag{6.48}$$

Predictions are affine in the current state and the d.o.f.

Summary: In effect, the CLP treats the embedded feedback as having beneficial properties to be preserved and thus the d.o.f. $\mathbf{c}_{k+i-1}$ capture how far one deviates away from this *desired/embedded* control law. This is a particularly useful insight, especially when it comes to constraint handling, or indeed the desire to ensure the MPC strategy acquires some pre-specified robustness properties [90]. Indeed, one could now hard wire this feedback, should that be desired.

The OLP and CLP are equivalent in the space covered but have different parameterisations of the d.o.f. In fact the CLP can be viewed as a means of normalising the space of optimisation variables which improves numerical conditioning.

6.6.2 The SOMPC/OMPC law with the closed-loop paradigm

Next the MPC implementation of the CLP is introduced in more detail. The main components of MPC are predictions, a performance index and d.o.f.. Hence there is a need to define each of these components for the CLP.

1. The d.o.f. are the perturbations $\mathbf{c}_{k+i}, i = 0, ..., n_c - 1$. It is usual to define these as:

$$\underset{\rightarrow k}{\mathbf{c}}^T = [\mathbf{c}_{k|k}^T, ..., \mathbf{c}_{k+n_c-1|k}^T]; \quad \mathbf{c}_{k+n_c+i|k} = 0, \ i \geq 0 \tag{6.49}$$

Note, it is implicit within the predictions that far future values of $\mathbf{c}_{k+i}$ are zero.

2. The optimisation of the performance index, for example (6.31), takes the normal format:

$$J = \sum_{i=0}^{n_c-1} \mathbf{x}_{k+i+1}^T Q \mathbf{x}_{k+i+1} + \mathbf{u}_{k+i}^T R \mathbf{u}_{k+i} \tag{6.50}$$
$$+ \sum_{i=0}^{\infty} \mathbf{x}_{k+n_c+i+1}^T Q \mathbf{x}_{k+n_c+i+1} + \mathbf{u}_{k+n_c+i}^T R \mathbf{u}_{k+n_c+i}$$

and hence J needs to be formulated in terms of the d.o.f. $\underset{\rightarrow k}{\mathbf{c}}$.

3. The predictions are given in (6.47, 6.48). There is a dependence on initial values in the loop and the assumptions on the perturbations $\underset{\rightarrow k}{\mathbf{c}}$.

4. It is noted from (6.34) that the part of the cost corresponding to mode 2 can be simplified to $J_2 = \mathbf{x}_{k+n_c|k}^T P \mathbf{x}_{k+n_c|k}$; $\Phi^T P \Phi = P - \Phi^T Q \Phi - K^T R K$.

5. Hence, following substituting (6.47, 6.48) into (6.50) gives the cost function as

$$
\begin{aligned}
J &= [P_{cl}\mathbf{x}_k + H_c \underset{\rightarrow k}{\mathbf{c}}]^T \mathrm{diag}(Q)[P_{cl}\mathbf{x}_k + H_c \underset{\rightarrow k}{\mathbf{c}}] \\
&\quad + [P_{clu}\mathbf{x}_k + H_{cu}\underset{\rightarrow k}{\mathbf{c}}]^T \mathrm{diag}(R)[P_{clu}\mathbf{x}_k + H_{cu}\underset{\rightarrow k}{\mathbf{c}}] \\
&\quad + [P_{cl2}\mathbf{x}_k + H_{c2}\underset{\rightarrow k}{\mathbf{c}}]^T P[P_{cl2}\mathbf{x}_k + H_{c2}\underset{\rightarrow k}{\mathbf{c}}] \\
&= \underset{\rightarrow k}{\mathbf{c}}^T S_c \underset{\rightarrow k}{\mathbf{c}} + 2\underset{\rightarrow}{\mathbf{c}}^T S_{cx}\mathbf{x}_k + w
\end{aligned}
\tag{6.51}
$$

where

$$
\begin{aligned}
S_c &= H_c^T \mathrm{diag}(Q)H_c + H_{cu}^T \mathrm{diag}(R)H_{cu} + H_{c2}^T P H_{c2} \\
S_{cx} &= H_c^T \mathrm{diag}(Q)P_{cl} + H_{cu}^T \mathrm{diag}(R)P_{clu} + H_{c2}^T P P_{cl2}
\end{aligned}
\tag{6.52}
$$

and w does not depend upon $\underset{\rightarrow k}{\mathbf{c}}$ and hence can be ignored.

NOTE: A numerically and algebraically easier method for defining S_c, S_{cx} is given in Section 6.6.4.1.

6. The CLP OMPC law is derived by minimising w.r.t. to the d.o.f., that is,

$$\min_{\underset{\rightarrow}{\mathbf{c}_k}} \; J = \underset{\rightarrow k}{\mathbf{c}}^T S_c \underset{\rightarrow k}{\mathbf{c}} + 2\underset{\rightarrow k}{\mathbf{c}}^T S_{cx}\mathbf{x}_k \tag{6.53}$$

which implies that the optimal unconstrained $\underset{\rightarrow k}{\mathbf{c}}$ is given from

$$\underset{\rightarrow k}{\mathbf{c}} = -S_c^{-1} S_{cx}\mathbf{x}_k \tag{6.54}$$

Notably this is a state feedback; that is, the optimal unconstrained perturbation is given by an additional feedback loop. The underlying control law is defined as $\mathbf{u}_k = -K\mathbf{x}_k + \mathbf{c}_k$ where $\underset{\rightarrow}{\mathbf{c}} = -S_c^{-1} S_{cx}\mathbf{x}_k$.

Summary: The CLP MPC control law in the unconstrained state space case is equivalent to a fixed state feedback:

$$\mathbf{u}_k = -\underbrace{[K + S_c^{-1} S_{cx}]}_{K_{SOMPC}}\mathbf{x}_k \tag{6.55}$$

Readers will note this is analogous to (6.40).

6.6.3 Properties of OMPC/SOMPC solved using the CLP

This section focusses on the advantages of the solution proposed in this section as opposed to the solution of Section 6.5.2. It is clear that both solve the same optimisation with the same flexibility within the predictions and thus, in theory will give identical control laws. However, the differing parameterisations of the d.o.f. can give significant differences to both numerical properties, conditioning and moreover, in sight.

6.6.3.1 Open-loop unstable systems

Readers are reminded of the findings in Section 5.8 where it was shown that basing computations on divergent predictions can lead to significant numerical ill-conditioning. It will be clear that the matrices in Section 6.5.2 suffer form this weakness and thus should not be used with open-loop unstable systems [138]. In such case the CLP is a simple and effective alternative.

6.6.3.2 Conditioning and structure of performance index J

Even for the case of open-loop stable models, one can argue that the conditioning of optimisation (6.53) could be preferred to (6.39). Evidence for this is given in the structure of S_c, S_{cx}.

Lemma 6.2 *For the case of OMPC, then $S_{cx} = 0$ and therefore the unconstrained optimal value for $\underset{\rightarrow k}{\mathbf{c}} = 0$.*

Proof: This follows from the definition of K as being optimal in which case one can state that $K_{OMPC} = K$ is a necessary outcome of optimisation (6.53) when $Q = Q_2, R = R_2$. This observation in combination with (6.55) proves that $S_{cx} = 0$. The consequence that $\underset{\rightarrow k}{\mathbf{c}} = 0$ is therefore automatic from (6.54). □

Lemma 6.3 *For the case of OMPC, then S_c is block diagonal of the form*

$$S_c = \mathrm{diag}([\Phi_1, \Phi_2, \cdots, \Phi_3])$$

That is, there is no interaction within the cost function between the different components $\mathbf{c}_k, \mathbf{c}_{k+1}, \ldots$.

Proof: One can consider the impact on the cost function of different perturbations, for example $\mathbf{c}_k, \mathbf{c}_{k+1}$. However, a key observation is whether the optimised choice for $\mathbf{c}_{k+1}$ depends, or not, on the previous choice of $\mathbf{c}_k$. From the previous lemma it is clear that in the unconstrained case that the optimum $\underset{\rightarrow k}{\mathbf{c}} = 0$, and therefore, irrespective of the choice of $\mathbf{c}_k$, the unconstrained optimal for $\mathbf{c}_{k+i}, i = 1, 2, \ldots$ must be zero. The diagonal structure of S_c is an obvious consequence of the lack of interdependence between these. □

Lemma 6.4 *For the case of OMPC, then the diagonal blocks in S_c will all be identical, that is:*

$$S_c = \mathrm{diag}([\Phi_1, \Phi_2, \cdots, \Phi_3]) = \mathrm{diag}([\Phi, \Phi, \cdots, \Phi])$$

Proof: From linearity it is clear that the change in expected performance due to an input perturbation over an infinite horizon cannot depend upon the sample when that perturbation occurred. Hence for perturbations of the same size it must be that:

$$\delta J = \mathbf{c}_k^T \Phi_i \mathbf{c}_k \equiv \mathbf{c}_k^T \Phi_j \mathbf{c}_k$$

The statement of the lemma then follows automatically as $\Phi_i \neq \Phi_j, i \neq j$ would be a contradiction of this.   □

Theorem 6.3 *The structure of optimisation (6.53) is ideal and well conditioned.*

Proof: This is an obvious consequence of the structure of S_c which is as close to an identity matrix as reasonably possible in addition to the observation that $S_{cx} = 0$. Indeed, for the scalar case then S_c is a simple multiple of the identity matrix.   □

> **Summary:** The use of the CLP offers improved conditioning compared to the open-loop paradigm. In many cases this improvement may be inconsequential, but at times it could be useful and also lead to optimisations that are easier to solve.
>
> Moreover, the optimisation gives a clear insight into the impact of constraints in that the optimised value for $\mathbf{c}_{\rightarrow k}$ in the unconstrained case should be zero, hence the larger the values the more constraints are causing a deviation away from the unconstrained optimal.

6.6.4   Using autonomous models in OMPC

When it comes to constraint handling, it will be important to have algebraically simple and clear mechanisms for forming both predictions and constraint inequalities. Although not strictly necessary in this chapter, here is a natural point to revisit the autonomous model formulation which is a natural way of capturing the dynamics of the various signals in play such as states, inputs, outputs, disturbances, references and perturbations $\mathbf{c}_k$.

The reader is reminded to review Section 2.9.5 which gives a concise overview of how to form predictions using an autonomous model with an augmented state capturing all core dynamics:

$$\mathbf{Z}_{k+1} = \underbrace{\begin{bmatrix} \Phi & [B,0,\ldots,0] \\ 0 & I_L \end{bmatrix}}_{\Psi} \mathbf{Z}_k; \quad \mathbf{Z}_k = \begin{bmatrix} \mathbf{x}_k \\ \mathbf{c}_{\rightarrow k} \end{bmatrix} \qquad (6.56)$$

where I_L is a block upper triangular matrix of identities. Define Γ, K_u such that

$$\mathbf{x}_k = \Gamma \mathbf{Z}_k; \quad \mathbf{u}_k = -K_u \mathbf{Z}_k \qquad (6.57)$$

6.6.4.1 Forming the predicted cost with an autonomous model

Let the cost function be given as

$$J = \sum_{i=0}^{\infty} \mathbf{x}_{k+i+1}^T Q \mathbf{x}_{k+i+1} + \mathbf{u}_{k+i-1}^T R \mathbf{u}_{k+i-1} \tag{6.58}$$

Substitute in from (6.56, 6.57), then

$$J = \sum_{i=0}^{\infty} \mathbf{Z}_{k+i+1}^T \Gamma^T Q \Gamma \mathbf{Z}_{k+i+1} + \mathbf{Z}_{k+i}^T K_u^T R K_u \mathbf{Z}_{k+i} \tag{6.59}$$

With some small rearrangement using $\mathbf{Z}_{k+1} = \Psi \mathbf{Z}_k$

$$J = \sum_{i=0}^{\infty} \mathbf{Z}_{k+i}^T [\Psi^T \Gamma^T Q \Gamma \Psi + K_u^T R K_u] \mathbf{Z}_{k+i} \tag{6.60}$$

or using $\mathbf{Z}_{k+i} = \Psi^i \mathbf{Z}_k$

$$J = \mathbf{Z}_k^T \underbrace{\left\{ \sum_{i=0}^{\infty} (\Psi^i)^T [\Psi^T \Gamma^T Q \Gamma \Psi + K_u^T R K_u] \Psi^i \right\}}_{S} \mathbf{Z}_k \tag{6.61}$$

Readers may notice immediately that S can be solved using a standard Lyapunov identity.

$$\Phi^T S \Phi = S - [\Psi^T \Gamma^T Q \Gamma \Psi + K_u^T R K_u] \tag{6.62}$$

And hence $J = \mathbf{Z}_k^T S \mathbf{Z}_k$ can be expanded as

$$J = \mathbf{Z}_k^T \begin{bmatrix} S_x & S_{cx} \\ S_{cx}^T & S_c \end{bmatrix} \mathbf{Z}_k \tag{6.63}$$

$$J = \mathbf{x}_k^T S_x \mathbf{x}_k + \underset{\rightarrow k}{\mathbf{c}}^T S_c \underset{\rightarrow k}{\mathbf{c}} + 2\mathbf{x}_k^T S_{cx} \underset{\rightarrow k}{\mathbf{c}}$$

The following section will analyse this formulation of J in detail and demonstrate the insights that can be derived from it.

Theorem 6.4 *The solution of (6.62) for a horizon of $n_c = 1$ is sufficient to determine the parameters for any horizon n_c for OMPC.*

  Proof: It is known that for OMPC, $S_{cx} = 0$ and the value of S_x is not needed, so need not be computed. Also it is known from Lemma 6.4 that S_c is block diagonal with all the blocks the same and independent of n_c, thus the value for $n_c = 1$ can be used to build S_c for any dimension. □

Summary: Using autonomous model formulations gives an algebraically convenient mechanism, that is (6.62), for determining the parameters of the cost function. In general this would be much more efficient than the conventional alternative of (6.52) and also has the advantage of extending easily to larger horizons.

6.6.4.2 Using autonomous models to support the definition of constraint matrices

In the case of constraint handling, it will be useful to expand the autonomous model formulation to capture more state information, and in particular variables such as the input, target and disturbance.

$$Z_{k+1} = \underbrace{\begin{bmatrix} \Phi & (I-\Phi)K_{xr} & [B,0,\ldots,0] & 0 \\ 0 & I & 0 & 0 \\ 0 & 0 & I_L & 0 \\ K & -K.K_{xr}-K_{ur} & [I,0,0,\ldots] & 0 \end{bmatrix}}_{\Psi} Z_k; \quad Z_k = \begin{bmatrix} \mathbf{x}_k \\ \mathbf{r}_{k+1}-\mathbf{d}_k \\ \underset{\rightarrow k}{\mathbf{c}} \\ \mathbf{u}_{k-1} \end{bmatrix}$$

(6.64)

where I_L is a block upper triangular matrix of identities. Define Γ, K_u such that

$$\mathbf{x}_k = \Gamma \mathbf{Z}_k; \quad \mathbf{u}_k = -K_u \mathbf{Z}_k \qquad (6.65)$$

In practice an SOMPC/OMPC cost function often makes use of deviation variables. These can be defined using Eqn.(2.24) and the expanded autonomous model above as follows (Note $K_{xr} = -K_{xd}, K_{ur} = -K_{ud}$):

$$\mathbf{x}_k - \mathbf{x}_{ss} = \underbrace{[I,-K_{xr},0,0]}_{K_{xss}} \mathbf{Z}_k; \quad \mathbf{u}_k - \mathbf{u}_{ss} = \underbrace{[-K,-K_{ur},0,0]}_{K_{uss}} \mathbf{Z}_k \qquad (6.66)$$

6.6.4.3 Forming the predicted cost with an expanded autonomous model

This section utilises the expanded autonomous model formulation of (6.64) alongside the use of deviation variables in the cost function and thus serves as a framework for a more realistic practical algorithm compared to the earlier parts of this chapter which deployed $\mathbf{x}_{ss} = \mathbf{u}_{ss} = 0$ for convenience.

Let the cost function be given as

$$J = \sum_{i=0}^{\infty} \tilde{\mathbf{x}}_{k+i+1}^T Q \tilde{\mathbf{x}}_{k+i+1} + \tilde{\mathbf{u}}_{k+i}^T R \tilde{\mathbf{u}}_{k+i} \qquad (6.67)$$

Substitute in from (6.56, 6.57, 6.66), then

$$J = \sum_{i=0}^{\infty} \mathbf{Z}_{k+i+1}^T K_{xss}^T \Gamma^T Q \Gamma K_{xss} \mathbf{Z}_{k+i+1} + \mathbf{Z}_{k+i}^T K_{uss}^T K_u^T R K_u K_{uss} \mathbf{Z}_{k+i} \qquad (6.68)$$

With some small re-arrangement using $\mathbf{Z}_{k+1} = \Psi \mathbf{Z}_k$

$$J = \sum_{i=0}^{\infty} \mathbf{Z}_{k+i}^T [\Psi^T K_{xss}^T \Gamma^T Q \Gamma K_{xss} \Psi + K_{uss}^T K_u^T R K_u K_{uss}] \mathbf{Z}_{k+i} \qquad (6.69)$$

or using $\mathbf{Z}_{k+i} = \Psi^i \mathbf{Z}_k$

$$J = \mathbf{Z}_k^T \underbrace{\left\{ \sum_{i=0}^{\infty} (\Psi^i)^T [\Psi^T K_{xss}^T \Gamma^T Q \Gamma K_{xss} \Psi + K_{uss}^T K_u^T R K_u K_{uss}] \Psi^i \right\}}_{S} \mathbf{Z}_k \qquad (6.70)$$

Readers may notice immediately that S can be solved using a standard Lyapunov identity.

$$\Phi^T S \Phi = S - [\Psi^T K_{xss}^T \Gamma^T Q \Gamma K_{xss} \Psi + K_{uss}^T K_u^T R K_u K_{uss}] \tag{6.71}$$

And hence $J = \mathbf{Z}_k^T S \mathbf{Z}_k$ can be expanded as

$$J = \mathbf{Z}_k^T \begin{bmatrix} S_x & S_{x,rd} & S_{cx} & S_{x,u} \\ S_{x,rd}^T & S_{rd,rd} & S_{c,rd} & S_{rd,u} \\ S_{cx}^T & S_{c,rd}^T & S_c & S_{cu} \\ S_{x,u}^T & S_{rd,u}^T & S_{cu}^T & S_u \end{bmatrix} \mathbf{Z}_k \tag{6.72}$$

In practice only the components depending on $\underset{\rightarrow k}{\mathbf{c}}$ need to be considered in any optimisation, so

$$J \equiv \underset{\rightarrow k}{\mathbf{c}}^T S_c \underset{\rightarrow k}{\mathbf{c}} + \underset{\rightarrow k}{\mathbf{c}}^T [S_{cx}\mathbf{x} + S_{c,rd}(\mathbf{r}_{k+1} - \mathbf{d}_k) + S_{cu}^T \mathbf{u}_{k-1}] \tag{6.73}$$

Remark 6.12 *As argued earlier, for OMPC the components $S_{cx}, S_{c,rd}, S_{cu}$ will be zero. However, this will not be the case for SOMPC.*

6.6.5 Advantages and disadvantages of the CLP over the open-loop predictions

In many cases whether one uses an OLP or CLP implementation of dual mode control, it will make little practical difference apart from the issue of insight. This is because if one minimises the same objective with the same d.o.f. (a re-parameterisation of the d.o.f. does not change the potential effects), then one must get the same answer. However, the reader will also be aware that the conditioning of different parameterisations can vary significantly. In general it is better to solve a well conditioned problem as opposed to a poorly conditioned problem. Herein lies one major potential benefit of the CLP; it changes the conditioning of the optimisation problem. In cases where the OLP is poorly conditioned the CLP may not be and hence it will give more reliable answers.

A contrary argument from some numerical specialists is linked to the potentially tailoring of the optimisation solvers. The CLP paradigm condenses a lot of information into the definition of the cost function (and later the constraint inequalities) and thus gives a dense formulation. However, at times formulations with far larger but sparse and strongly structured matrices can be easier to handle efficiently with a tailored solver and thus a dense formulation may not always be advantageous.

Summary: Whether one derives the SOMPC/OMPC law using the CLP or open-loop predictions, it is nevertheless clear that the algebra is far more complicated than for GPC and needs careful handling. The choice of which to use will ultimately be linked to personal preference and secondary issues such as numerical conditioning, choice of solver and constraint handling.

6.7   Numerical illustrations of OMPC and SOMPC

This section will use numerical (MATLAB based) illustrations to reinforce the findings of this chapter.

1. If K is chosen using (6.25), with $Q = Q_2, R = R_2$ (that is OMPC), what is the structure of the optimisation problem?

2. If K is chosen using (6.25), with $Q \neq Q_2, R \neq R_2$ (that is SOMPC), what is the structure of the optimisation problem?

3. Is there evidence of consistent decision making from one sample to the next for OMPC and SOMPC?

4. Is the optimised cost function J_k Lypanov as expected?

5. In addition, a number of MATLAB files are provided so that readers can explore the behaviour of OMPC/SOMPC across a range of scenarios and models (e.g., set point changes, rejection of output disturbances, with parameter uncertainty, with and without and observer to estimate the states).

As in the previous chapter, for clarity of illustration, primarily SISO illustrations are used. With MIMO examples the number of interacting issues and input/output signals mitigates against clear illustrations on paper. However, the code provided will cater for MIMO processes so that readers can undertake their own investigations.

6.7.1   Illustrations of the impact of Q, R on OMPC

It is assumed that the reader recognises that the weights Q, R effect the priorities in the performance index between different states and inputs. However, as noted in the previous chapter, the links between the actual values in the weights and the performance that results is not simple in general and is likely to need some degree of trial and error. Consequently, this topic is not presented in any detail beyond a few simple illustrations. Readers are invited to use the supplied code to assess such impacts for themselves.

> **ILLUSTRATION: Impact of changing Q, R on OMPC**
>
> Consider the model given in **ompcsimulate_impact_qr.m**. Three alternative choices of Q, R are used.

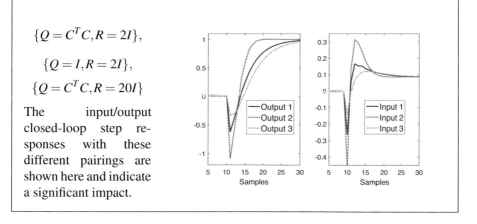

$\{Q = C^T C, R = 2I\}$,

$\{Q = I, R = 2I\}$,

$\{Q = C^T C, R = 20I\}$

The input/output closed-loop step responses with these different pairings are shown here and indicate a significant impact.

Remark 6.13 *Readers should note that the files used to support the illustrations in the next few subsections are* **ompc_simulate.m** *and* **opmc_simulateb.m**. *These files are designed to illustrate key points, but also, to be easy for users to read and edit, and thus they have very simple coding and consequently make simplifying assumptions such as states being fully measureable and disturbance values being known.*

For more complete (still without constraint handling) simulations, see the alternative files outlined in Section 6.7.7.

STUDENT PROBLEM

For a number of examples of your choice, investigate the impact of changes in Q, R in the closed-loop step response behaviour. To what extent are Q, R easy parameters to choose to achieve a desired trade-off between output tracking and input activity?

6.7.2 Structure of cost function matrices for OMPC

The data in **ompccost_example1.m** is based on the model:

$$A = \begin{bmatrix} 0.9 & 1 \\ 0.2 & 0.1 \end{bmatrix}; \quad B = \begin{bmatrix} 0.1 \\ 0.8 \end{bmatrix}; \quad C = \begin{bmatrix} 1.9 & -1 \end{bmatrix}; \quad Q = C^T C, R = I \quad (6.74)$$

and is based on OMPC so uses $Q_2 = Q, R_2 = R$.

An indicative output for S_c for $n_c = 1, 2, 3$ is:

$$\begin{bmatrix} 1.4673 \end{bmatrix}; \quad \begin{bmatrix} 1.4673 & 0 \\ 0 & 1.4673 \end{bmatrix}; \quad \begin{bmatrix} 1.4673 & 0 & 0 \\ 0 & 1.4673 & 0 \\ 0 & 0 & 1.4673 \end{bmatrix}; \quad (6.75)$$

This shows, as expected, that the diagonal elements of S_c do not change as n_c changes. Readers are encouraged to run the file for themselves as the matrices are rather large to show in general. It will be obvious that (to within computer accuracy), as expected S_c has a block diagonal structure with all the blocks identical. Also $S_{cx} = 0, \forall n_c$.

The data in **ompccost_example2.m** is based on the MIMO power generation model (see Appendix A). Readers are encouraged to run the file for themselves as the matrices are rather large to show in print for all n_c. It will be obvious that (to within computer accuracy), as expected S_c has a block diagonal structure with all the blocks identical and with values that do not vary with n_c. Also $S_{cx} = 0$.

6.7.3 Structure of cost function matrices for SOMPC

The data in **sompccost_example1.m** is based on the model (6.74) but the terminal mode control is such that $Q_2 \neq Q, R_2 \neq R$ so that $K \neq K_{OMPC}$.

$$K = [0.000098\ 0.599]; \quad K_{OMPC} = [0.006\ 0.041]$$

An indicative output for S_c for $n_c = 1, 2, 3$ is:

$$\begin{bmatrix} 1.469 \end{bmatrix}; \quad \begin{bmatrix} 1.469 & 0.048 \\ 0.048 & 1.469 \end{bmatrix}; \quad \begin{bmatrix} 1.469 & 0.048 & 0.007 \\ 0.048 & 1.469 & 0.048 \\ 0.007 & 0.048 & 1.469 \end{bmatrix}; \ S_{cx} \neq 0 \quad (6.76)$$

This shows, as expected, that the off-diagonal elements of S_c are no longer zero. Moreover, the matrices $S_{cx} \neq 0$ and are not negligible so that

$$\|S_c^{-1} S_{cx}\| \gg 0 \quad \Rightarrow \quad K_{SOMPC} \neq K_{OMPC}$$

Similar insights can be obtained from the power generation model example in **sompccost_example2.m**.

STUDENT PROBLEM

For a number of examples of your choice, validate that for OMPC, the matrix $S_{cx} = 0$ for any n_c whereas for SOMPC $S_{cx} \neq 0$ and thus, in general $K_{OMPC} = K$ whereas $K_{SOMPC} \neq K$.

Also validate that for OMPC the matrix S_c has a block diagonal structure, with each block identical. You can use the m-files **ompccost_example1,2.m sompccost_example1,2.m** as templates to save time.

6.7.4 Demonstration that the parameters of K_{SOMPC} vary with n_c

The variability in S_c and S_{cx} implies that K_{SOMPC} will vary in the case of SOMPC (but not for OMPC where $K_{OMPC} = 0$). This section gives a single example of how the parameters change with n_c.

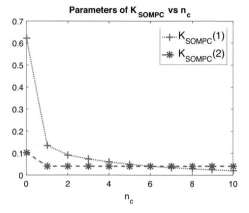

The file **monotonic_example2.m** is used to generate this figure of the two parameters of K_{SOMPC} and how these vary for small values of n_c and eventually settle at $K_{OMPC} = [0.006, 0.409]$ for large n_c.

STUDENT PROBLEM

For a number of examples of your choice, validate that for SOMPC the parameters K_{SOMPC} depend upon n_c but moreover, converge to K_{OMPC} for large n_c. You can use the files **monotonic_example1,2.m** as templates for this.

6.7.5 Demonstration that J_k is a Lyapunov function with OMPC and SOMPC

This section illustrates that, in the nominal case, the cost function J_k is monotonically decreasing as expected whenever the target and disturbance are constant. This property holds true for both OMPC and SOMPC.

The state space model and choice of Q, R, n_c are arbitrary in example **monotonic_example1.m** and hence readers are encouraged to open this file and vary these to validate that the same insights follow. It is clear that, for a fixed target, the cost function J_k decreases monotonically as expected. **Moreover**, the optimal value of $c_k = 0$.

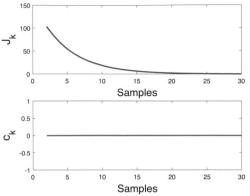

Remark 6.14 *In the case of OMPC for the unconstrained case, the optimum* $\mathbf{c}_k = 0$ *by definition as $S_{cx} = 0$.*

The state space model and choice of Q, R are arbitrary in example **monotonic_example2.m** but $n_c = 4$. Readers are encouraged to open this file and vary the parameters to validate that the same insights follow. It is clear that, for a fixed target, the cost function J_k decreases monotonically as expected.

The power station model is used in **monotonic_example3.m** with $n_c = 5$ and the simulation has a target change and a disturbance change around samples 25, 40, respectively. It is clear that, for a fixed scenario the cost function J_k decreases monotonically as expected.

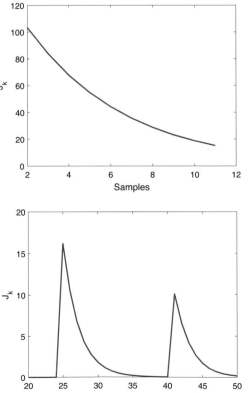

6.7.6 Demonstration that with SOMPC the optimum decision changes each sample

In the unconstrained case, OMPC produces $\underset{\rightarrow k}{\mathbf{c}} = 0$ and thus is consistent from one sample to the next. Conversely, SOMPC produces non-zero values and thus an interesting question is to consider the extent to which the proposed *optimum* strategy is consistent from one sample to the next. This is analogous to the discussions in the previous chapter on GPC and the preference for algorithms which do not keep changing their mind on what constitutes the optimum strategy. The examples in this section demonstrate that *the tail* of the optimised $\underset{\rightarrow k}{\mathbf{c}}$ is different from the next optimum, that is $\underset{\rightarrow k+1,tail}{\mathbf{c}} \neq \underset{\rightarrow k+1|k+1}{\mathbf{c}}$ and thus there is some inconsistency in the decision making.

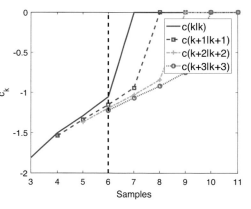

The example used in **monotonic_example2.m** with $n_c = 4$ shows the optimum predicted trajectories for $\mathbf{c}_k$ taken at several neighbouring sampling instants. It is clear that there is inconsistency, for example:

$$\mathbf{c}_{6|3} \neq \mathbf{c}_{6|4} \neq \mathbf{c}_{6|5} \neq \mathbf{c}_{6|6}$$

The example used in **mono-tonic_example3.m** with $n_c = 5$ shows the optimum predicted trajectories for $\mathbf{c}_k$ taken at several neighbouring sampling instants. It is clear that there is inconsistency, for example:

$$\mathbf{c}_{30|27} \neq \mathbf{c}_{30|28} \neq \mathbf{c}_{30|29} \neq \mathbf{c}_{30|30}$$

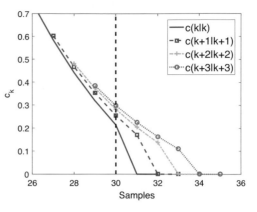

6.7.7   List of available illustrations

So far the numerical illustrations have focussed on demonstrating some underpinning concepts and insights into OMPC/SOMPC. This section focuses more on the actual closed-loop behaviour for a range of scenarios and thus lists the resources students can use to create these for themselves, such as:

1. Script files demonstrating how to form SOMPC and OMPC closed-loop responses.

2. Evidence that OMPC/SOMPC deals with uncertainty (output disturbances and parameter uncertainty) effectively.

3. How much difference is there between OMPC based on an independent model (**opmc_simulatec.m**) and one where the states are estimated via an observer (**opmc_simulated.m**)?

Students should run the corresponding m-files to reproduce the relevant data and note that the underlying model parameters, weights, choices of targets and disturbances signals and so forth should be self-evident in the script file and therefore easy for readers to edit to simulate their own examples.

All these examples are in the constraint free case; constraints are introduced in the next chapter.

1. **ompcsimulate_example1.m** Demonstrates the OMPC algorithm with no uncertainty and full state measurement but includes an output disturbance. This can be edited by the user for different models.

2. **ompcsimulate_example2.m** Demonstrates the SOMPC algorithm with no uncertainty and full state measurement but includes an output disturbance. This can be edited by the user for different models.

3. **ompcsimulate_example3.m** Demonstrates the OMPC/SOMPC algorithm with disturbance uncertainty. Uses an independent model structure (e.g., Section 2.7) for state and disturbance estimation. This can be edited by the user for different models.

4. **ompcsimulate_example4.m** Demonstrates the OMPC/SOMPC algorithm with disturbance uncertainty. Uses an independent model structure (e.g., Section 2.7) for state and disturbance estimation. **This example is a MIMO case.** This can be edited by the user for different models.

5. **ompcsimulate_example5.m** Demonstrates the OMPC/SOMPC algorithm with model parameter uncertainty and disturbance uncertainty. Uses an independent model structure for state and disturbance estimation. This can be editted by the user for different models.

6. **ompcsimulate_example6.m** Demonstrates the OMPC/SOMPC algorithm with model parameter uncertainty and disturbance uncertainty. Uses an independent model structure for state and disturbance estimation. **This example is a MIMO case.** This can be edited by the user for different models.

7. **ompcsimulate_example7.m** Demonstrates the OMPC/SOMPC algorithm with model parameter uncertainty and disturbance uncertainty. Uses an observer for state and disturbance estimation. This can be edited by the user for different models.

8. **ompcsimulate_example7b.m** Demonstrates the OMPC/SOMPC algorithm with disturbance uncertainty. Uses an observer for state and disturbance estimation. **This example is a MIMO case.** This can be edited by the user for different models.

9. **ompcsimulate_example8.m** Compares the disturbance rejection and measurement noise rejection properties of OMPC/SOMPC where state estimation comes from either an independent model formulation or an observer.

10. **ompcsimulate_example9.m** Compares the disturbance rejection and parameter uncertainty sensitivity properties of OMPC/SOMPC where state estimation comes from either an independent model formulation or an observer. **This example is a MIMO case.**

STUDENT PROBLEM

Using the code provided or otherwise, compare the behaviour of OMPC and SOMPC for a number of different models, choices of weights, targets and disturbances and with parameter uncertainty.

6.8   Motivation for SOMPC: Different choices for mode 2 of dual-mode control

Readers might be wondering why we have bothered to give so much space to SOMPC when it seems obvious that OMPC is a better posed optimisation and in terms of the defined cost function is expected to perform better. The answer is twofold.

- Historical developments in MPC went through several variations of SOMPC before settling on OMPC [75, 103, 116, 126, 170].

- There are still a number of scenarios where SOMPC could have advantages compared to OMPC [31].

The fundamental discussion is based around the trade-off between a desire for good performance such as fast tracking and a contrary need to meet constraints such as hard limits on input rates. Constraints are discussed in more detail in the next chapter, so this section will simply give some insight into the issues.

First we will give a brief overview of the different forms of dual-mode control that have been popularised in the literature, followed by some comments on their respective strengths and weaknesses. Some terminology is needed:

1. **Terminal control law or terminal feedback:** Sometimes denoted terminal conditions, this is the implied control law used in mode 2 of the predictions.

2. **Terminal region or target set S_T:** This will be defined in Chapter 8. It comprises a constraint for the mode 2 predictions so that in essence one requires $\mathbf{x}_{k+n_c|k} \in S_T$ in order for the dual-mode algorithm to be properly defined.

It is the terminal region S_T which fundamentally drives the user towards SOMPC as opposed to OMPC.

1. Where the terminal control law is focussed on high bandwidth/fast responses, there is a good likelihood that the associated terminal region S_T will be small, compared to an alternative choice for the terminal control law.

2. Small terminal regions carry the danger of the associated MPC algorithm having a relatively small operating range where it is defined effectively.

In practice the user needs to find a compromise between large operating regions and good performance. This trade-off may be supported by allowing a relatively detuned control law in mode 2, but an emphasis in the performance index on better tuning, that is SOMPC!

Remark 6.15 *If n_c is large, say around the system settling time, then the choice of terminal control law will have negligible effect on the unconstrained control law*

(6.55), as all the transients will occur in mode 1 and mode 2 will comprise of only small numbers. In other words

$$\lim_{n_c \to \infty} [K + S_c^{-1} S_{cx}] = K_{opt} \tag{6.77}$$

where K_{opt} is the optimal controller minimising (6.6); this can be derived using optimal control algorithms.

However, it is usually assumed that for computational reasons n_c is small. In this case the choice of terminal control law has a significant effect as it shapes much of the transients which are now in mode 2 of the predictions.

Summary: The choice of terminal control law has a significant impact on unconstrained and constrained performance for small n_c. Moreover, it has a significant impact on the operating region for which a dual-mode control law is well defined.

6.8.1   Dead-beat terminal conditions

The dead beat choice was popularised in the MPC field by [26, 75, 103] but actually known earlier (e.g., [68, 84]). Historically these were the first insightful suggestions of how to guarantee, a priori, the stability of MPC algorithms.

The key idea was to use a dead-beat form of terminal mode, i.e., force the predicted output to be identically equal to the set point for n steps (n the model order). Hence, for the nominal case, the predicted output would remain at the set point thereafter, assuming that the corresponding predicted $\Delta \mathbf{u}$ were also zero. However, this approach may not be recommended in general.

1. Dead-beat control is known to use very active input signals. Hence the terminal region S_T may be very small.

2. As S_T is small, the use of small n_c will give rise to small operating regions. To improve the volume of feasibility regions one would need a large n_c with a consequent increase in the on-line computational load.

3. Dead-beat control would tend to give poor performance, measured by way of typical quadratic performance indices such as (6.6).

4. Dead-beat control usually has poor robustness; some of this poor sensitivity will inevitably be inherited by the associated MPC law.

Summary: While interesting from a historical perspective, few authors these days would recommend the use of dead-beat terminal conditions except for a few exceptional scenarios and thus details are not provided here.

6.8.2   A zero terminal control law

The choice of no terminal control in effect means the choice:

$$\{\mathbf{u}_{k+i} - \mathbf{u}_{ss} = 0 \quad \text{or} \quad \Delta\mathbf{u}_{k+i} = 0\} \quad i \geq n_c \tag{6.78}$$

Readers will immediately realise that these are exactly the two terminal conditions deployed by GPC depending on the choice of performance index.

GPC is a variant of SOMPC so could also be considered as dual mode

SOMPC was defined in fairly loose terms as having two modes, the second of which was determined by a fixed linear feedback law. Consideration of (6.78) makes it clear GPC does in fact also have a dual-mode structure. The only difference is that GPC insists that the input is constant in mode 2 and this is analogous to a very specific choice of feedback, that is, effectively zero.

However, as highlighted in the previous chapter, here are significant possible weaknesses in using what is in effect a zero terminal control law.

1. Mode 2 is given by open-loop behaviour. In many cases this may imply poor performance and significant prediction mismatch.

2. For small values of n_c, it may be difficult to gain good closed-loop performance.

3. It is not appropriate to unstable open-loop systems, as the implied terminal cost (6.30) with open-loop behaviour is infinite (e.g., see Section 5.8).

4. The combination of an infinite output horizon and small n_c will give open-loop behaviour, essentially equivalent to low gain integral control (see Section 5.5.3).

Nevertheless a possible significant advantage of having a zero terminal mode is the associated terminal constraint S_T may be much larger which makes constraint handling easier overall. This will become more apparent in the following chapters which consider constraints.

> **Summary:** Using a zero terminal control law for mode 2 in effect reduces to a GPC/DMC type of algorithm with the corresponding weaknesses. Nevertheless, this may still be popular in practice because the constraint handling may be far easier and feasible regions larger in this case; compared to dual-mode strategies with a non-zero feedback in mode 2.

6.8.3 Input parameterisations to eliminate unstable modes in the prediction

It was noted that a GPC type of algorithm is poorly posed in general for systems with open-loop unstable poles. One alternative to deploying a stabilising terminal feedback within mode 2 of the predictions is to parameterise the predictions in such a way that the predictions are guaranteed convergent (and to the correct steady-state). Such methods were developed independently by several authors [51, 52, 132, 135, 116]. Interestingly, all these early works deployed a zero terminal control law of the type in (6.78), but they do not really fit into the OMPC/SOMPC framework because the parameterisation of the future input trajectory is much more constrained.

The basic proposal was to use systematic analysis of the predictions in order to separate these into convergent parts and divergent components. Then, to find the dependence of the divergent components on past measured data as well as the future control moves. This dependence can be exploited to ensure those components are zero and in essence it reduces to a solution of the following form for suitable P_1, P_2:

$$\Delta \underset{\rightarrow k}{\mathbf{u}} = P_1 \begin{bmatrix} \Delta \underset{\leftarrow k-1}{\mathbf{u}} \\ \underset{\leftarrow k}{\mathbf{y}} \\ \mathbf{r}_k \end{bmatrix} + P_2 \underset{\rightarrow k}{\mathbf{c}} \tag{6.79}$$

As long as $\Delta \underset{\rightarrow k}{\mathbf{u}}$ is parameterised in this form, then irrespective of the choice of $\underset{\rightarrow k}{\mathbf{c}}$, the associated output predictions are convergent to the correct asymptotic output.

This book will not go into details of how P_1, P_2 might be computed as largely this type of approach has been superseded by SOMPC (dual-mode) approaches but interested readers will find more details in the references and [161]. There may be some specific scenarios where such an approach has benefits and thus awareness of this as a possibility is likely to be useful.

Summary: Although alternatives to the dual-mode structure (6.18) exist in the literature, largely these have not been pursued in recent years with the possible exception of GPC/DMC which is still very popular in industry, perhaps because the implementation and constraint handling is somewhat easier than for a dual-mode approach.

6.8.4 Reflections and comparison of GPC and OMPC

Readers may be wondering why this book spent so much time on finite horizon MPC control laws, especially given the weaknesses these contain. Why not focus on infinite horizon (dual-mode) approaches from the outset? The answer involves two main observations.

1. Most algorithms deployed in industry are of a GPC/DMC type and thus a reader needs to be familiar with these to be able to understand current

practice and indeed, to have insights into effective tuning and management of these implementations.

2. OMPC has strong theoretical foundations but these come at a price. As will be seen in the following chapters, the most obvious price is that the constraint handling is more cumbersome and numerically demanding in general and moreover, it is more likely to encounter feasibility difficulties (that is conflicts between constraints). In an industrial scenario, these weaknesses might outweigh the benefits of a better posed optimisation.

6.8.4.1 DMC/GPC is practically effective

Given all the recent academic developments in MPC, it is rather odd to note that industrial applications very rarely use state space or transfer function models. Moreover, they very rarely use algorithms that can be related closely to optimal control and hence with a priori stability results. Why is this?

The most popular algorithm is essentially of the DMC form (Section 4.6) using the following guidelines:

- n_y is large and certainly greater than the system settling time.

- n_u is small, often just one.

- Models are usually step responses (e.g., DMC).

It should be emphasised (as argued in the previous chapter) that these guidelines can ensure effective control, especially where the open-loop dynamics are reasonably benign.

6.8.4.2 The potential role of dual-mode algorithms

The basic MPC algorithm using finite horizons can give good performance, notwithstanding arguments on prediction mismatch given in Chapter 5. Hence one might wonder why bother with dual-mode implementations. The simple answer is that there is no need for many cases, but:

1. The development of dual-mode paradigms/infinite horizon algorithms solved a theoretical problem for the academics and hence gave MPC some analytic rigor. The insight gained is invaluable in understanding better the strengths and weaknesses of more typical finite horizon MPC implementations.

2. When dealing with unstable open-loop processes or those with quite complex transient behaviour, a dual-mode type of algorithm is likely to perform well with minimal tuning as the necessary insight is built-in.

Summary: In practice the DMC/GPC algorithm is good enough to handle most industrial problems. Recent advances have given a better understanding of why this is so and also provided the tools to deal with more challenging scenarios.

6.9 Chapter summary

This chapter has introduced the concept of dual-mode predictive control and focussed on two alternatives denoted as: (i) Suboptimal MPC (SOMPC) and (ii) Optimal (OMPC); the former includes GPC/DMC as a subset.

Dual-mode approaches are a mechanism for facilitating the use of infinite input and output prediction horizons while simultaneously keeping the optimisation complexity down to tractable levels. Indeed, one could argue that with the use of the appropriate Lyapunov equations (e.g., Section 6.6.4.1), a dual-mode approach with a sensible number of d.o.f. is actually computationally less demanding than GPC with a sensible output horizon; this is because one can capture the impact of the predictions from n_c to infinity on the cost function very efficiently and thus avoid much of the prediction algebra from Chapter 2.

The use of infinite horizons also enables much stronger statements about algorithm properties such as a priori *guarantees of stability* in the nominal case and the potential to ensure consistency of decision making from one sample to the next. These properties will be invaluable when moving to the industrially critical constraint handling case.

Finally the reader will have noticed that in the constraint free case, OMPC reduces to the well known optimal control solution and thus may wonder why bother with MPC at all. They may also wonder why bother with SOMPC which seems to embed some inconsistency which is contrary to many of the arguments in this book. The rationale for both of these will become clear in the following chapters on constraint handling.

6.10 MATLAB files in support of this chapter

6.10.1 Files in support of OMPC/SOMPC simulations

The reader should note that the focus of these files is on simplicity and transparency so none of these include constraint handling or error catching. You can use the illustration script files as templates for how to call these correctly. Most of the files will default to an SOMPC implementation; to get OMPC ensure the weights for the terminal mode and the cost function are selected to be identical.

1. **ompc_cost.m, sompc_cost.m**: Forms the parameters of the cost function for OMPC and SOMPC.

2. **opmc_observer.m** Forms an observer for states and the disturbance term used to ensure offset-free tracking.

3. **ompc_simulate.m** Assumes full state measurement. Nominal case only. Caters for non-zero initial state but target is zero and no disturbances are included. Simplest code so easiest for a beginner to read!

4. **ompc_simulateb.m** Assumes full state measurement. Nominal case only. Allows for non-zero targets and disturbances.

5. **ompc_simulatec.m** Assumes an independent model structure for prediction and disturbance estimation. Also allows for parameter uncertainty.

6. **ompc_simulated.m** Uses an observer for state and disturbance estimation. Also allows for parameter uncertainty. Readers might like to note that the choice of the observer is *arbitrary* and they may wish to replace this with a different design.

6.10.2 MATLAB files producing numerical illustrations from Section 6.7

These files are written as script files and in such a way that readers should easily be able to modify and thus enter examples of their own choosing.

1. **ompcsimulate_impact_qr.m** Allows the user to compare nominal performance with different choices of the weighting matrices Q, R.

2. **monotonic_example1,2,3.m** Demonstrates that the cost function is Lyapunov wherever the disturbance/set point is constant. Also produces plots of the parameters of K_{SOMPC} as a function of n_c and shows how the optimised trajectory $\underset{\to k}{c}$ changes with each sample.

3. **ompccost_example1,2.m** Demonstrates the structure of the OMPC cost function parameters using the CLP.

4. **sompccost_example1,2.m** Demonstrates the structure of the SOMPC cost function parameters using the CLP.

5. **ompcsimulate_example1.m** - **ompcsimulate_example9.m** Demonstrates the OMPC/SOMPC algorithms across a range of scenarios (see Section 6.7.7).

7

Constraint handling in GPC/finite horizon predictive control

CONTENTS

7.1 Introduction

One of the main reasons for the popularity of MPC in industry is the ability to incorporate constraint handling in a systematic as opposed to an ad hoc manner. Nevertheless, this issue has been left until last because constraints have a significant impact on closed-loop behaviour and it was important that this impact did not distract from the reader understanding the foundational principles and good design in MPC.

It so happens that the inclusion of constraints is relatively straightforward in principle, that is, the algebra and algorithm formulation is very little changed from the earlier chapters. The real challenge is in the associated online optimisation and implementation [27] which becomes far more complicated to code. Nevertheless, such issues would occupy an entire book in themselves and require a different set of skills and hence this book focusses on the principles, algebra and illustrations and leaves optimisation theory and practice to other authors. For readers happy to use MATLAB®, suitable optimisation routines are readily available and easy to code with.

For organisation purposes and because some of the concepts and derivations are different, this chapter deals with finite horizon algorithms and the next chapter with dual-mode algorithms.

7.2 Guidance for the lecturer

The main content of this chapter is relatively straightforward, that is, how constraints are represented by linear inequalities, and thus all the content could be included in an introductory course in predictive control. However, the author would recommend that students are required to build constraint inequalities for just one prediction model scenario (e.g., transfer function or state space or independent model) as the principles are similar and such a focus within a module is an easy way to avoid over-load/duplication. The chapter also gives some focus on why constraint handling is important and a large number of numerical illustrations which are relatively straight-forward for a reader to engage with.

Typical learning outcomes for an examination assessment: Within written examinations students can easily demonstrate how inequalities representing constraints are constructed and then embedded into the computation of an MPC control law. They can also be asked to discuss and illustrate (without algebra) the importance of systematic constraint handling using both common sense and/or based on some mathematical interpretations. Some discussion is provided in sections 7.3 and 7.6 to support this.

Typical learning outcomes for an assignment/coursework: Assignments should focus on students providing numerical examples of effective constraint handling and the impact on behaviour. As in the rest of this book, the author suggests students are provided with code and thus they can focus on the applications and demonstrating an understanding of scenarios where constraint handling is important.

7.3   Introduction

The first priority is to demonstrate that systematic constraint handling is important and often improves behaviour considerably compared to alternative methods. Nevertheless, there is a significant offline and online cost to deploying systematic constraint handling, and this expense needs to be properly justified.

One could use rather trivial examples from everyday life and contrast the outcomes from *anticipating constraints* to *saturation policies*.

7.3.1   Definition of a saturation policy and limitations

In simple terms, a saturation policy **is defined here** solely on the current input so that, if the unconstrained optimised input $\mathbf{u}_k$ violates an input constraint, then reset $\mathbf{u}_k$ to the nearest value such that input limits and input rates are satisfied.

Saturation policies cannot easily deal with output constraints in a systematic fashion and indeed may also give no consideration to future predicted values such as $\mathbf{u}_{k+i|k}, i > 0$. Thus, for the purpose of this book, the presence of implied violations in far future input predictions is not considered in the definition of $\mathbf{u}_k$ in a simple saturation algorithm.

7.3.2   Limitations of a saturation policy

Saturation policies bring a number of obvious limitations, some of which are emphasised in the following simple illustrations.

1. A car driver has limited braking and steering capacity. If their policy (control law design) is to go as fast as possible assuming unlimited braking and steering, it is clear that the entry speed to a corner would often be too fast. Applying maximum braking and steering at the point of need may be insufficient to avoid a crash and even if it were, the corresponding behaviour is likely to be poor. This is an example of a saturation policy as

the optimisation takes no account of the need to meet this constraint in the planning.

2. The operator of an interactive MIMO system plans an operating point change based on a linear control law which, in general, manages the interaction well. However, assuming linear behaviour throughout, halfway through this change one of the inputs reaches a maximum. If this maximum is beyond the input limit, then this input would saturate for a length of time well after the target change had been initiated. In consequence, the underlying control would now differ from the linear design, and hence the corresponding behaviour could become very poor.

3. A nominal control law for controlling the level in a tank allows some oscillation in the level. What is the consequence of this control causing the level to go beyond the depth of the tank. Clearly there will be some spillage.

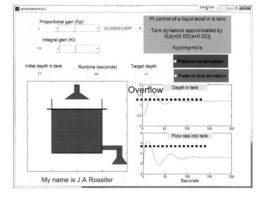

STUDENT PROBLEM

Give convincing arguments for a number of different scenarios where saturation control is likely to lead to ineffective control. That is, demonstrate how saturation control may lead to undesirable behaviour compared to what might be achievable with a more systematic approach.

ILLUSTRATION: Saturation control in GPC seems fine

Consider the system and constraints.

$$y(z) = \frac{z^{-1} + 0.3z^{-2}}{1 - 1.2z^{-1} + 0.32z^{-2}} u(z); \tag{7.1}$$

$$\|u_k\| \le 1; \quad \|\Delta u_k\| \le 1$$

It can be seen that the input saturates at its upper limit during transients, but otherwise the behaviour is still fine. In this case, it so happens that saturation of the current input coincides with the optimum choice, and indeed that may often be the case.

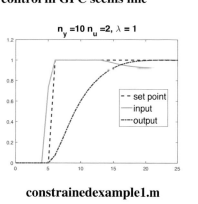

$n_y = 10 \; n_u = 2, \; \lambda = 1$

set point
input
output

constrainedexample1.m

ILLUSTRATION: Impact of saturation control on closed-loop performance with PID

The simplest constraint handling policy will simply saturate the input if it violates input limits. However, even this is fraught with potential dangers. Compare the impact of saturating the output of a PID (input/output 2) with an alternative which saturates the internal PID variables (input/output 1). Clearly in this case a simple change makes a huge difference to behaviour and this emphasises the potential weaknesses and dangers of simple saturation; it is not always clear how effective this will be!

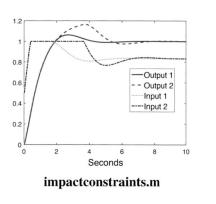

Output 1
Output 2
Input 1
Input 2

Seconds

impactconstraints.m

ILLUSTRATION: Saturation control in GPC fails

Consider the following system and constraints.

$$y(z) = \frac{z^{-1} - 1.4z^{-2}}{1 - 1.7z^{-1} + 1.2z^{-2}} u(z); \quad \|u_k\| \le 1; \quad \|\Delta u_k\| \le 0.2 \tag{7.2}$$

It can be seen that using a saturation policy, the input rate saturates during transients and thus limits performance and indeed leads to an oscillatory output, a bigger non-minimum phase characteristic and very unsatisfactory behaviour. Conversely, the algorithm which is optimised subject to constraints over the entire prediction, produces a smoother output (albeit with an unavoidable offset due to the input being saturated). It achieves

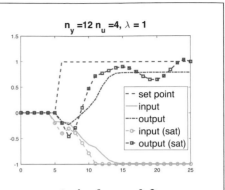

constrainedexample2.m

this by backing off the input slightly during initial transients, that is, not going all the way up to the rate constraint.

7.3.3   Summary of constraint handling needs

Having established that systematic constraint handling is beneficial, it remains to demonstrate how that is achieved with MPC. This chapter will first discuss how potential constraint violations are detected and described mathematically using linear inequalities. Next, it will show how these inequalities can be incorporated into an MPC algorithm to ensure that constraints are not violated and that the predicted performance is optimised subject to constraint satisfaction.

Summary: Good performance in the constraint free case need not imply good performance in the constrained case and often saturation control is ineffective. There is a need for effective constraint handling algorithms. Unsurprisingly, optimum constrained performance could be significantly different from the optimum unconstrained feedback law performance. Finally, the word *feasibility* is used to capture the concept that constraints are, or can be, satisfied.

DEFINITION - FEASIBILITY IN MPC

It is useful at this point to clarify a word often used in the MPC literature and how it is interpreted. A prediction (or class of predictions) is said to be feasible if it satisfies (or can satisfy) all constraints and infeasible if it does not.

Clearly, an infeasible prediction, one which violates constraints, cannot exist in general and thus, its use would invalidate any stability and performance assurances as well as render meaningless any optimisation using this prediction.

7.4 Description of typical constraints and linking to GPC

The constraints may occur on any variables in a feedback loop. The desire is that none of the constraints are violated by the optimal predictions as any violation can have undesirable consequences, although violation of some constraints (for example so-called softer constraints or constraints on prediction components far into the future) may be less critical than for others. The reader will note that this statement includes an important fundamental concept.

We need to check that the entire prediction (in principle for an infinite horizon) satisfies constraints. If any part of a prediction proposed by an MPC algorithm violates a constraint, then that prediction is potentially meaningless in that it may not be implementable! In such a case it certainly cannot be considered optimal!

In practice, constraints are checked over the horizons deployed by the relevant MPC algorithm:

- For GPC the output predictions are checked against constraints over the first n_y samples and input predictions are checked over the first n_u samples.

- For OMPC both input and output predictions should be checked against constraints over an infinite horizon.

The potential weakness of the checking taking place only over limited horizons within GPC is often ameliorated by the good practice tuning guidance discussed in Chapter 5.

STUDENT PROBLEM

Give some examples where failure to consider constraints over a long enough prediction horizon can lead to significant negative impacts on behaviour.

Remark 7.1 *The use of long output horizons is critical for constraint handling scenarios (Section 5.5.6), as with short output horizons, the part of the output prediction* *which is unobserved (see illustrations in Section 5.5.2) could be violating constraints (is infeasible) and thus the associated prediction is meaningless/invalid with the consequence that the associated optimisation is also meaningless (the solution is not optimal/valid).*

The following subsections show how to set up inequalities which ensure the predictions satisfy constraints (are feasible). It should also be noted that the constraint equations should be expressed in terms of the d.o.f. within the predictions so that there is transparency on what the user can change to ensure the relevant inequalities are satisfied.

> **Summary:** Constraints must be tested over long horizons (until signal convergence) to ensure the validity/feasibility of the underlying MPC optimisation.

7.4.1   Input rate constraints with a finite horizon (GPC predictions)

This section will assume the typical GPC assumption that within the predictions:

$$\Delta \mathbf{u}_{k+i|k} = 0, \quad i \geq n_u \tag{7.3}$$

Dual-mode prediction structures will be considered in the following chapter.

Take upper and lower limits on the input rate to be:

$$\underline{\Delta \mathbf{u}} \leq \Delta \mathbf{u}_{k+i|k} \leq \overline{\Delta \mathbf{u}}, \quad \forall i \geq 0 \tag{7.4}$$

Given that the input increments are predicted to be zero beyond the horizon n_u, one can check the constraints up to then with the following matrix/vector inequalities:

$$\underbrace{\begin{bmatrix} \underline{\Delta \mathbf{u}} \\ \underline{\Delta \mathbf{u}} \\ \vdots \\ \underline{\Delta \mathbf{u}} \end{bmatrix}}_{L\underline{\Delta \mathbf{u}}} \leq \underbrace{\begin{bmatrix} \Delta \mathbf{u}_k \\ \Delta \mathbf{u}_{k+1} \\ \vdots \\ \Delta \mathbf{u}_{k+n_u-1} \end{bmatrix}}_{\Delta \underset{\rightarrow k}{\mathbf{u}}} \leq \underbrace{\begin{bmatrix} \overline{\Delta \mathbf{u}} \\ \overline{\Delta \mathbf{u}} \\ \vdots \\ \overline{\Delta \mathbf{u}} \end{bmatrix}}_{L\overline{\Delta \mathbf{u}}}; \quad L = \begin{bmatrix} I \\ I \\ \vdots \\ I \end{bmatrix} \tag{7.5}$$

or in simpler terms

$$L\underline{\Delta \mathbf{u}} \leq \Delta \underset{\rightarrow k}{\mathbf{u}} \leq L\overline{\Delta \mathbf{u}} \tag{7.6}$$

where the reader is reminded of the flexible definition of L given for example in Remark 2.1. A more conventional representation for (7.5) is in terms of a single set of linear inequalities such as:

$$\underbrace{\begin{bmatrix} I \\ -I \end{bmatrix}}_{C_{\Delta u}} \Delta \underset{\rightarrow k}{\mathbf{u}} - \underbrace{\begin{bmatrix} L\overline{\Delta \mathbf{u}} \\ -L\underline{\Delta \mathbf{u}} \end{bmatrix}}_{\mathbf{d}_{\Delta u}} \leq 0 \tag{7.7}$$

where I represents a suitable dimension identity matrix. In simpler form:

$$C_{\Delta u}\Delta\underset{\rightarrow k}{\mathbf{u}} \leq \mathbf{d}_{\Delta u} \tag{7.8}$$

where matrices $C_{\Delta u}, \mathbf{d}_{\Delta u}$ are time invariant in this case.

> **Summary:** Satisfaction of rate constraints by the predicted inputs can be tested using a simple matrix/vector inequality expression such as (7.8).

7.4.2 Input constraints with a finite horizon (GPC predictions)

This section will deploy the same assumption about input predictions as in the previous section given in Eqn. (7.3) and using a similar mechanism derive the inequalities to represent the satisfaction of absolute input constraints by input predictions. However, first one must express the future inputs in terms of the d.o.f. $\Delta\underset{\rightarrow k}{\mathbf{u}}$ as optimisation will ultimately require everything to be defined with respect to the same d.o.f..

Given $\mathbf{u}_{k+i} = \mathbf{u}_{k-1} + \Delta\mathbf{u}_k + \cdots + \Delta\mathbf{u}_{k+i}$, it is easy to see that:

$$\underset{\rightarrow k}{\mathbf{u}} = C_{I/\Delta}\Delta\underset{\rightarrow k}{\mathbf{u}} + \underbrace{\begin{bmatrix} I \\ I \\ \vdots \\ I \end{bmatrix}}_{L}\mathbf{u}_{k-1}; \quad C_{I/\Delta} = \begin{bmatrix} I & 0 & \cdots & 0 \\ I & I & \cdots & 0 \\ \vdots & \vdots & \ddots & \vdots \\ I & I & \cdots & I \end{bmatrix} \tag{7.9}$$

where $C_{I/\Delta}$ is a Toeplitz matrix based on $I/(1-z^{-1})$ (see Chapter 2) and hence is block lower triangular with identity matrices filling the lower triangular part.

Define upper and lower limits on the input predictions

$$\underline{\mathbf{u}} \leq \mathbf{u}_{k+i|k} \leq \overline{\mathbf{u}}, \quad \forall i \geq 0 \tag{7.10}$$

One need only test for satisfaction of these constraints over the prediction horizon n_u as this automatically gives satisfaction thereafter from the implied assumption that:

$$\mathbf{u}_{k+n_u+i|k} = \mathbf{u}_{k+n_u-1|k}, \quad \forall i \geq 0$$

Hence, absolute input constraints can be represented with the following:

$$\underbrace{\begin{bmatrix} \underline{\mathbf{u}} \\ \underline{\mathbf{u}} \\ \vdots \\ \underline{\mathbf{u}} \end{bmatrix}}_{L\underline{\mathbf{u}}} \leq \underbrace{\begin{bmatrix} \mathbf{u}_k \\ \mathbf{u}_{k+1} \\ \vdots \\ \mathbf{u}_{k+n_u-1} \end{bmatrix}}_{\underset{\rightarrow k}{\mathbf{u}}} \leq \underbrace{\begin{bmatrix} \overline{\mathbf{u}} \\ \overline{\mathbf{u}} \\ \vdots \\ \overline{\mathbf{u}} \end{bmatrix}}_{L\overline{\mathbf{u}}} \tag{7.11}$$

or in simpler terms:

$$L\underline{u} \leq \underset{\rightarrow k}{u} \leq L\overline{u} \tag{7.12}$$

Hence, substituting from (7.9)

$$L\underline{u} \leq C_{I/\Delta}\underset{\rightarrow k}{\Delta u} + Lu_{k-1} \leq L\overline{u} \tag{7.13}$$

Re-arranging into a more conventional form gives:

$$\underbrace{\begin{bmatrix} C_{I/\Delta} \\ -C_{I/\Delta} \end{bmatrix}}_{C_u} \underset{\rightarrow k}{\Delta u} + \underbrace{\begin{bmatrix} L \\ -L \end{bmatrix}}_{\mathbf{d}_{Lu}} u_{k-1} \leq \underbrace{\begin{bmatrix} L\overline{u} \\ -L\underline{u} \end{bmatrix}}_{\mathbf{d}_u} \tag{7.14}$$

Finally this can be expressed in compact form analogous to (7.8) as:

$$C_u\underset{\rightarrow k}{\Delta u} \leq \mathbf{d}_u - \mathbf{d}_{Lu}u_{k-1} \tag{7.15}$$

where matrices $C_u, \mathbf{d}_u, \mathbf{d}_{Lu}$ are time invariant in this case.

> **Summary:** Satisfaction of input constraints by the predicted inputs can be tested using a simple matrix/vector inequality expression of the form of (7.15). The reader will notice that these inequalities are time varying due to the dependence upon u_{k-1}.

7.4.3 Output constraints with a finite horizon (GPC predictions)

Output constraints can be set up analogously to input constraints by defining: (i) the output prediction dependence on the d.o.f. and (ii) the output constraint limits.

First define the dependence of the output predictions on the d.o.f. as in for example (2.49):

$$\underset{\rightarrow k+1}{y} = H\underset{\rightarrow k}{\Delta u} + P\underset{\leftarrow k-1}{\Delta u} + Q\underset{\leftarrow k}{y} \tag{7.16}$$

Define upper and lower limits on the output predictions

$$\underline{y} \leq y_{k+i|k} \leq \overline{y}, \quad \forall i \geq 0 \tag{7.17}$$

One chooses to only test for satisfaction of these constraints over the prediction horizon n_y and assume satisfaction thereafter (this is a vulnerability of GPC). Hence, output constraints can be represented with the following:

$$\underbrace{\begin{bmatrix} \underline{y} \\ \underline{y} \\ \vdots \\ \underline{y} \end{bmatrix}}_{L\underline{y}} \leq \underbrace{\begin{bmatrix} y_{k+1} \\ y_{k+2} \\ \vdots \\ y_{k+n_y} \end{bmatrix}}_{\underset{\rightarrow k+1}{y}} \leq \underbrace{\begin{bmatrix} \overline{y} \\ \overline{y} \\ \vdots \\ \overline{y} \end{bmatrix}}_{L\overline{y}} \tag{7.18}$$

or in simpler terms:

$$L\underline{\mathbf{y}} \le \underset{\rightarrow k+1}{\mathbf{y}} \le L\bar{\mathbf{y}} \tag{7.19}$$

Hence, substituting from (7.16):

$$\underbrace{\begin{bmatrix} H \\ -H \end{bmatrix}}_{C_y} \underset{\rightarrow k}{\Delta\mathbf{u}} + \underbrace{\begin{bmatrix} P \\ -P \end{bmatrix}}_{\mathbf{d}_{yu}} \underset{\leftarrow k-1}{\Delta\mathbf{u}} + \underbrace{\begin{bmatrix} Q \\ -Q \end{bmatrix}}_{\mathbf{d}_{yy}} \underset{\leftarrow k}{\mathbf{y}} \le \underbrace{\begin{bmatrix} L\bar{\mathbf{y}} \\ -L\underline{\mathbf{y}} \end{bmatrix}}_{\mathbf{d}_y} \tag{7.20}$$

Finally this can be expressed in compact form analogous to (7.8) as:

$$C_y \underset{\rightarrow k}{\Delta\mathbf{u}} \le \mathbf{d}_y - \mathbf{d}_{yu} \underset{\leftarrow k-1}{\Delta\mathbf{u}} - \mathbf{d}_{yy} \underset{\leftarrow k}{\mathbf{y}} \tag{7.21}$$

where matrices $C_y, \mathbf{d}_y, \mathbf{d}_{yu}, \mathbf{d}_{yy}$ are time invariant in this case.

Summary: Satisfaction of output constraints by the predicted outputs can be tested using a simple matrix/vector inequality expression such as (7.21). The reader will notice that these inequalities are time varying due to the dependence upon $\underset{\leftarrow k-1}{\Delta\mathbf{u}}$, $\underset{\leftarrow k}{\mathbf{y}}$.

Remark 7.2 *Ideally one should take the output constraint horizon to be at least equal to the setting time, even if this is bigger than n_y. Non-inclusion in the cost function J of a term $\mathbf{y}_{k+m|k}$, $m > n_y$ may not imply that the outputs are no longer changing beyond sample m. It could be that the unconsidered part of the output prediction, beyond the horizon n_y, does violate a constraint. Failure to account for this may not be resolved simply by the usual receding horizon arguments (in effect the stability of the nominal unconstrained feedback loop) because linearity is lost if a constraint is enforced unexpectedly.*

7.4.4 Summary

All the constraints, that is, input rates, inputs and outputs, must be satisfied simultaneously; hence one can combine inequalities (7.8, 7.15, 7.21) into a single set of linear inequalities of the form:

$$C_c \underset{\rightarrow k}{\Delta\mathbf{u}} \le \mathbf{d}_c - \mathbf{d}_{cu}\mathbf{u}_{k-1} - \mathbf{d}_{c\Delta u} \underset{\leftarrow k-1}{\Delta\mathbf{u}} - \mathbf{d}_{cy} \underset{\leftarrow k}{\mathbf{y}} \tag{7.22}$$

where

$$C_c = \begin{bmatrix} C_{\Delta u} \\ C_u \\ C_y \end{bmatrix}; \; \mathbf{d}_c = \begin{bmatrix} \mathbf{d}_{\Delta u} \\ \mathbf{d}_u \\ \mathbf{d}_y \end{bmatrix}; \; \mathbf{d}_{cu} = \begin{bmatrix} 0 \\ \mathbf{d}_{Lu} \\ 0 \end{bmatrix}; \; \mathbf{d}_{c\Delta u} = \begin{bmatrix} 0 \\ 0 \\ \mathbf{d}_{yu} \end{bmatrix}; \; \mathbf{d}_{cy} = \begin{bmatrix} 0 \\ 0 \\ \mathbf{d}_{yy} \end{bmatrix}$$

where $C_c, \mathbf{d}_c$ are time invariant and the other terms depend upon the current state (past input and output information).

Remark 7.3 *The reader will note that even with a relatively small n_u and n_y, nevertheless there can be a very large number of linear inequalities. Let the system be MIMO of dimension $n \times n$ so that each limit (e.g., $\overline{u}, \overline{y}$) is of dimension n. Then the constraints above imply p inequalities where*

$$p = 2n(2n_u + n_y) \tag{7.23}$$

STUDENT PROBLEM

For simplicity and brevity (to avoid unnecessary duplication), this section has considered inputs and outputs based on an MFD model and thus predictions of the form (7.16). However, in Chapter 2, it was noted that GPC could be formulated with:

1. MFD models with a T-filter (2.59).
2. Independent model predictions, e.g., (2.76).
3. State space based predictions such as (2.14, 2.34) in absolute terms or linked to deviation variables (2.30).
4. Step response models (2.100).
5. Closed-loop predictions (2.112).

Construct inequalities to represent the satisfaction of inputs, input rates and output constraints with several different types of modelling/prediction structures.

HINT: In cases where the prediction model uses u_k directly, the implementation of rate constraints will require a modification from the matrices used in Section 7.4.1. The final structure will also be slightly different from (7.22), for example it may depend on state x_k and/or a disturbance estimate $y_p - y_m$.

Summary: Constraints at each sampling instant can be represented by a set of simple linear inequalities (7.22) that depend upon the d.o.f. and the current state (past inputs and outputs). Even low dimensional systems can require large numbers of inequalities to be handled, especially in the case where there are output constraints.

7.4.5   Using MATLAB to build constraint inequalities

This section gives a number of examples of constraint inequalities and demonstrates how simple MATLAB code can be used to construct the vectors and matrices required in (7.22).

ILLUSTRATION: Input limits and input rate limits

Consider the SISO case with no output constraints and input limits

$$-2 \leq u_k \leq 3; \quad \|\Delta u_k\| \leq 1 \quad (7.24)$$

Take an input horizon of $n_u = 3$.
The relevant matrices $C_c, \mathbf{d}_c, \mathbf{d}_{cu}$ of Eqn.(7.22) are shown here.
The structure of C_c clearly matches that expected from (7.7,7.14), the values in $\mathbf{d}_c, \mathbf{d}_{cu}$ are taken from the limits (7.24).

File: **inputlimitsexample1.m**

```
[Cc,dc,dcu]

ans =

    1    0    0    1    0
    0    1    0    1    0
    0    0    1    1    0
   -1    0    0    1    0
    0   -1    0    1    0
    0    0   -1    1    0
    1    0    0    3   -1
    1    1    0    3   -1
    1    1    1    3   -1
   -1    0    0    2    1
   -1   -1    0    2    1
   -1   -1   -1    2    1
```

STUDENT PROBLEM

The file **inputlimitsexample1.m** also contains code for a MIMO example with $n_u = 2$ and the following input constraints:

$$\begin{bmatrix} -2 \\ -0.5 \end{bmatrix} \leq \mathbf{u}_k \leq \begin{bmatrix} 1.3 \\ 0.9 \end{bmatrix}; \quad \Delta \mathbf{u}_k \leq \begin{bmatrix} 1 \\ 0.5 \end{bmatrix} \quad (7.25)$$

Form the matrices $C_c, \mathbf{d}_c, \mathbf{d}_{cu}$ by hand and compare to the results from the code.
Investigate the matrices you obtain with various dimensions of systems, various choices of n_u and different limits.

ILLUSTRATION: Input limits, input rate limits and output limits

Consider the SISO case $y_{k+1} - 0.8y_k = 0.4u_k$ and with constraints:

$$-2 \leq u_k \leq 3; \quad \|\Delta u_k\| \leq 1; \quad -0.4 \leq y_k \leq 4 \quad (7.26)$$

Take an input horizon of $n_u = 3$ and an output horizon $n_y = 5$. The relevant matrices $C_c, \mathbf{d}_c, \mathbf{d}_{cu}, \mathbf{d}_{c\Delta u}, \mathbf{d}_{cy}$ are shown here. The structure of C_c clearly matches that expected from (7.7, 7.14, 7.20, 7.22).

Readers can replicate this using the file **inputlimitsexample2.m**.

```
>> [Cc,dc,dcu,dcdu,dcy]

ans =

    1.0000        0        0    1.0000        0        0        0        0
         0   1.0000        0    1.0000        0        0        0        0
         0        0   1.0000    1.0000        0        0        0        0
   -1.0000        0        0    1.0000        0        0        0        0
         0  -1.0000        0    1.0000        0        0        0        0
         0        0  -1.0000    1.0000        0        0        0        0
    1.0000        0        0    3.0000  -1.0000        0        0        0
    1.0000   1.0000        0    3.0000  -1.0000        0        0        0
    1.0000   1.0000   1.0000    3.0000  -1.0000        0        0        0
   -1.0000        0        0    2.0000   1.0000        0        0        0
   -1.0000  -1.0000        0    2.0000   1.0000        0        0        0
   -1.0000  -1.0000  -1.0000    2.0000   1.0000        0        0        0
    0.4000        0        0    4.0000        0        0  -1.8000   0.8000
    0.7200   0.4000        0    4.0000        0        0  -2.4400   1.4400
    0.9760   0.7200   0.4000    4.0000        0        0  -2.9520   1.9520
    1.1808   0.9760   0.7200    4.0000        0        0  -3.3616   2.3616
    1.3446   1.1808   0.9760    4.0000        0        0  -3.6893   2.6893
   -0.4000        0        0    0.4000        0        0   1.8000  -0.8000
   -0.7200  -0.4000        0    0.4000        0        0   2.4400  -1.4400
   -0.9760  -0.7200  -0.4000    0.4000        0        0   2.9520  -1.9520
   -1.1808  -0.9760  -0.7200    0.4000        0        0   3.3616  -2.3616
   -1.3446  -1.1808  -0.9760    0.4000        0        0   3.6893  -2.6893
```

A higher-order system example is not presented as this increases the column dimension of P, Q and thus the matrices/vectors would be too large to display.

STUDENT PROBLEM

The file **inputlimitsexample3.m** also contains code for a MIMO example based on an MFD model. Use this example to compare the results you would get for the inequalities on pen and paper with the output from the file (the matrices are too big to display here).

Investigate the matrices you obtain with various dimensions and orders of systems, various choices of n_u, n_y and different limits.

Summary: Readers should be comfortable forming matrix inequalities which represent GPC predictions satisfying constraints. The inequalities typically depend upon fixed information and time varying information (e.g., past inputs and outputs) and thus may include a number of separate components which need to be clearly defined and updated each sample, as in (7.22).

7.5 Constrained GPC

The previous section established that GPC predictions will satisfy constraints as long as a suitable set of inequalities is satisfied, such as:

$$C_c \Delta \underset{\rightarrow k}{\mathbf{u}} \leq \mathbf{d}_c - \mathbf{d}_{cu} \mathbf{u}_{k-1} - \mathbf{d}_{c\Delta u} \Delta \underset{\leftarrow k-1}{\mathbf{u}} - \mathbf{d}_{cy} \underset{\leftarrow k}{\mathbf{y}} \qquad (7.27)$$

This form allows for input limits, input rate limits and output limits so it may take a simpler form where only a subset of those limits apply. In general, it is worth simplifying this to take the following format:

$$C_c \Delta \underset{\rightarrow k}{\mathbf{u}} \leq \mathbf{f}_k; \quad \mathbf{f}_k = \mathbf{d}_c - \mathbf{d}_{cu} \mathbf{u}_{k-1} - \mathbf{d}_{c\Delta u} \Delta \underset{\leftarrow k-1}{\mathbf{u}} - \mathbf{d}_{cy} \underset{\leftarrow k}{\mathbf{y}} \qquad (7.28)$$

where $\mathbf{f}_k$ is time varying, but C_c is fixed. The time varying nature of the inequalities is a key obstacle to ad hoc solutions and as will be seen is also a reason why, in general, simple saturation policies can be ineffective.

Next, we ask the question, how do such constraints impact on the computation of a GPC control law?

7.5.1 Quadratic programming in GPC

In simple terms, the aim of an MPC algorithm is to optimise predicted behaviour, subject to those predictions being implementable (or feasible), which means subject to those predictions satisfying constraints. This statement can be reduced to a very simple format.

Consider a typical performance objective such as given in Section 4.5.3.

$$\min_{\Delta \underset{\rightarrow k}{\mathbf{u}}} \Delta \underset{\rightarrow k}{\mathbf{u}}^T \underbrace{(H^T H + \lambda I)}_{S} \Delta \underset{\rightarrow k}{\mathbf{u}} + \Delta \underset{\rightarrow k}{\mathbf{u}}^T \underbrace{2H^T [P \Delta \underset{\leftarrow k-1}{\mathbf{u}} + Q \underset{\leftarrow k}{\mathbf{y}} - \underset{\rightarrow k+1}{\mathbf{r}}]}_{\mathbf{p}} \qquad (7.29)$$

or in simple terms:

$$\min_{\Delta \underset{\rightarrow k}{\mathbf{u}}} \Delta \underset{\rightarrow k}{\mathbf{u}}^T S \Delta \underset{\rightarrow k}{\mathbf{u}} + \Delta \underset{\rightarrow k}{\mathbf{u}}^T \mathbf{p}; \quad S > 0 \qquad (7.30)$$

We can add constraints to this problem just by stating them alongside the objective, that is:

$$\min_{\Delta \underset{\rightarrow k}{\mathbf{u}}} \Delta \underset{\rightarrow k}{\mathbf{u}}^T S \Delta \underset{\rightarrow k}{\mathbf{u}} + \Delta \underset{\rightarrow k}{\mathbf{u}}^T \mathbf{p} \quad \text{s.t.} \quad C_c \Delta \underset{\rightarrow k}{\mathbf{u}} \leq \mathbf{f}_k \qquad (7.31)$$

This is called a quadratic programming (QP) problem:

- The objective to be minimised is quadratic in form.

- The constraints (that is, constraints (7.28)) are linear in the d.o.f.

7.5.2 Implementing constrained GPC in practice

The solution of optimisation (7.31) is non-trivial in general and rarely amenable to simple solutions. This means that the control engineer needs to rely on optimisation toolboxes which are likely to have been created by someone else with the required mathematical and coding skills base.

A discussion of the alternative algorithms for solving (7.31) is far beyond the remit of this book, being a research area in its own right, and hence some simple statements only are made here.

1. The solution of a QP optimisation is considered straightforward in general and many standard solvers are available.

2. Where the constraints are feasible (that is not mutually incompatible), the QP algorithm is guaranteed to converge to the optimum for problem (7.31).

3. While theoretical upper limits on the convergence times of these solvers can be high, in practice, QP solvers are fast and except for very large problems with tens or hundreds of d.o.f. and thousands of inequalities, the computation time is not a problem for systems with sample times of seconds or slower [27].

4. This book will use the solver provided by MATLAB. The relevant file is named **quadprog.m** and has some options to allow the user to select the solving algorithm of their choice. A typical call statement would take a form using the parameters $S, \mathbf{p}, C_c, \mathbf{f}_k$ from (7.31):

$$opt = \textbf{quadprog}(S, \mathbf{p}, C_c, \mathbf{f}_k) \; ; \quad opt \equiv \Delta \underset{\rightarrow k}{\mathbf{u}}$$

Summary: Most practical MPC algorithms dealing with constraints rely on a quadratic programming optimisation which therefore must be available on the relevant hardware/software environment.

7.5.3 Stability of constrained GPC

Although one can define an MPC algorithm that does constraint handling, it is less straightforward to establish if that algorithm will be stabilising. The stability of the underlying linear controller (e.g., 4.31, 4.90) does not guarantee the stability of the constrained controller arising from the repeated use of the solution to (7.31).

- With finite horizon algorithms, both a priori and a posteriori stability results are difficult to derive in general for the constrained case and exist only for special cases.

- Guarantees of recursive feasibility (to be discussed in the next chapter) can be given for the nominal case under some mild conditions such as: (i) the use of large output horizons within the constraint inequalities and (ii) convergent predictions, but these results are not necessarily robust to any uncertainty which is present in all real scenarios.

There is no easy answer to this failing; however, the lack of a rigorous mathematical guarantee of stability does not imply instability nor need it imply poor performance. In practice, if sensible guidelines (use a large enough output and constraint horizon) are followed and one avoids infeasibility, then the constrained loop will have a similar performance to the unconstrained loop.

> **Summary:** Practical industrial algorithms with constraint handling do not have stability guarantees, or rigorous guarantees of recursive feasibility, but this is rarely a problem in practice when long output horizons are used.

7.5.4 Illustrations of constrained GPC with MATLAB

This section will provide a number of simulation examples of the use of GPC with the constraint handling facility enabled. To demonstrate the importance of this, the illustrations will also include the corresponding behaviours from a saturation policy (as defined in Section 7.3.1). The illustrations will consider a combination of input limits, input rate limits and output limits. From a user's perspective, how many of these are included does not really change the complexity of the optimisation format; however, it may have a significant impact on the complexity of the code to construct the constraint inequality matrices. Consequently, code is provided with varying levels of complexity/capability so that the code is as transparent to the reader as is possible.

7.5.4.1 ILLUSTRATION: Constrained GPC with only input constraints

The figures given below overlay unconstrained control (which could not occur in practice) with the behaviour from applying GPC with constraint handling. Here only absolute inputs constraints and input rate constraints are used.

gpcconstraints1.m:
Compares unconstrained and constrained solutions (at each sampling instant) for a simple example. In this case one could argue that saturation control would likely give equivalent behaviour, but of course the QP has handled the constraints effectively.

gpcconstraints2.m:
Compares unconstrained and constrained solutions (at each sample) for a simple example. In this case it is clear that saturation control is not a good solution and differs significantly from the optimised QP solution.

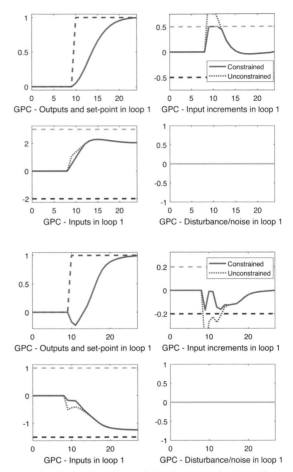

gpcconstraints3.m: Provides equivalent plots for a 3 by 3 example (one for each loop). Readers are encouraged to run this themselves and will notice that in this case, partially due to the MIMO nature of the problem, saturation control is not always the same as the optimal constrained solution.

7.5.4.2 ILLUSTRATION: Constrained GPC with output constraints

The examples in this section indicate clearly that including output constraints is much more challenging and moreover may imply inputs remain far of their limits. In other words, the optimised trajectories are often far away from what would arise with saturation control.

gpcconstraints4.m:
In this illustration the output constraint is still violated and inspection of the *command window* will show that the QP was infeasible. The disturbance was so large that it was impossible to satisfy all the constraints simultaneously! Feasibility is a genuine challenge for constrained GPC.

gpcconstraints5.m:
Demonstrates how an output constraint is satisfied by deviating the inputs from the unconstrained optimal. Here the potential violation was caused by a large output disturbance.

gpcconstraints6.m: A very challenging output constraint is satisfied for a non-minimum phase system. Of course the consequence is a significant reduction in the overall response time, although this is still far better than that achievable by the *saturation* alternative.

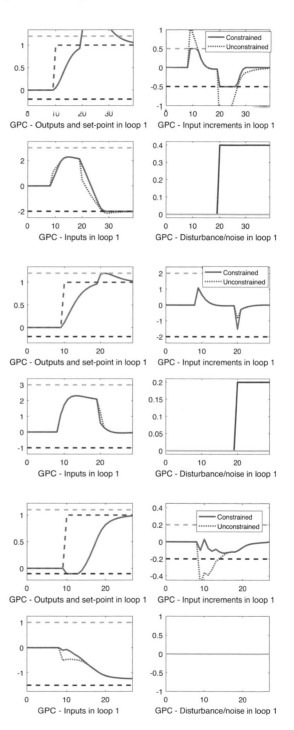

gpcconstraints7.m: A MIMO example where the impact on behaviour of enforcing the output constraint is significant.

7.5.4.3 ILLUSTRATIONS: Constrained GPC with a T-filter

gpcconstraints8,9,10.m: Provides code for input constraints only but with predictions and constraint inequalities based on the use of a T-filter.
gpcconstraints11,12.m: Provides code for input and output constraints with predictions and constraint inequalities based on the use of a T-filter.
gpcconstraints13,14.m: Demonstrates that constraint handling with a noisy measurement benefits from the inclusion of a T-filter in a similar way to the unconstrained case, but disturbance rejection may be slightly worse. [MATLAB figures 1-3 are with no T-filter and figures 4-6 with a T-filter.]

7.5.4.4 ILLUSTRATION: Constrained GPC with an independent model

gpcconstraints15.m: Provides code for input constraints only but with predictions and constraint inequalities based on the use of an independent transfer function model.
gpcconstraints16.m: Provides code for input constraints and output constraints with predictions and constraint inequalities based on the use of an independent transfer function model.
gpcconstraints17.m: Provides code for constrained GPC with a noisy output and produces figures for comparison using a T-filter and also for use of an independent model.
gpcconstraints18.m: MIMO example with a MIMO independent transfer function model.
gpcconstraints19.m: Independent model GPC based on a state space model and a performance index linked to deviation variables (that is $u - u_{ss}$) rather than Δu.
gpcconstraints20,21.m: Independent model GPC based on a state space model and a performance index linked to Δu.

STUDENT PROBLEMS

Use models and constraints of your choice to investigate the following scenarios with a GPC algorithm.

1. The differences between unconstrained and constrained solutions.

2. The impact of the different types of constraints (inputs, input rates and outputs).

3. The impact on the closed-loop behaviour of MIMO systems.

4. The potential for creating infeasible optimisations (the

MATLAB code should give a warning message in the command window).

5. The impact of using different types of predictions such as those based on a T-filter and an independent model formulation.

You can use the different m-files above and in Section 7.8 to produce simulations corresponding to different scenarios/assumptions.

7.6 Understanding a quadratic programming optimisation

This section is a little bit of an aside but is critically important in helping readers to understand what is going on during a QP optimisation and in particular, why certain events are observed.

1. Can a QP optimisation be visualised and make intuitive sense?
2. Why is saturation control often suboptimal and potentially very poor compared to an optimal constrained solution?
3. How does infeasibility occur and are there simple fixes?

7.6.1 Generic quadratic function optimisation

For simplicity define a generic QP optimisation as follows where it is assumed that the constraints are feasible.

$$\min_{\mathbf{x}} \underbrace{\mathbf{x}^T S \mathbf{x} + \mathbf{x}^T \mathbf{p} + c}_{J} \quad \text{s.t.} \quad C\mathbf{x} \leq \mathbf{f} \tag{7.32}$$

In the unconstrained case the optimum is given by:

$$\mathbf{x} = -0.5S^{-1}\mathbf{p} \tag{7.33}$$

First it is useful to understand the shape of the quadratic function J and how this depends upon the d.o.f. $\mathbf{x}$.

ILLUSTRATION: Contour curves for quadratic function (7.32)

Consider the following function and plot contour curves for J and their dependence on d.o.f. $\mathbf{x}$.

$$J = \mathbf{x}^T \begin{bmatrix} 1.2 & -0.2 \\ -0.2 & 0.8 \end{bmatrix} \mathbf{x} + \mathbf{x}^T \begin{bmatrix} 1 \\ 5 \end{bmatrix} + 2 \qquad (7.34)$$

The contours (ellipsoids) correspond to values of $\mathbf{x}$ which give the same value of J. The smallest value of J is given from *the center* or (7.33) as $\mathbf{x} = [-0.98, -3.37]^T$. The key observation is that the position of the unconstrained minimum is *obvious*!

Contours of fixed J

File: **contoursplot1.m**

7.6.2 Impact of linear constraints on minimum of a quadratic function

Next, we consider what happens when a linear constraint is added to a quadratic optimisation; what happens to the minimum? The key point to note is that the unconstrained optimum of (7.33) may be inconsistent with the constraint, and therefore the constrained optimum will be in a different place.

7.6.2.1 ILLUSTRATION: Contour curves, constraints and minima for a quadratic function (7.32)

Consider the function of (7.34) and add the following linear constraint $\mathbf{x} \geq 1.5$.

The contours (ellipsoids) are overlaid with this constraint (dashed line) and also, the new constrained optimum is marked. Notably, the constrained optimum (marked with a cross) which is the smallest value of *J*, or smallest contour, which also meets the constraint **MUST LIE** on the constraint as to not do so would be a contradiction of this being a minimum.

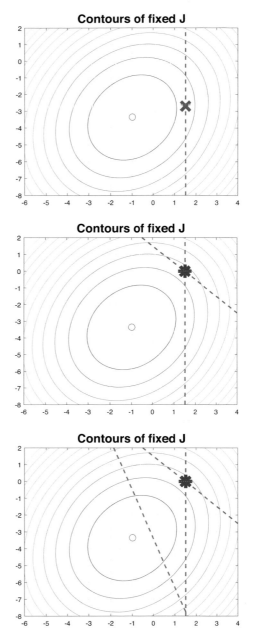

Next add a second constraint that $[1 \quad 1]\mathbf{x} \geq 1$. Adding this to the contour plot, it is clear that now only the top right-hand part of the figure satisfies both constraints. This time the constrained optimum (the contour with the smallest value), lies on the intersection of these two constraints.

Finally add a third constraint that $[1 \quad 0.2]\mathbf{x} \geq 1$. Adding this to the contour plot, it is clear this has no affect on the constrained optimum as the constrained optimum from the previous two constraints also satisfies this new constraint. This is an example of a redundant constraint; it plays no part in the computation of the constrained optimum for this choice of *J*.

File: **contoursplot2.m**

It is useful at this point to emphasise some terminology.

- A constraint is active if its inclusion affects the constrained optimum.

- A constraint is inactive, or redundant, if its inclusion does not affect the optimum.

- The constraints which are active will depend upon the position of the unconstrained optimum (and shape).

This latter point is particularly important: the inactive/active constraints may change as the cost function J and associated unconstrained minimum changes and thus to find the constrained minimum we first must identify the active constraints. Within the context of MPC, it is only the linear component of J which changes each sample (S is constant but $\mathbf{p}$ changes). The following illustrations demonstrate how the active set (constraints on which the constrained optimum lies) change as $\mathbf{p}$ changes.

7.6.2.2 ILLUSTRATION: Impact of change of linear term in quadratic function (7.34) on constrained minimum

Consider a quadratic function with the same S as in (7.34) and three linear constraints (the top right corner of the figure is the feasible region) but vary the value of $\mathbf{p}$. As $\mathbf{p}$ changes, the unconstrained optimum (7.33) changes and with it the centre of the ellipsoid. With $\mathbf{p} = [5 \ 0]^T$ the constrained optimum (marked with a cross) lies on a single constraint, say the 1st constraint.

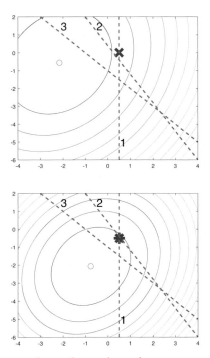

With $\mathbf{p} = [1 \ 3]^T$ the constrained optimum (marked with an asterisk) lies on the corner of two constraints (say constraints 1 and 2).

[File: **contoursplot3.m**]

The next figure is particulary interesting as the optimum is no longer on $x_1 = -0.5$ so this is a case where a simple saturation strategy could be misleading. Choosing $x_1 = -0.5$ (constraint 1) would restrict the USER to higher values of J and thus be suboptimal.

With $\mathbf{p} = [4\ 6]^T$ the constrained optimum (marked with a plus) lies on just constraint 2.

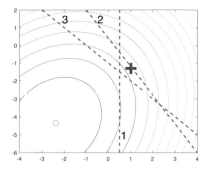

In general it can be seen that the constrained optimum could involve any single one of the 3 constraints, any pair or indeed none as the unconstrained optimum $-S^{-1}\mathbf{p}$ moves.

7.6.3 Illustrations of MATLAB for solving QP optimisations

This section gives some illustrations of how to use MATLAB to solve a QP optimisation along with figures which give the reader confidence that the file is doing the expected job.

MATLAB call statement for QP

MATLAB assumes the function is given in the format

$$\min_{\mathbf{x}}\ 0.5\mathbf{x}^T S\mathbf{x} + \mathbf{x}^T\mathbf{p} + c \quad \text{s.t.} \quad C\mathbf{x} \leq \mathbf{f} \tag{7.35}$$

where the reader should note in particular the **0.5** on the quadratic term. The constrained optimum is therefore given as $\mathbf{x} = -S^{-1}\mathbf{p}$. To find the constrained optimum, use the call statement:

xopt=quadprog(S,p,C,f,[],[],[],[],[],opt)

The blank variables and variable *opt* can be omitted and readers should use *help quadprog* to learn more about their role, along with possible error flags, for example to indicate infeasibility.

The following illustrations show MATLAB solutions for different choices of *J* and different constraints. They also overlay the contour plots of *J* to aid insight.

ILLUSTRATION: Using MATLAB to solve QP problems

With cost function and constraints:

$$J = \frac{\mathbf{x}^T \begin{bmatrix} 1.2 & -0.2 \\ -0.2 & 0.8 \end{bmatrix} \mathbf{x}}{2} + \mathbf{x}^T \begin{bmatrix} 1 \\ 1 \end{bmatrix};$$

$$\begin{bmatrix} -1 & 0 \\ 0 & -1 \\ -1 & -1 \end{bmatrix} x \le \begin{bmatrix} -1 \\ -1 \\ -3 \end{bmatrix}$$

File: **matlabqpex1.m**

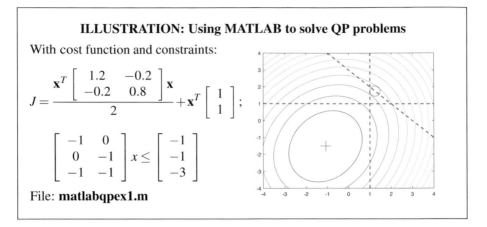

ILLUSTRATION: Using MATLAB to solve QP problems

With cost function and constraints:

$$J = 0.5 x^T \begin{bmatrix} 0.8 & 0.6 \\ 0.6 & 1.8 \end{bmatrix} + x^T \begin{bmatrix} 1 \\ 1 \end{bmatrix} + 3;$$

$$\begin{bmatrix} -1 & 0 \\ 0 & -1 \\ -1 & -1 \end{bmatrix} x \le \begin{bmatrix} -1 \\ -1 \\ -3 \end{bmatrix}$$

File: **matlabqpex2.m**

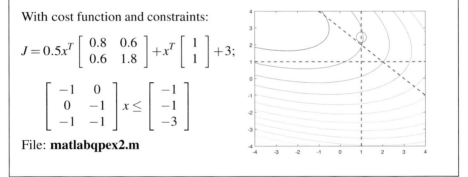

7.6.4 Constrained optimals may be counter-intuitive: saturation control can be poor

It was indicated in the introduction to this chapter that a failure to properly consider constraints when planning a strategy can result in severely suboptimal behaviour. This section gives a single example to give some mathematical evidence of this using the insights gained from contour plots overlaid with inequalities. In simple terms, consideration of all the figures in this section, and the following example, indicate that it is not atypical for saturation control (that is selecting the value of $\mathbf{u}_k$ at its constraint value) will often lead to worse or even far worse, performance (a bigger contour J), than selecting an alternative value which may be much smaller. This is unsurprising in qualitative terms as sometimes we need to back off the current input to ensure we have the freedom of movement (ability to move the input both ways) desired further into the future.

ILLUSTRATION: Counter-intuitive results for a QP

With cost function and box constraints:

$$J = 0.5x^T \begin{bmatrix} 0.8 & 0.8 \\ 0.8 & 5.0 \end{bmatrix} + x^T \begin{bmatrix} 1.5 \\ 20 \end{bmatrix} + 3;$$

$$\begin{bmatrix} -1 & 0 \\ 0 & -1 \\ 1 & 0 \\ 0 & 1 \end{bmatrix} x \leq \begin{bmatrix} 1 \\ 1 \\ 1 \\ 1 \end{bmatrix}$$

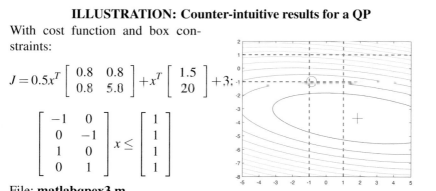

File: **matlabqpex3.m**

In this case one can see that the constrained optimal (which must lie inside the square formed by the constraints) has selected $x_1 = -1$ whereas a simple saturation policy would use the closest limit to the unconstrained optimal, that is, it would select $x_1 = 1$. **The constrained optimal is the opposite of the result given by simple saturation!** In this case, simple saturation would be a very poor choice indeed.

Summary: A failure to take proper account of constraints during a feedback design can have several undesirable consequences. It has been demonstrated that the inclusion of linear constraints changes the optimum, and not always as we expect.

7.7 Chapter summary

This chapter has demonstrated how to introduce systematic constraint handling into the GPC algorithm. Specifically several illustrations emphasise the importance of incorporating constraints within the performance optimisation rather than using a posteriori saturation approaches; the constrained optimum trajectory can be very different and much better than the saturation trajectory.

Systematic constraint handling requires the use of a quadratic programming optimisation. This chapter has shown that building the linear inequalities to represent satisfaction of constraints by predictions is straightforward for GPC and involves relatively minimal computation and coding. Nevertheless, the solution of the QP optimisation itself is less straightforward to code and readers are recommended to use routines that are already available. In general these algorithms have good reliability and in terms of modern computing, can be solved in well under a second for typical MPC problems.

7.8   MATLAB code supporting constrained MPC simulation

This code is provided so that students can learn immediately by trial and error, entering their own parameters directly. Readers should note that this code is not *professional quality* and is deliberately simple, short and free of comprehensive *error catching* so that it is easy to understand and edit by users for their own scenarios. Please feel free to edit as you please.

Readers may wonder why such a diversity of files is provided when many are subsumed in others. The rationale is to keep any given file as simple as possible so that readers who wish to read and edit a file are not overwhelmed by unnecessary complexity and conditional expressions.

7.8.1   Miscellaneous MATLAB files used in illustrations

inputlimitsexample1-3.m	Demonstrates the creation of the matrix inequalities representing constraint handling for a prediction model (2.49).
matlabqpex1-2.m	Overlays contour plots, lines representing inequalities and the constrained optimal to give clear insight into the action of a 2-dimensional QP optimisation. Users can change the cost function and linear constraints to a limited extent but would need to update the plotting of the lines representing the inequalities.
contourplots1-2.m	Produces contour plots for a specified 2-dimensional cost function. Users can change the cost function to a limited extent.
impactconstraints.m	Demonstrates weakness of saturation control in Section 7.3.2.
constrainedexample1,2.m	Demonstrates weakness of saturation control in Section 7.3.2.

7.8.2 MATLAB code for supporting GPC based on an MFD model

mpc_simulate_noconstraints.m	Performs GPC simulations for an MFD model ignoring constraints altogether.
mpc_simulate_noconstrainthandling.m	Performs GPC optimisations without constraint handling, but enforces input saturation as required.
mpc_simulate.m	Performs GPC optimisations with input and input rate constraint handling.
mpc_simulate_outputconstraints.m	Performs GPC optimisations with input, input rate and output constraints within the online optimisation.
mpc_simulate_overlay.m	Overlays constrained and unconstrained solutions from each sample so that the impact of constraints on behaviour can be observed. Implements the constrained solution.
mpc_simulate_tfilt_noconstraints.m	Performs GPC simulations with a T-filter for an MFD model ignoring constraints altogether.
mpc_simulate_tfilt.m	Performs GPC optimisations with input and input rate constraint handling and a T-filter.
mpc_simulate_tfiltoutputconstraints.m	Performs GPC with a T-filter and optimises with input, input rate and output constraints within the online optimisation.
gpcconstraints1-14.m	Files used to generate illustrations for a variety of scenarios. Useful as starting points for how to enter data and call main files for readers who wish to create their own examples. See Section 7.5.4 for more details.
mpc_predmat.m, **mpc_law.m,** **mpc_constraints.m,** **mpc_constraints2.m**	Utility files used to build prediction matrices, control law parameters and constraint matrices for the MFD case.
mpc_predtfilt.m	Additional utility file used to build prediction matrices for GPC with a T-filter.

7.8.3 MATLAB code for supporting GPC based on an independent model

impc_tf_simulate_constraints.m	Performs GPC simulations for an independent MFD model with input and output constraints.
impc_simulate.m	Performs GPC simulations for an independent state space model with input constraints only and a performance index weighting $\mathbf{u} - \mathbf{u}_{ss}$.
impc_simulate_Du.m	Performs GPC simulations for an independent state space model with input and output constraints and a performance index weighting $\Delta\mathbf{u}$.
gpcconstraints15-21.m	Files used to generate illustrations for a variety of scenarios using independent models. Useful as starting points for how to enter data and call main files for readers who wish to create their own examples. See Section 7.5.4 for more details.
imgpc_tf_predmat.m, imgpc_tf_law.m, imgpc_constraints_tfmodel.m	Utility files used to build prediction matrices, control law parameters and constraint matrices for the independent model MFD case.
imgpc_costfunction.m, imgpc_predmat.m, imgpc_constraints.m	Utility files used to build prediction matrices, control law parameters and constraint matrices for the independent model state space case based on a performance index with $\mathbf{u} - \mathbf{u}_{ss}$.
imgpc_costfunctionDu.m, imgpc_predmatDu.m, imgpc_constraintsDu.m	Utility files used to build prediction matrices, control law parameters and constraint matrices for the independent model state space case based on a performance index with $\Delta\mathbf{u}$.

8

Constraint handling in dual-mode predictive control

CONTENTS

8.1 Introduction

This chapter will assume that the reader has gone through Chapter 7 and will build directly from there. Including constraints in dual-mode algorithms is altogether more difficult both algebraically and conceptually than for finite horizon algorithms, and thus merits a separate discussion. Of particular note is the need to extend constraint handling over an infinite horizon. Naturally one cannot build infinite dimensional matrices and thus mechanisms are needed to deal with this apparent dichotomy in a rigorous and computationally efficient manner.

8.2 Guidance for the lecturer/reader and introduction

The author would advise that this content is not included in a first course in predictive control as it requires yet more new concepts to be understood by the students and this would cause a danger of serious overload. Moreover, while interesting and potentially valuable for reinforcing core principles, this is a luxury best afforded by those doing a second course or pursuing research and/or development which involves MPC.

In view of this, the chapter will begin somewhat abruptly and assume familiarity with the previous two chapters. It begins by showing briefly how similar techniques to the previous chapter can be used for constraint handling, but in general, although easy to deploy, flexible and often effective, these have fundamental theoretical weaknesses for dual-mode algorithms. Hence some new concepts are introduced around set representations of inequalities, admissible sets, terminal sets and invariant sets. These concepts are required to extend the stability guarantees of unconstrained OMPC/SOMPC to the constrained case. Having dealt with concepts using indicative and simple algebra, the chapter then moves on to developing precise algebra for some specific cases and MATLAB® implementations. Finally, despite the improvement in rigour, it is demonstrated that the use of terminal sets and invariant sets can actually be an impediment to effective use in practice and ironically, the user may prefer the more simplistic but less rigorous approaches.

8.3 Background and assumptions

Readers may wonder why the discussion of constraints has been separated for dual-mode approaches, but this will now quickly become apparent.

1. Dual-mode approaches deploy notably different classes of predictions and

this has a significant impact on the construction of the relevant inequalities for constraint handling.

2. Dual-mode approaches deploy a terminal control law and have input predictions that do not satisfy (7.3), that is, do not move to a constant in finite time. This subtlety needs careful handling.

There are two conceptual approaches that can be used to capture constraint information, that is satisfaction of sample constraints such as:

$$
\left\{
\begin{array}{c}
\underline{\Delta \mathbf{u}} \le \Delta \mathbf{u}_{k+i|k} \le \overline{\Delta \mathbf{u}} \\
\underline{\mathbf{u}} \le \mathbf{u}_{k+i|k} \le \overline{\mathbf{u}} \\
\underline{\mathbf{y}} \le \mathbf{y}_{k+i|k} \le \overline{\mathbf{y}} \\
\underline{\mathbf{x}} \le K_x \mathbf{x}_{k+i|k} \le \overline{\mathbf{x}}
\end{array}
\right\}, \quad \forall i \ge 0
\tag{8.1}
$$

1. Only check constraints (8.1) for the first N_c samples and assume that any violations thereafter are at most minimal and thus will have a negligible impact.

2. Do some off-line analysis to establish the existence and value of an N_c such that satisfying constraints for the first N_c samples necessarily implies satisfaction thereafter.

While these two may seem very similar in principle, there is a significant difference.

An arbitrary, albeit large, choice of N_c may not be large enough and this would invalidate any guarantees of recursive feasibility (although in practice a sensible choice is likely to be large enough). However, the coding of inequalities to capture constraints (8.1) for N_c steps is moderately straightforward as illustrated in Section 7.4 and the following. Consequently this option may be favoured by early users and thus is presented first in the following section.

Conversely, the computation of a large enough N_c to ensure recursive feasibility requires more insight, an understanding and use of recursive linear programming and potentially substantial off-line analysis. Nevertheless, the resulting constraint set is likely to be far more compact and thus one reaps the benefits in the online implementation. This approach is introduced in Section 8.5.6.

Remark 8.1 *Using algebra analogous to that given in Section 7.4, one can ensure the satisfaction of state constraints from (8.1) with an expression of the form:*

$$
L\underline{\mathbf{x}} \le D_{K_x} \mathbf{x}_{\rightarrow k+1} \le L\overline{\mathbf{x}}
\tag{8.2}
$$

where D_{K_x} is pseudo-block diagonal with blocks taking the values K_x.

Summary: A simplistic, but often effective, constraint handling policy, ensures sample constraints (8.1) are satisfied for N_c steps into the future.

8.4 Description of simple or finite horizon constraint handling approach for dual-mode algorithms

8.4.1 Simple illustration of using finite horizon inequalities or constraint handling with dual mode predictions

A conceptually simple, but approximate approach to constraint handling for dual-mode predictions has strong analogies with the previous chapter on finite horizon methods.

1. Choose a horizon N_c which is greater than the settling time.
2. Build inequalities equivalent to (8.1) for the given horizon N_c.
3. Substitute in the relevant predictions (such as (6.18)) to determine a compact form equivalent to (7.22).

The ease with which one can form inequalities to test constraints over a finite horizon was apparent in Chapter 7 and thus the following illustration uses an identical process, but with dual-mode predictions.

ILLUSTRATION: Constraint inequalities with dual-mode predictions

Consider the input predictions for an asymptotic target of the origin:

$$\{\mathbf{u}_{k+i} \text{ a d.o.f., } i < n_c\}, \quad \{\mathbf{u}_{k+i} = -K\mathbf{x}_{k+i}, i \ge n_c\}, \tag{8.3}$$

alongside model $\mathbf{x}_{k+i} = A\mathbf{x}_{k+i-1} + B\mathbf{u}_{k+i-1}$. Combining these and the prediction algorithms in Chapter 2 or for example Eqn.(6.32), one can form predictions for any horizon N_c we choose of the form:

$$\underset{\rightarrow k+1}{\mathbf{x}} = P_x \mathbf{x}_k + H_x \underset{\rightarrow k}{\mathbf{u}}; \quad \underset{\rightarrow k}{\mathbf{U}} = P_{xu}\mathbf{x}_k + H_{xu}\underset{\rightarrow k}{\mathbf{u}} \tag{8.4}$$

where $\underset{\rightarrow k}{\mathbf{U}}$ comprises the input predictions for a large horizon N_c whereas $\underset{\rightarrow k}{\mathbf{u}}$ comprises only the first n_c values which constitute the d.o.f. Combining predictions (8.4) with input and state constraints only from (8.1) and noting (8.2) gives:

$$\underbrace{\begin{bmatrix} D_{K_X}P_x \\ -D_{K_x}P_x \\ P_{xu} \\ -P_{xu} \end{bmatrix}}_{C_c}\mathbf{x}_k + \underbrace{\begin{bmatrix} H_x \\ -H_x \\ H_{xu} \\ -H_{xu} \end{bmatrix}}_{\mathbf{d}_{xu}}\underset{\rightarrow k}{\mathbf{u}} \le \underbrace{\begin{bmatrix} L\bar{\mathbf{x}} \\ -L\underline{\mathbf{x}} \\ L\bar{\mathbf{u}} \\ -L\underline{\mathbf{u}} \end{bmatrix}}_{\mathbf{d}_c} \tag{8.5}$$

NOTE: For simplicity this illustration has neither included the details required for unbiased prediction nor input rate constraints.

Practically speaking, such a set of inequalities is likely to be sufficient, but it will suffer from two notable weaknesses: (i) any guarantees associated to the use of dual-mode are invalidated by the use of a finite horizon for the constraint handling so that one cannot give assurances that the associated predictions are feasible and (ii) in practice the associated definitions for the inequalities for a suitable N_c are likely to contain a very large number of redundant constraints and thus could be highly inefficient to use.

> **Summary:** Using finite horizon predictions based on a dual-mode prediction
> structure, the development of inequalities to represent constraint satisfaction is
> very similar to that deployed by GPC.

8.4.2 Using finite horizon inequalities or constraint handling with unbiased dual-mode predictions

In practice, it is necessary to use unbiased predictions and allow for non-zero steady-state targets and disturbances, and so the algebra becomes slightly more involved. As shown in Chapter 2, define the expected steady-state values as follows:

$$\begin{bmatrix} \mathbf{x}_{ss} \\ \mathbf{u}_{ss} \end{bmatrix} = \begin{bmatrix} K_{xr} \\ K_{ur} \end{bmatrix} \begin{bmatrix} \mathbf{r}_{k+1} - \mathbf{d}_k \end{bmatrix}; \quad \mathbf{d}_k = \mathbf{y}_p(k) - C\mathbf{x}(k) \tag{8.6}$$

and replace the predicted control moves during the first n_c prediction steps by the equivalent CLP form of:

$$\underbrace{\mathbf{u}_{k+i} - \mathbf{u}_{ss}}_{\tilde{\mathbf{u}}_{k+i}} = -K\underbrace{[\mathbf{x}_{k+i} - \mathbf{x}_{ss}]}_{\tilde{\mathbf{x}}_{k+i}} + \mathbf{c}_{k+i}; \quad i = 0,1,\cdots,n_c - 1$$

So in effect, with $\Phi = A - BK$, the system dynamics within the first n_c steps can be represented as:

$$\tilde{\mathbf{x}}_{k+i} = \Phi\tilde{\mathbf{x}}_{k+i-1} + B\mathbf{c}_{k+i-1} ; \quad i = 0,1,\cdots,n_c - 1 \tag{8.7}$$

As indicated in Section 2.9.4, the associated predictions take a form equivalent to Eqn.(8.4) where notation $\underset{\rightarrow k}{\mathbf{U}}$ indicates a long horizon for the future inputs:

$$\underset{\rightarrow k+1}{\tilde{\mathbf{x}}} = P_{cl}\tilde{\mathbf{x}}_k + H_c\underset{\rightarrow k}{\mathbf{c}}; \quad \underset{\rightarrow k}{\tilde{\mathbf{U}}} = P_{clu}\tilde{\mathbf{x}}_k + H_{cu}\underset{\rightarrow k}{\mathbf{c}} \tag{8.8}$$

Substituting in from (8.6), the predictions will be affine in the current state, current target, current disturbance and future control values.

All the constraints, that is, input rates, inputs and outputs, must be satisfied simultaneously. Hence, using methods analogous to Section 7.4, one can combine inequalities (8.1) with prediction Equations (8.8) defined for a given horizon N_c to determine a single set of linear inequalities:

- of the following form for predictions using the first n_c inputs as d.o.f. in the prediction equation:

$$C_c \underset{\rightarrow k}{\mathbf{u}} \leq \mathbf{d}_c - d_{cu}\mathbf{u}_{k-1} - d_{cx}\mathbf{x}_k - d_{cr}\underset{\rightarrow k+1}{\mathbf{r}} - d_{cd}\mathbf{d}_k \qquad (8.9)$$

- and the following form for predictions using the input perturbation $\mathbf{c}_k$ as the d.o.f.:

$$C_c \underset{\rightarrow k}{\mathbf{c}} \leq \mathbf{d}_c - d_{cu}\mathbf{u}_{k-1} - d_{cx}\mathbf{x}_k - d_{cr}\underset{\rightarrow k+1}{\mathbf{r}} - d_{cd}\mathbf{d}_k \qquad (8.10)$$

where $C_c, \mathbf{d}_c$ are time invariant and the other terms depend upon the current state (past input, state information, the target and the disturbance estimate). Depending on the constraints and model predictions deployed, the predictions (and therefore inequalities) may or may not include the terms associated to $\mathbf{u}_{k-1}$.

In the dual-mode case it is likely that the required horizon for the constraint handling N_c will be large (say 30 or more) and therefore the total number of inequalities in (8.9,8.10) can quickly run into the thousands even for a moderate size of system.

Summary: One can include constraints for a dual-mode algorithm by testing the predictions only up to N_c samples (N_c greater than the settling time). In many scenarios, for example where numerical efficiency and rigorous proofs are not needed, this is effective and simple to code. Constraint satisfaction can be represented by a set of linear inequalities (8.9, 8.10) that depend upon the d.o.f. and the current state, past inputs, disturbance estimate and the desired target, which thus the inequalities must be updated each sample.

The inclusion of the target information in the constraint inequalities is a notable difference from those required for GPC (see Eqn.(7.22)) and arises due to the terminal mode control law.

STUDENT PROBLEMS

1. Form the constraint inequalities equivalent to (8.9, 8.10) for unbiased dual-mode predictions of the form given in Eqn.(8.7) for a specified horizon N_c.

2. How would you extend the inequalities to include input rate constraints and output constraints?

3. Construct inequalities to represent the satisfaction of inputs, input rates and output constraints with several different types of dual-mode modelling/prediction structures.

4. Create MATLAB code to build constraint matrices of the form given in (8.9,8.10) for different dual-mode prediction models and constraints of your choice and to allow for non-zero steady-states and disturbances. Use the prediction code created for Chapter 2.

8.4.3 MATLAB code for constraint inequalities with dual-mode predictions and SOMPC/OMPC simulation examples

The code provided here is based on update model (8.7) (see Section 6.6) in conjunction with constraints (8.1). The core commands building the constraint inequalities use the following predictions which are unbiased in the steady-state in the presence of both non-zero set points and disturbances:

$$\underset{\rightarrow k+1}{\mathbf{x}} = P_{cl}\mathbf{z}_k + H_c \underset{\rightarrow k}{\mathbf{c}} + P_r \mathbf{r}_{k+1}; \quad \underset{\rightarrow k}{\mathbf{u}} = P_{clu}\mathbf{z}_k + H_{cu}\underset{\rightarrow k}{\mathbf{c}} + P_{ru}\mathbf{r}_{k+1} \tag{8.11}$$

where $\mathbf{z}_k = [\mathbf{x}_k^T, \mathbf{d}_k^T]^T$ is a state which includes the most upto date disturbance estimate $\mathbf{d}_k = \mathbf{y}_p(k) - \mathbf{y}_m(k)$. For simplicity of handling, the inequalities (8.10) can be reduced to take the form:

$$N\underset{\rightarrow k}{\mathbf{c}} \le \mathbf{d}_{fixed} + \underbrace{[M,P]}_{Mp}\begin{bmatrix} \mathbf{z}_k \\ \mathbf{r}_{k+1} \end{bmatrix}; \quad \mathbf{z}_k = \begin{bmatrix} \mathbf{x}_k \\ \mathbf{d}_k \end{bmatrix} \tag{8.12}$$

A summary of core files is:

- **sompc_data.m**, amongst other things, builds the closed-predictions (8.11) for a specified horizon N_c and uses these to build the constraint inequalities (8.12).

- **sompc_constraints.m** is used by **sompc_data.m** and takes the prediction matrices and combines with the inequalities.

- For convenience students can see the use of these files within a SOMPC simulation using **sompc_simulate.m**.

The data presented next can be reproduced using the MATLAB code in the files **sompc_constraints_example1-3.m**. By finite horizon sets, we mean using a technique such as indicated in Eqn.(8.5). That is, form predictions of the input/state over a specified horizon N_c and ensure those predictions meet constraints.

The examples given next include absolute input and state constraints but not input rate constraints. The figures demonstrate that the constraints are handled effectively as expected and the perturbation signal $\mathbf{c}_k$ is non-zero only during transients. Once close enough to the target, the unconstrained optimal satisfies constraints and thus $\mathbf{c}_k$ is not required.

8.4.3.1 ILLUSTRATION: Constraint handling in SISO OMPC

Consider the model given in **sompc_constraints_example1.m** which features two set point changes and a step change in the output disturbance. It is clear from the following figure that during each transition, non-zero values of $\mathbf{c}_k$ are required, but these quickly return to zero and also offset-free tracking is achieved.

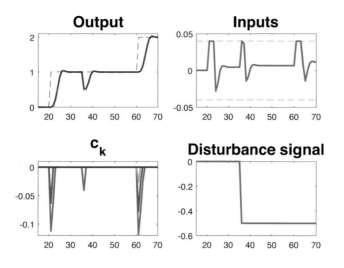

8.4.3.2 ILLUSTRATION: Constraint handling in MIMO OMPC

Consider the highly interactive MIMO example **sompc_constraints_example2.m**

which features two set point changes and a step change in the output disturbance. It is clear from the following figure that during some transitions, non-zero values of c_k are required, but these quickly return to zero and also offset-free tracking is achieved.

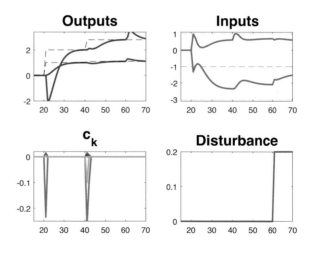

STUDENT PROBLEMS

1. Look at the simulations in the example files **sompc_constraints_example1-3.m** and explore the impact of changing the parameters n_c, Q, R, Q_2, R_2, n_p and the constraints.

2. Extend the code in **sompc_data.m** and **sompc_simulate.m** to cater for input rate constraints.

3. Do some simulations on systems and constraints of your choice (it is straightforward to edit these in the files provided).

Summary: Despite the use of finite horizons for constraint handling, the OMPC/SOMPC algorithm has performed well and the lack of a formal guarantee of recursive feasibility has not caused any problems. The total number of constraints is manageable and the failure to remove redundant constraints has not created any obvious difficulties.

8.4.4 Remarks on using finite horizon inequalities in dual-mode MPC

This section and the MATLAB code deployed, although relatively straightforward to code and implement, could be highly inefficient for on-line implementation. There are a number of obvious weaknesses.

1. For a conservative choice of N_c, many, quite possibly the majority, of the inequalities will be redundant and ideally should be removed off-line.

2. The arbitrary choice of horizon N_c for constraint handling could, in some cases, imply a constraint violation by the predictions beyond N_c samples. This would negate any convergence/stability assurances.

Nevertheless, the simplicity of coding required to produce inequalities such as (8.5) means that this approach is likely to be popular in practice and moreover it has the significant advantage that the impact of changes in the target $\mathbf{r}_k$ and disturbance estimate $\mathbf{d}_k$ are automatically included in (8.9, 8.10); this will also facilitate the use of reference governor types of approach [43] where required to retain feasibility.

Summary: The key advantage of the finite horizon approach is simplicity of coding and the explicit inclusion of target information as a separate state in the resulting inequalities, which means the inequalities can easily be updated for different values of target/disturbance. Nevertheless, there is a lack of rigour which is tackled in the following sections.

8.4.5 Block diagram representation of constraint handling in dual-mode predictive control

It is useful to have a block diagram representation of the impact of constraint handling within OMPC as shown here.

It is clear that constraint handling acts a little like a second feedback loop. However, the dotted lines are used here to represent the fact that this loop is not active much of the time and hence, in effect, often $\mathbf{c}_k = 0$. Non-zero values are needed only when unconstrained predictions violate constraints and thus need modification.

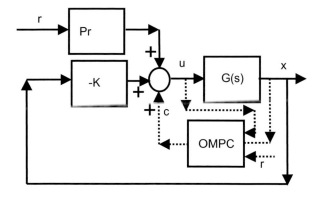

Figure: Block diagram representation of OMPC with constraint handling.

This block diagram representation also makes clear that the embedded/terminal feedback K need not be designed via optimal control or indeed any other specified method and can actually be selected however the user wishes.

8.5 Concepts of redundancy, recursive feasibility, admissible sets and autonomous models

Before we move onto rigorous approaches to constraint handling with dual-mode predictions, first it is helpful to define some useful concepts. These are useful background before proceeding with more popular/rigorous constraint handling approaches for dual-mode algorithms in the following section. The specific concepts to be introduced are:

• What do we mean by constraint redundancy and how is this useful?

• What do we mean by recursive feasibility and how is this useful in giving rigour to stability analysis?

• What is a terminal region and how does this help us formulate and gain insight into a constrained dual-mode algorithm?

• How do we use autonomous model formulations for dual-mode predictions?

• The concepts of invariance and the maximal admissible set are introduced.

• What are the links between invariance, terminal regions and constraint handling with a dual-mode MPC algorithm?

8.5.1   Identifying redundant constraints in a set of inequalities

In practice many of the inequalities in a set of inequalities such as (8.10) are redundant, that is, they cannot be violated while all the other constraints are satisfied. For computational efficiency, these should be removed in any off-line analysis. This section demonstrates a simple algorithm for determining whether a given inequality is redundant or not.

Take the full set of inequalities to be defined as below and also define the ith inequality (or ith row):

$$H\mathbf{x} \le \mathbf{g} \quad \equiv \quad h_i\mathbf{x} \le g_i, \forall i \tag{8.13}$$

Find the maximum value the ith row can take subject to the other inequalities being satisfied:

$$\hat{h}_i = \max_{\mathbf{x}} \; [h_i\mathbf{x} - g_i] \;\; s.t. \; \{h_j\mathbf{x} \le g_j, \forall j \ne i\} \tag{8.14}$$

The ith row is redundant if:

$$\hat{h}_i \le 0 \tag{8.15}$$

In general, if the constraint set (8.13) does not change from one sample to the next (that is $\mathbf{g}$ is fixed), then it is reasonable to remove all redundant constraints off-line so that any on-line optimisation is as simple as possible. However, readers will note from Eqn.(8.9) that typically the constraints depend upon variables that change each sample.

Remark 8.2 *Removing redundant constraints from (8.9, 8.10) is not as straightforward as one might hope. Rearranged as in Eqn.(8.12), the constraint set of (8.1) puts no limits on the augmented states of $\mathbf{r}_k, \mathbf{d}_k$ and thus, in principle these states could be unbounded which hence corrupts optimisation (8.14). In practice there is likely to be a need to add additional limits on $\mathbf{r}_k, \mathbf{d}_k$ for any significant reductions in the complexity of set (8.12) through identifying and removing finding redundant constraints.*

It is commonplace in the literature to assume $\mathbf{r}_k, \mathbf{d}_k$ are fixed to avoid this complication, but of course such an assumption limits the flexibility of the associated algorithm.

> **Summary:** It is straightforward using a linear programme to identify redundant inequalities. These should then be removed during an off-line phase hence reducing the total number of inequalities being used on-line.
>
> It is not the purpose of this book to define efficient algorithms for handling this scenario. Moreover, as this is an off-line task, except in extreme circumstances, efficiency is less important than for on-line tasks.

8.5.2   The need for recursive feasibility

It was noted in Section 7.5.3 that for finite horizon MPC, it is difficult to derive guarantees of closed-loop stability in the constrained case. This is largely because one cannot fall back on the linear loop analysis and implied closed-loop poles as the control law becomes non-linear during constraint handling. Herein there is a need to understand more precisely the possible impact of constraints and the most important characteristic here is so called *feasibility* and more specifically the terminology *recursive feasibility*.

Feasibility in the context of MPC

Feasibility means that it is possible to solve the inequalities implicit in the MPC optimisation, for example (8.10). Infeasibility means that there does not exist a suitable choice for the d.o.f. such that constraints can be satisfied. In general the feasible region is defined, mathematically, as follows. Consider the inequalities equivalent to (8.12):

$$N\underset{\rightarrow k}{\mathbf{c}} \le \mathbf{d}_{fixed} + M_P \underbrace{\left[\begin{array}{c} \mathbf{z}_k \\ \mathbf{r}_{k+1} \end{array} \right]}_{\mathbf{w}_k} \tag{8.16}$$

where $\mathbf{w}_k$ is an augmented parametric state including information about the asymptotic target (and steady-state) which implicitly depends upon both $\mathbf{r}_k$ and $\mathbf{d}_k$. The feasible region S_w for augmented state $\mathbf{w}_k$ is defined from:

$$S_w = \{\mathbf{w}_k : \exists \underset{\rightarrow k}{\mathbf{c}} \text{ s.t. } N\underset{\rightarrow k}{\mathbf{c}} \le \mathbf{d}_{fixed} + M_P \mathbf{w}_k\} \tag{8.17}$$

Feasibility regions might be time varying

Unsurprisingly, the corresponding *feasible region* for just state $\mathbf{x}_k$ is dependent on $\mathbf{r}_k$ and $\mathbf{d}_k$ as $\mathbf{w}_k = [\mathbf{x}_k^T, \mathbf{d}_k^T, \mathbf{r}_k^T]^T$ and thus is time varying.

$$S_x = \{\mathbf{x}_k : \exists \underset{\rightarrow k}{\mathbf{c}} \text{ s.t. } N\underset{\rightarrow k}{\mathbf{c}} \le \mathbf{d}_{fixed} + M_P[\mathbf{x}_k^T, \mathbf{d}_k^T, \mathbf{r}_k^T]^T\} \tag{8.18}$$

Due to the time varying nature of the feasible region for $\mathbf{x}_k$, we need a language and terminology to cope with this; hence the notation of recursive feasibility.

Recursive feasibility with constant targets

Recursive feasibility means that given a MPC optimisation (more precisely the associated constraint inequalities) is feasible at the current sample, that is $\mathbf{w}_k \in S_w$ or equivalently $\mathbf{x}_k \in S_x$, we can guarantee that the MPC optimisation will be feasible at the next sampling instant, and thus implicitly for all future samples. It so happens that with dual-mode MPC, recursive feasibility is automatic in the nominal case **in the absence of changes in the target/disturbance**. This is because it is implicit that one has checked the predictions over an infinite horizon and also the potential to utilise *the tail* (see Section 6.3.1) means the same prediction (already known to be feasible) can be used at the subsequent sampling instant should one choose to.

8.5.3 Links between recursive feasibility and closed-loop stability

In the case of dual-mode algorithms as seen in Chapter 6, strong stability results are possible due to the recursive feasibility properties although conversely, these may come at the price of some conservatism in terms of operating regions and performance.

- For dual-mode algorithms, the nominal guarantee of stability given in Chapter 6 is valid under the assumption that feasibility can be guaranteed. In essence one needs assurances of recursive feasibility.

- Guarantees of recursive feasibility can be given for the nominal case alongside the use of dual-mode predictions, but these results are not necessarily robust to any uncertainty which is present in all real scenarios and also, often not valid during changes in target/disturbance.

The basic argument is that the Lypaunov stability proof outlined in Section 6.5 applies during constraint handling and thus is equally applicable in the constrained case. However, this is subject to **feasibility**, that is the assumption that the associated predictions meet any constraints. The specific challenge is that recursive feasibility is guaranteed if and only if the target steady-state does not change. Any change in the target implies a change in the implied inequalities (8.12) and associated feasible region (8.17,8.18). A large enough change in r_k or d_k can imply $w_k \notin S_w$ which consequently implies infeasibility. In general, guaranteeing recursive feasibility may be a significant challenge for dual-mode approaches where the implied asymptotic target is too ambitious [43, 119], or indeed during large disturbances [114].

A loss of recursive feasibility, that is infeasibility, typically occurs due to two causes:

1. at the initialisation stage, that is the initial state x_k and/or initial target r_k are incompatible with (8.12).

2. following target and/or disturbance changes. In this case feasibility could be recovered by reducing or removing the implied change [43].

The most straightforward recursive feasibility proofs are typically predicated on the assumption of no changes to r_k, d_k, thus infeasibility may occur following a large target change or a large disturbance. In the literature this issue tends to be dealt with using hierarchical methods and/or reference governing [43] and/or tubes [114] and is not a topic pursued in this book.

Summary: Dual-mode algorithms such as OMPC/SOMPC have a guarantee of stability in the constrained case, if one can give a guarantee of initial feasibility and recursive feasibility. The guarantee of recursive feasibility is automatic if no target/disturbance changes occur, but is vulnerable to any changes and thus may imply significant conservatism (small target changes, small disturbances, etc.).

8.5.4 Terminal regions and impacts

It is useful at this point to define the concept of a terminal region. A dual-mode prediction defaults to a linear control law such as (6.17) during mode 2. The associated behaviour in mode 2 is feasible if it satisfies (8.1) and this can be guaranteed to be true if mode 2 lies entirely inside a feasible terminal region.

In effect one could determine the associated set (feasible values of $\mathbf{x}_{k+n_c}$) from (8.12) by setting $\mathbf{c}_k = 0$ and hence the mode 2 predictions are feasible if:

$$\mathbf{x}_{k+n_c|k} \in S_T; \quad S_T = \{\mathbf{x}_k : 0 \leq \mathbf{d}_{fixed} + M_P[\mathbf{x}_k^T, \mathbf{d}_k^T, \mathbf{r}_k^T]^T\} \tag{8.19}$$

where S_T is a shorthand to represent terminal region. Again, it is obvious that the shape and volume of the terminal region is affected by values of $\mathbf{d}_k, \mathbf{r}_k$.

The significance and computation of a terminal region is revisited in sections 8.5.6.3 and 8.8. It will be shown that the volume and shape of this set can be critical to the efficacy of the associated dual-mode MPC algorithm. More specifically, in general terms one should seek to make the terminal region *large* enough [34], which has an impact on the choice of the associated terminal control law in (6.17).

Summary: A terminal region (8.19) is the region within which the mode 2 predictions, or equivalently the behaviour associated to the unconstrained control law, are guaranteed to satisfy constraints.

8.5.5 Autonomous model formulations for dual-mode predictions

This section briefly revisits the definitions of autonomous prediction models initially presented in Section 2.9.5. These will prove useful formulations for handling the algebra required to give rigorous recursive feasibility assurances and also convenient and compact computations of terminal regions.

8.5.5.1   Unbiased prediction and constraints with autonomous models

We will assume the standard offset correction format for the predictions, and thus assume a model with the following update equations:

$$\mathbf{x}_{k+i} = A\mathbf{x}_{k+i-1} + B\mathbf{u}_{k+i-1}; \quad \mathbf{u}_{k+i-1} - \mathbf{u}_{ss} = -K[\mathbf{x}_{k+i-1} - \mathbf{x}_{ss}] + \mathbf{c}_{k+i-1}; \quad i = 1,\dots,n_c$$
$$\mathbf{x}_{k+i} = A\mathbf{x}_{k+i-1} + B\mathbf{u}_{k+i-1}; \quad \mathbf{u}_{k+i-1} - \mathbf{u}_{ss} = -K[\mathbf{x}_{k+i-1} - \mathbf{x}_{ss}]; \qquad\qquad\quad i > n_c$$

$$(8.20)$$

where $\Phi = A - BK$ and also the model and process outputs have a difference $\mathbf{d}_k = \mathbf{y}_p(k) - C\mathbf{x}(k)$ used to eliminate bias. The steady-state estimates are derived from (2.24), that is:

$$\begin{bmatrix} \mathbf{x}_{ss} \\ \mathbf{u}_{ss} \end{bmatrix} = \begin{bmatrix} K_{xr} & K_{xd} \\ K_{ur} & K_{ud} \end{bmatrix} \begin{bmatrix} \mathbf{r}_{k+1} \\ \mathbf{d}_k \end{bmatrix} \tag{8.21}$$

Hence, an autonomous prediction model, which also enables computation of the rate $\Delta\mathbf{u}_k$, can be derived as in Section 6.6.4.2:

$$\mathbf{Z}_{k+1} = \underbrace{\begin{bmatrix} \Phi & (I-\Phi)K_{xr} & [B,0,\dots,0] & 0 \\ 0 & I & 0 & 0 \\ 0 & 0 & I_L & 0 \\ K & -K.K_{xr} - K_{ur} & [I,0,0,\dots] & 0 \end{bmatrix}}_{\Psi} \mathbf{Z}_k; \quad \mathbf{Z}_k = \begin{bmatrix} \mathbf{x}_k \\ \mathbf{r}_{k+1} - \mathbf{d}_k \\ \underset{\rightarrow k}{\mathbf{c}} \\ \mathbf{u}_{k-1} \end{bmatrix}$$

$$(8.22)$$

and deviation variables are:

$$\mathbf{x}_k - \mathbf{x}_{ss} = \underbrace{[I,-K_{xr},0,0]}_{K_{xss}}\mathbf{Z}_k; \quad \mathbf{u}_k - \mathbf{u}_{ss} = \underbrace{[-K,-K_{ur},0,I]}_{K_{uss}}\mathbf{Z}_k; \tag{8.23}$$

$$\Delta\mathbf{u}_k = \underbrace{[-K,-K_{ur},0,-I]}_{K_{Duss}}\mathbf{Z}_k;$$

The associated performance index based on deviation variables would be:

$$J = \sum_{i=0}^{\infty} [\mathbf{x}_{k+i+1} - \mathbf{x}_{ss}]^T Q[\mathbf{x}_{k+i+1} - \mathbf{x}_{ss}] + [\mathbf{u}_{k+i}^T - \mathbf{u}_{ss}]R[\mathbf{u}_{k+i} - \mathbf{u}_{ss}] \tag{8.24}$$

In effect this is the same as (6.67) and thus the performance index to be minimised reduces to (6.70) and then (6.73) (assuming $S_{cx} = 0$, $S_{c,rd} = 0$, $S_{cu} = 0$):

$$J \equiv \underset{\rightarrow k}{\mathbf{c}}^T S_c \underset{\rightarrow k}{\mathbf{c}} \tag{8.25}$$

For convenience the target information has been included in the autonomous model as a state. This is logical in that the predictions necessarily depend on the values $(\mathbf{r}_{k+1} - \mathbf{d}_k)$ via (8.21).

The constraints to be handled at every sample, namely (8.1), can also be captured very compactly using the associated augmented state $\mathbf{Z}_k$.

$$\underbrace{\begin{bmatrix} K_{uss} \\ -K_{uss} \\ K_{Duss} \\ -K_{Duss} \\ K_{xss} \\ -K_{xss} \end{bmatrix}}_{G\Psi} \mathbf{Z}_k \leq \underbrace{\begin{bmatrix} \overline{\mathbf{u}} \\ -\underline{\mathbf{u}} \\ \overline{\Delta \mathbf{u}} \\ -\underline{\Delta \mathbf{u}} \\ \overline{\mathbf{x}} \\ -\underline{\mathbf{x}} \end{bmatrix}}_{f\Psi} \tag{8.26}$$

or in effect, $G\Psi \mathbf{Z}_k \leq \mathbf{f}\Psi, \ \forall k$.

Summary: The autonomous model formulation of this section is useful to capture the prediction equations for all the important states $\mathbf{x}_k, \mathbf{u}_k, \Delta\mathbf{u}_k$ and the dependence on the d.o.f. $\underset{\rightarrow k}{\mathbf{c}}$ and target $\mathbf{r}_k$ in a single simple prediction model (8.22, 8.26).

$$\mathbf{Z}_{k+1} = \Psi\mathbf{Z}_k; \quad G\Psi\mathbf{Z}_k \leq \mathbf{f}\Psi, \ \forall k \tag{8.27}$$

This model is also useful for expressing the various inequalities, at each sample, required for constraint handling in a simple expression.

8.5.5.2 Constraint inequalities and autonomous models with deviation variables

A standard assumption in the admissible set algorithm was that no constraints were active at steady-state and a standard mechanism for achieving unbiased prediction is with the associated deviation variables.

$$\begin{aligned} \tilde{\mathbf{x}}_{k+i} &= \Phi\tilde{\mathbf{x}}_{k+i-1} + Bc_{k+i-1}; & \tilde{\mathbf{u}}_{k+i-1} &= -K\tilde{\mathbf{x}}_{k+i-1} + \mathbf{c}_{k+i}; & i = 1,2,...,n_c \\ \tilde{\mathbf{x}}_{k+i} &= \Phi\tilde{\mathbf{x}}_{k+i-1}; & \tilde{\mathbf{u}}_{k+i-1} &= -K\tilde{\mathbf{x}}_{k+i-1}; & i > n_c \end{aligned} \tag{8.28}$$

where $\tilde{\mathbf{x}}_k = \mathbf{x}_k - \mathbf{x}_{ss}, \tilde{\mathbf{u}}_k = \mathbf{u}_k - \mathbf{u}_{ss}$.

The corresponding autonomous model is:

$$\mathbf{Z}_{k+1} = \underbrace{\begin{bmatrix} \Phi & [B,0,...,0] & 0 \\ 0 & I_L & 0 \\ K & [I,0,0,...] & 0 \end{bmatrix}}_{\Psi} \mathbf{Z}_k; \quad \mathbf{Z}_k = \begin{bmatrix} \tilde{\mathbf{x}}_k \\ \underset{\rightarrow k}{\mathbf{c}} \\ \tilde{\mathbf{u}}_{k-1} \end{bmatrix} \tag{8.29}$$

The constraint inequalities (8.26) can be updated likewise but noting a suitable shift to the sample constraints, for example:

$$\begin{aligned} \underline{\mathbf{u}} - \mathbf{u}_{ss} &\leq \tilde{\mathbf{u}}_k \leq \overline{\mathbf{u}} - \mathbf{u}_{ss} \\ \underline{\mathbf{x}} - \mathbf{x}_{ss} &\leq \tilde{\mathbf{x}}_k \leq \overline{\mathbf{x}} - \mathbf{x}_{ss} \end{aligned} \tag{8.30}$$

Summary: The autonomous prediction model (8.29) using deviation variables is more compact than the one (8.22) which includes the target/disturbance information explicitly. This is helpful when the target is fixed and indeed is the most common assumption in the mainstream literature to allow simplicity of notation. However, it is a disadvantage and inappropriate where such an assumption is not correct, which indeed could be the case whenever the disturbance estimate is changing.

STUDENT PROBLEMS

Use the autonomous models of (8.22) and (8.29) alongside inequalities such as (8.26) to build suitable inequalities which represent the likely constraint inequalities (to within practical accuracy) using $N_c = 20$.

Also, use MATLAB to compare the efficiency of this with the use of an admissible set algorithm (to be discussed in the following sections).

8.5.5.3   Core insights with dual-mode predictions and constraint handling

The preceding subsections have shown that one can capture dual-mode predictions in an equivalent single mode autonomous model formulation. Moreover, the constraints at each sample can be represented using the augmented state linked to this autonomous model (e.g., see (8.26)). Hence, predictions and constraints can be represented in a compact form which will ease the algebraic and coding requirements of implementing OMPC.

$$\mathbf{Z}_{k+1} = \Psi \mathbf{Z}_k; \quad G_\Psi \mathbf{Z}_k \leq \mathbf{f}_\Psi, \ \forall k \tag{8.31}$$

The following section introduces efficient algorithms for ensuring the predictions associated to (8.31) meet the associated constraints. Specifically, the reader will note that the effort used to put dual-mode predictions into the format of (8.31) is repaid many times by the simplicity of coding and additional rigour thereafter for constraint handling as compared to the developments in Section 8.4.2.

8.5.6   Constraint handling with maximal admissible sets and invariance

The earlier section indicated that a simplistic approach to determining the required constraint inequalities is to build these from the dual-mode predictions in a manner analogous to that used by GPC. However, that approach fails to properly consider how large a horizon N_c is required and moreover, is likely to include large numbers of redundant constraints which could make on-line implementation very inefficient.

This section introduces approaches which find the required N_c systematically and moreover, in parallel, remove redundant inequalities and thus ensure the final constraint set is minimal in the sense that it has the fewest inequalities possible. Nevertheless, one downside of these approaches is that the off-line computation is far more demanding (by several orders of magnitude) and moreover, as they require off-line computations, they are less flexible when it comes to handling target and disturbance changes.

A side benefit of using complete sets is that the useful property of invariance drops out. Hence this section will also define what is meant by invariance in the context of MPC and demonstrate how this property can be used to add rigour to convergence and stability analysis.

This section will first use a very simple example based around maximal admissible sets (MAS) [42] to ground the core concepts, before moving to show how these concepts can be exploited within SOMPC/OMPC using some small changes. Finally, a more efficient approach is presented [112] which is conceptually similar to the standard MAS approach, but exploits the structure in a cleverer way.

Summary: A polyhedral invariant set which captures the constraint handling over an infinite horizon can be constructed in a straightforward manner and moreover may be denoted MAXIMAL in that it includes all necessary inequalities, if computed appropriately.

8.5.6.1 Maximal admissible set

This section gives a concise summary of results available in [42]. For simplicity of presentation initially we will assume that the asymptotic target is the origin.

Maximal admissible sets (MAS) are based on a simple update model and constraint structure as follows:

$$\{\mathbf{x}_{k+i+1} = \Phi\mathbf{x}_{k+i}, \quad G\mathbf{x}_{k+i} \leq \mathbf{f}\} \quad \forall i \geq 0; \quad \mathbf{f} > 0 \tag{8.32}$$

For convenience we assume $\mathbf{f} > 0$ (this ensures $\mathbf{x}_k = 0$ meets constraints), and indeed this is normal for constraints of the type in (8.5) and regulation problems where the implied target is zero and no constraints are active in steady-state.

One can ensure the constraints $G\mathbf{x}_{k+i} \leq \mathbf{f}, \ \forall i$ are satisfied for a specified initial condition and for a given horizon N using a simple set of inequalities as follows:

$$\underbrace{\begin{bmatrix} G\Phi \\ G\Phi^2 \\ \vdots \\ G\Phi^N \end{bmatrix}}_{G_N} \mathbf{x}_k \leq \underbrace{\begin{bmatrix} \mathbf{f} \\ \mathbf{f} \\ \vdots \\ \mathbf{f} \end{bmatrix}}_{L_N\mathbf{f}} \tag{8.33}$$

so in essence $G_N\mathbf{x} \leq L_N\mathbf{f}$. Before discussing the MAS algorithm, it is convenient to define the sets:

$$S_N = \{\mathbf{x}_k: \ G_N\mathbf{x}_k \leq L_N\mathbf{f}\} \tag{8.34}$$

Lemma 8.1 *Predictions from a given initial value $\mathbf{x}_k$ are bounded by a known set $S_\infty = \{\mathbf{x}: G_\infty\mathbf{x} \leq L_\infty\mathbf{f}\}$ if and only if the associated dynamics (8.32) are convergent.*

Proof: If the eigenvalues of Φ are not inside the unit circle, then the associated predictions can be divergent and thus are not bounded. We will not deal with the special case of eigenvalues on the unit circle and for now assume the eigenvalues are all strictly inside the unit circle. In consequence

$$\lim_{i \to \infty} \Phi^i\mathbf{x}_k = 0 \tag{8.35}$$

and the trajectory can be bounded based on the values in G_∞ where implicitly therefore the inequalities in (8.33) imply a set restriction to S_∞. □

Lemma 8.2 *There exists a finite value N_c such that the constraints $G\Phi^{N_c+i}\mathbf{x}_k \leq \mathbf{f}$, $i > 0$ are redundant, that is, cannot be violated assuming (8.33) holds for $N = N_c$.*

Proof: This follows directly from (8.35) in combination with the assumption $\mathbf{f} > 0$ which means that eventually, for large enough N_c, the largest values in $\Phi^{N_c+i}\mathbf{x}_k$ where $\mathbf{x}_k \in S_{N_c}$ become so small that they are negligible compared to $\mathbf{f}$. It is assumed here that $\mathbf{x}_k \in S_{N_c}$ implies $\mathbf{x}_k$ is bounded (e.g., via (8.1)). □

Remark 8.3 *An equivalent statement of the previous lemma is that there exists an $N \geq N_c$ which implies that:*

$$\mathbf{x} \in S_N \ \Rightarrow \ \mathbf{x} \in S_{N+i}, \quad \forall i > 0 \tag{8.36}$$

so that the set $S_{N_c} = S_{N_c+1} = \cdots = S_\infty$.

Theorem 8.1 *If all the additional constraints associated to set S_{N+1} which are not in set S_N are redundant, then N is an upper bound on the required N_c such that satisfying (8.32) for $N = N_c$ implies satisfying (8.32) for $N = \infty$.*

Proof: Redundancy (see Section 8.5.1) of constraints $\Phi^{N+1}\mathbf{x}_k \leq \mathbf{f}$ means that

$$\mathbf{x}_k \in S_N \Rightarrow \max_{\mathbf{x}_k} \mathbf{e}_j^T G\Phi^{N+1}\mathbf{x}_k \leq \mathbf{f}_j, \forall j \tag{8.37}$$

where $\mathbf{e}_j$ are the standard basis vectors and for all j means for each inequality in $G\mathbf{x}_k \leq \mathbf{f}$. That is, it is impossible to violate any of the constraints $G\Phi^{N+1}\mathbf{x}_k \leq \mathbf{f}$ while simultaneously lying inside S_N, thus the additional constraints do not change the set S_N in any way and so (8.37) implies that $S_N = S_{N+1}$. It is implicit that the argument is then recursive so

$$\{S_{N+1} = S_N \ \Rightarrow \ S_{N+i} = S_N, \forall i > 0\} \ \Rightarrow \ N_c \leq N$$

□

ILLUSTRATION: Convergence and constraint redundancy for $N > N_c$

Consider the simple first order model in **finitehorizone-nough1.m**. A plot of the state trajectories from two different initial conditions demonstrates that in this case, a horizon $N_c = 1$ would be sufficient.

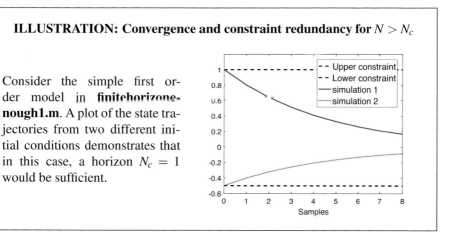

We are now in a position to define the Maximal Admissible Set.

Maximal Admissible Set (MAS)

The maximal admissible set is defined as S_{N_c} as this is the largest set within which $\mathbf{x}_k$ can lie and still satisfy both system dynamics and constraints of (8.32). The converse also applies, that is a set $S_N, N < N_c$ is larger than the MAS, but states within such an S_N include some whose associated trajectories would violate a constraint.

Algorithm 8.1 *[DEFINING THE MAS] An iteration is used as follows: Initialise $i = 1$ and define S_i using (8.34)*

1. Perform the sequence of linear programmes in (8.37) and deduce whether the set of proposed inequalities $G\Phi^{i+1}\mathbf{x}_k \leq \mathbf{f}$ is redundant.

2. If the inequalities are not redundant, define S_{i+1} using (8.34), augment i by one and go back to step 1.

3. If the inequalities are redundant, then $N_c = i$ and the MAS is defined as $S_\infty = S_{N_c}$.

Readers may note that the algorithm as presented here is conceptual rather than numerically efficient. More efficient algorithms are possible [112].

ILLUSTRATION: Convergence and constraint redundancy for $N > N_c$

Consider a second order model in **finitehorizonenough2.m**. A plot of the state trajectories from four different initial conditions on the edges of the sample constraints $Gx \leq f$ (marked with a dashed lines) demonstrates that in this case,

the trajectories (marked with '+,*' etc., for each sample) initially go outside the sample constraints. Simple inspection of the trajectories suggests after about 10 samples, the trajectories remain within the constraints and therefore that N_c will be around 10, as for larger values of N the trajectories will clearly satisfy the sample constraints thereafter.

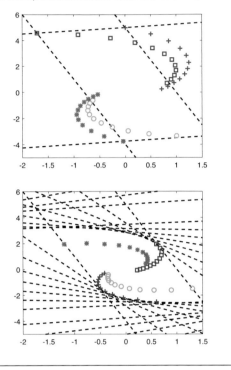

Next, plot all the inequalities for $N = N_c$ (dashed lines). The set satisfying these is much smaller than that satisfying just sample constraints. A plot of the state trajectories from four different initial conditions on the edges of S_{N_c} demonstrates that in this case, the trajectories all remain inside S_{N_c}, so S_{N_c} is invariant.

ILLUSTRATION: MAS is often much smaller than notional state constraints

Consider a simple under damped second order system with box state constraints $|x_i| \leq 10$, $i = 1, 2$. Plot the state trajectories from different points inside the state constraints and also overlay the MAS. It is clear here that the MAS is much smaller than the box but moreover, trajectories originating within the MAS remain within the MAS, whereas trajectories originating outside the MAS at some point violate the box constraints.

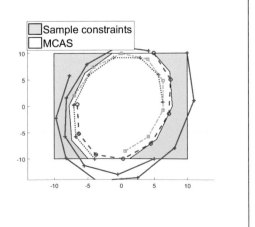

ILLUSTRATION: MAS may be very different from sample constraints

A critical point for the reader to note is that often the MAS is a much smaller region in the state space than perhaps indicated by the sample constraints (8.1). These figures show the sample constraints on the left and the MAS on the right for a simple 3-state example,

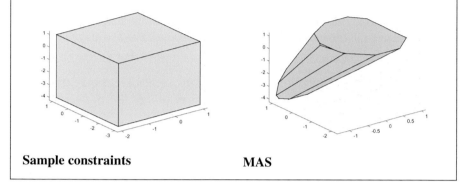

Sample constraints   **MAS**

ILLUSTRATION: MATLAB code for finding the MAS

A simple piece of code is provided for finding the MAS for systems/constraint pairs which meet the structure of (8.32). The corresponding call statement takes the form:

$$[G_n, \mathbf{f}_N] = \textbf{findmas}(\Phi, G, \mathbf{f})$$

It is assumed that (8.35) is satisfied but there is no error checking or removal of redundant constraints.

Summary: Constraint handling over an infinite horizon can be captured with the finite number of inequalities implicit in S_{N_c} as long as:

1. Predictions and constraints can be represented in the form of (8.32).
2. It is implicit that $\mathbf{f} > 0$ so that no constraint is active in steady-state.
3. The eigenvalues of Φ are strictly inside the unit circle.

For convenience, the terminology MAS [42] is commonly adopted.

8.5.6.2   Concepts of invariance and links to the MAS

A set is invariant [6] if, once a state enters that set, it can no longer leave. So, for instance, a set $\mathscr{S}$ is invariant iff:

$$\mathbf{x}_k \in \mathscr{S} \Rightarrow \mathbf{x}_{k+1} \in \mathscr{S} \tag{8.38}$$

The reader may like to note that condition (8.38) implies that $\mathbf{x}_{k+i} \in \mathscr{S}$, $\forall i > 0$.

Invariance is useful for giving bounded stability results in that, once inside a state can never leave and therefore it cannot diverge. This condition can also be made even stronger.

Lemma 8.3 *If an invariant set $\mathscr{S}$ is a subset of the terminal region S_T (see (8.19)), then within that set, the associated predictions from control law (6.17) are guaranteed to satisfy constraints.*

$$\{\mathscr{S} \subset S_T \,\&\, \mathbf{x}_k \in \mathscr{S}\} \ \Rightarrow \ \{G\mathbf{x}_{k+i} \leq \mathbf{f}, \ \forall i > 0\} \tag{8.39}$$

Proof: This is self-evident in that $\mathbf{x}_k \in S_T$ implies satisfaction of constraints and $\mathbf{x}_k \in \mathscr{S}$ implies $\mathbf{x}_k \in S_T$, $\forall i > 0$. The invariance of $\mathscr{S}$ ensures the result can be used recursively.   □

An even stronger and very useful result also exists which is a consequence of this lemma and the definitions in Section 8.5.6.1 which mean that, in effect, the terminal region S_T can be defined using the MAS approach.

Theorem 8.2 *A maximal admissible set as defined in Algorithm 8.1 is invariant.*

Proof: The definition of S_{N_c} around state predictions (see (8.32)) means that this is automatically an invariant set in that implicitly it meets condition (8.38) given $\mathbf{x}_{k+1} = \Phi\mathbf{x}_k$.   □

STUDENT PROBLEMS

> Using the code provided or otherwise, produce some examples of your own satisfying (8.32) to demonstrate that finding the MAS is straightforward in general and moreover, ensuring that $\mathbf{x}_k \in S_{N_c}$ implies that the state remains within S_{N_c} (that is S_{N_c} is invariant) and moreover, satisfies all constraints. Include some examples with more than 2 states.

There are various types of invariance, e.g., [6, 69], but this book will focus mainly on controlled invariance, that is, the invariance that arises in the closed-loop system deploying typical MPC dual-mode predictions. The MAS is a useful building block towards that.

Summary: The maximal admissible set as defined in Algorithm 8.1 is invariant and thus membership of the MAS is sufficient to guarantee satisfaction of constraints for the entire future (that is recursive feasibility).

8.5.6.3 Efficient definition of terminal sets using a MAS

The biggest headache with dual-mode approaches was the terminal mode where the input was subject to a standard feedback law and thus evolved over an infinite horizon rather than becoming fixed after a finite horizon. Defining the associated terminal region S_T was done with an arbitrary choice of N_c in the earlier Section 8.5.4. However, the MAS provides a neater solution to constraint handling for this mode.

Lemma 8.4 *If one can represent the predictions of mode 2 using a formulation equivalent to (8.32), then one can ensure constraint satisfaction of the mode 2 predictions with a simple set membership test of the form:*

$$\mathbf{x}_{k+n_c} \in S_{N_c}; \quad S_{N_c} \equiv S_{\infty} \tag{8.40}$$

Proof: This follows directly from the definition of the MAS so that

$$\mathbf{x}_{k+n_c} \in S_{N_c} \;\Rightarrow\; \mathbf{x}_{k+n_c+i} \in S_{N_c}, \;\; \forall i > 0$$

as long as the predictions/constraints are indeed represented by (8.32). □

Theorem 8.3 *A MAS can be used to capture the constraint handling of mode 2 predictions of a dual-mode MPC algorithm and thus, satisfaction of constraints over an infinite horizon can be represented by a finite number of inequalities within the appropriate S_{N_c}, as long as the steady-state satisfies $\mathbf{Gx}_{ss} < \mathbf{f}$ (readers will note a strict inequality here which means the steady-state is not on a constraint in steady-state).*

Proof: The summary in Equation (8.27) demonstrates the known convergent dual-mode predictions can indeed be represented in the format of (8.32). It suffices then to demonstrate that the autonomous model formulation meets the steady-state requirements that:

$$\lim_{k \to \infty} \mathbf{Z}_k = \mathbf{Z}_{\infty} \;\Rightarrow\; G\mathbf{Z}_{\infty} < \mathbf{f} \tag{8.41}$$

The asymptotic requirement reduces to the steady-state not lying on a constraint limit as this assumption was embedded to ensure the strict convergence of the MAS algorithm. □

Summary: Mode 2 of a dual-mode prediction is feasible if it meets constraints. A terminal set S_T (or terminal region) **should be** defined as an invariant and feasible set for the given mode 2 dynamics so that, $\mathbf{x}_{k+n_c} \in S_T$ implies that the mode 2 predictions are both convergent and satisfy constraints. In practice terminal regions are defined using an MAS for convenience and hence with the following approach.

1. Define the system predictions in the from of (8.32) but with $\mathbf{c}_k = 0$.

2. Find the corresponding MAS $S_{N_c} = S_{MAS}$ and define $S_T = S_{MAS}$.

Within an MPC control law, satisfaction of constraints by mode 2 predictions is then captured with the requirement that $\mathbf{x}_{k+n_c} \in S_{MAS}$.

8.5.6.4 Maximal controlled admissible set

The previous section focussed on constraint handling for mode 2 predictions but did not discuss constraint handling for mode 1 part of the predictions. In fact, the use of autonomous model formulations such as (8.27), which effectively combine both modes into a single prediction model, mean that we can use an MAS algorithm to capture the mode 1 and mode 2 predictions and constraint handling requirements simultaneously.

The overall dual-mode prediction class is feasible if and only if:

- Mode 1 predictions satisfy constraints.

- Mode 2 predictions satisfy constraints, that is $\mathbf{x}_{k+n_c} \in S_{MAS}$.

The maximal controlled admissible set (MCAS) is a shorthand description of the set of inequalities which ensure that both of these conditions are satisfied. It is denoted MCAS under the assumption that the mode 2 conditions are based on the MAS.

Algorithm 8.2 *The MCAS for a dual-mode algorithm can be determined conceptually in a few simple steps.*

> *1. Combine the mode 1 and 2 predictions into an autonomous model formulation such as (8.27).*
>
> *2. Apply the MAS Algorithm 8.1 to this prediction/constraint pair to find a suitable N.*
>
> *3. The resulting inequalities defining the MCAS will take the form:*

$$\{G_N \mathbf{Z}_k \le \mathbf{f}_N\} \equiv \{N \underset{\rightarrow k}{\mathbf{c}} + M_p \mathbf{w}_k \le \mathbf{d}_{fixed}\} \qquad (8.42)$$

WARNING: To ensure convergence, it will be necessary to add an additional constraint on $\mathbf{r}_k - \mathbf{d}_k$ which enforces the implied steady-state $\mathbf{x}_{ss}, \mathbf{u}_{ss}$ to lie strictly inside any of the input and state constraint limits.

Of particular note here is that the terminal region S_T does not need to be computed separately and in fact the terminal region is implicit in (8.42) by setting $\underset{\rightarrow k}{\mathbf{c}} = 0$.

ILLUSTRATION: MCAS changes size and shape with n_c

Consider a simple second-order system

$$A = \begin{bmatrix} 0.8 & 0.1 \\ -0.2 & 0.9 \end{bmatrix}; \quad B = \begin{bmatrix} 0.1 \\ 0.8 \end{bmatrix}; \quad C = [1.9 \; -1]; \qquad (8.43)$$

with feedback $K = [-1.4105 \; 0.9414]$. Find the MCAS for various values of n_c and assuming that $\mathbf{r}_k = 1$. Two observations are striking:

- The MCAS are much smaller than the region defined solely by the state constraints (dashed lines).

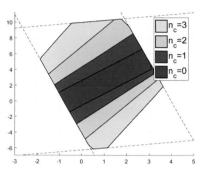

- The MCAS grows, possibly substantially, as n_c increases.

Using a small n_c could be seriously detrimental to overall feasibility of OMPC!

It is also clear that the MAS could be very small and far away from the state constraints and therefore unconstrained control is often not a realistic option.

ILLUSTRATION: The MCAS changes size and shape with the choice of terminal mode

Consider the same example as in (8.43), but change the terminal feedback (but not the objective function) to $K = [-1.3863 \; 0.9416]$ which in essence is equivalent to allowing the use of SOMPC. Again find the MCAS for various values of n_c and assuming that $\mathbf{r}_k = 1$. Two observations are striking:

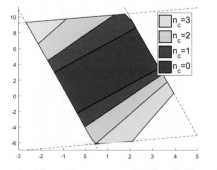

- The MCAS and MAS are larger than in the previous illustration.

- The MCAS grows, possibly substantially, as n_c increases.

The choice of terminal mode can have significant impact on the feasibility of OMPC/SOMPC and hence there could be strong arguments for using SOMPC, to ensure good feasibility while allowing the performance index to be focussed towards a different (better tuned) feedback where this is feasible.

Summary: Maximal admissible sets give rigour to constraint handling over infinite horizons in that they facilitate a guarantee that a given horizon N is large enough. Expressing dual-mode predictions using autonomous models such as (8.27) gives transparency to how an MAS algorithm can be used for dual-mode predictions. Finally, implicit within the use of an MCAS is that, the set is invariant which therein brings a guarantee of recursive feasibility (in the absence of target/disturbance changes).

8.5.6.5 Properties of invariant sets

The reason why invariance is so popular as a tool is that an invariance condition is equivalent to a bounded stability test. Once the state is inside $\mathscr{S}$ it is guaranteed to remain inside and thus, in the worst case, the system is bounded output stable. In fact it is commonplace to determine stronger results which also enable convergence to be established. A second key factor is that invariance is invaluable in the case of constraint handling as, assuming sets are defined to include constraint information (that is $\mathbf{x}_k \in \mathscr{S}$ implies constraints are satisfied), then these sets also enable a bounded output guarantee in the presence of constraints.

Assuming that a system is subject to feedback, the shape and volume of the invariant set depends upon several factors:

1. The open-loop system dynamics (or model).

$$\mathbf{x}_{k+1} = A\mathbf{x}_k + B\mathbf{u}_k \tag{8.44}$$

2. The embedded feedback law K and associated mode 2 dynamics.

$$\mathbf{x}_{k+1} = A\mathbf{x}_k + B\mathbf{u}_k, \quad \mathbf{u}_k = -K\mathbf{x}_k, \quad \mathbf{x}_{k+1} = \Phi\mathbf{x}_k; \quad \Phi = A - BK \tag{8.45}$$

3. The number and positioning of the d.o.f. within the predictions. Typically the d.o.f. are taken to be the choices of perturbations $\mathbf{c}_{k+i}$ for the first n_c control moves.

$$\{\mathbf{x}_{k+i+1} = A\mathbf{x}_{k+i} + B\mathbf{u}_{k+i}; \ \mathbf{u}_{k+i} = -K\mathbf{x}_{k+i} + \mathbf{c}_{k+i}\} \ i = 0, 1, ..., n_c - 1 \tag{8.46}$$

4. The constraints (8.1).

Summary: The existence of an invariant set may be equivalent to the existence of a Lyapunonv function and hence equivalent to a stability test.

An invariant set is one which once entered cannot be left. In the context of MPC, its definition/shape/volume depends upon the model, constraints and the mode 2 control law. However, it is difficult to make a priori judgements about the link between properties and choices, especially with polyhedral sets such as the MAS, so systematic choices are difficult in general.

Under the assumptions of convergent dynamics within the predictions and the steady-state not on the boundary, the invariant set will be finitely determined. Hence constraint satisfaction over an infinite horizon can be ensured by checking a finite number of inequalities.

Remark 8.4 *Alternative definitions of invariant sets do exist (e.g., limited complexity polyhedra and ellipsoids) but in general these can be very suboptimal in volume and therefore their use is restricted to problems where such a reduction in complexity is critically important. [Some brief discussion is in the appendices.]*

8.6 The OMPC/SOMPC algorithm using an MCAS to represent constraint handling

We have now developed enough concepts to propose a rigorous approach to dual-mode MPC, that is to define an algorithm with several valuable properties:

- A guarantee of recursive feasibility.

- A guarantee of stability for the nominal case and in the presence of constraints.

- Constraint handling over an infinite prediction horizon can be captured with a known finite number of inequalities.

This section will also make a link with the associated MATLAB code provided to the reader for implementing OMPC/SOMPC using the inequalities based on an MCAS.

8.6.1 Constraint inequalities for OMPC using an MCAS

The dynamics and constraints have now been expressed in a form exactly equivalent to (8.32) and thus a standard MAS algorithm can be used to find the appropriate admissible set in the form $G_N \mathbf{Z}_k \leq \mathbf{f}_N$. For convenience it is better to unpack $\mathbf{Z}_k$ to see the explicit dependence of the inequalities on different states.

$$MCAS = \{\mathbf{Z}_k : G_N \mathbf{Z}_k \leq \mathbf{f}_N\}; \qquad G_N = [M,P,N,Q]$$
$$\text{or} \tag{8.47}$$
$$MCAS = \{\mathbf{x}_k : \exists \underset{\rightarrow k}{\mathbf{c}} \ s.t. \ M\mathbf{x}_k + N\underset{\rightarrow k}{\mathbf{c}} + P(\mathbf{r}_{k+1} - \mathbf{d}_k) + Q\mathbf{u}_{k-1} \leq \mathbf{f}_N\}$$

For convenience the target information has been included in the autonomous model as a state. This is logical in that the MCAS (8.47) necessarily depends on the values $(\mathbf{r}_{k+1} - \mathbf{d}_k)$. However, these values have associated modes with eigenvalues of one which affects convergence of a standard admissible set algorithm and thus, in practice a slight modification to the MAS algorithm may be needed to ensure convergence.

The reader could also deploy the autonomous model formulation of (8.29) based on deviation variables. However, there is a warning in this approach. The new set will now be computed based on a specific assumption about the values of $\mathbf{u}_{ss}, \mathbf{x}_{ss}$ and thus will not be suitable for other values. The set will take the form:

$$MCAS = \{\mathbf{Z}_k : G_N \mathbf{Z}_k \leq \mathbf{f}_N\}; \qquad G_N = [M,N,Q]$$
$$\text{or} \tag{8.48}$$
$$MCAS = \{\tilde{\mathbf{x}}_k : \exists \underset{\rightarrow k}{\mathbf{c}} \ s.t. \ M\tilde{\mathbf{x}}_k + N\underset{\rightarrow k}{\mathbf{c}} + Q\tilde{\mathbf{u}}_{k-1} \leq \mathbf{f}_N\}$$

Summary: The autonomous model formulation is useful to capture the entire picture in a simple update model (8.27) and sample constraints (8.26) and allows a compact representation which can be deployed in an admissible set algorithm to find the MCAS. It is useful to unpack these sets to note the specific dependence of the MCAS on the various components. The explicit inclusion of the target as a state in the autonomous model ensures the corresponding MCAS is defined flexibly to allow for changes in the target, although the associated dynamics include constants, that is states $(\mathbf{r}_{k+1} - \mathbf{d}_k)$, and so, without limits on the target and other *tailoring*, the admissible set algorithm will not converge.

One can ensure convergence of the autonomous model by removing the state linked to the target information. However, the corresponding set is only guaranteed to be the MCAS for single specified $\mathbf{u}_{ss}, \mathbf{x}_{ss}$ and may not be correct for other values. Small but important details such as these are rarely discussed in the academic literature and in consequence, a practical approach is likely to use (8.33) with a suitably large value of N or some other tailoring.

STUDENT PROBLEMS

Use the autonomous models of (8.22) and (8.29) to build suitable inequalities which represent the likely MCAS (to within practical accuracy) using the set definition of (8.33) with $N = 20$.

Use MATLAB to compare the efficiency of this with the use of an admissible set algorithm.

8.6.2 The proposed OMPC/SOMPC algorithm

The OMPC/SOMPC algorithm is defined by combining inequalities (8.47,8.48) with a standard CLP performance index and thus:

$$\min_{\underset{\to k}{\mathbf{c}}} \; \mathbf{c}_{\to k}^T S_c \mathbf{c}_{\to k} + \mathbf{p}_k^T \mathbf{c}_{\to k} \quad \text{s.t.} \quad M\mathbf{x}_k + N\mathbf{c}_{\to k} + P(\mathbf{r}_{k+1} - \mathbf{d}_k) + Q\mathbf{u}_{k-1} \leq \mathbf{f}_N \quad (8.49)$$

where a suitable value of $\mathbf{p}_k$ is time varying as indicated in Equation (6.73); $\mathbf{p}_k = 0$ for OMPC. The MATLAB code in the following section assumes this formulation.

Summary: Readers will note that in principle this algorithm allows for changes in the target, the disturbance and also implicitly uses unbiased prediction and therefore achieves offset-free tracking in the presence of some uncertainty. In practice, one would remove any redundant inequalities off-line so that the inequalities comprise a minimum complexity set.

8.6.3 Illustrations of the dual-mode prediction structure in OMPC/SOMPC

Readers may find it useful to have a pictorial representation of feasibility/infeasibility and also a representation of dual-mode predictions in a phase diagram. Here we can use the same example as in Section 8.5.6.4. The figure shows the OMPC state evolution from an initial condition of $\mathbf{x}_0 = [-2.2 \;\; -4]^T$ to the

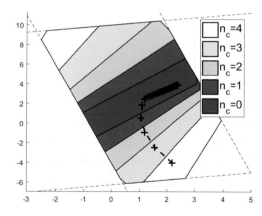

final steady-state of $\mathbf{x}_{ss}$ corresponding to $\mathbf{r} = 1$. It is clear that the trajectory begins in the MCAS associated to $n_c = 4$ and outside the MCAS associated to $n_c = 3$ and thus the prediction takes 4 samples before entering the MAS (the region with $n_c = 0$).

Infeasibility in OMPC/SOMPC

For the illustration above, if $\mathbf{x}_o \notin MCAS$, then the OMPC/SOMPC algorithm is undefined and the implied inequalities will be inconsistent. Infeasibility can easily occur due to a poor choice of n_c, for example in the figure above, a choice of $n_c \leq 3$ with the initial condition given would be infeasible as this state does not lie within the MCAS associated to $n_c = 3$.

Summary: Readers may note in the literature the concept of an n-step set, that is, the set from which within n steps the trajectory can enter the MAS while satisfying constraints. In effect, the n-step set corresponds to the MCAS with $n = n_c$. If a state lies outside the n-step set and $n_c \leq n$, then a feasible solution cannot be determined by OMPC/SOMPC.

8.7 Numerical examples of the SOMPC/OMPMC approach with constraint handling

This section uses the algorithm of (8.49) and gives a number of simulation examples to demonstrate the efficacy of the algorithm.

The examples are separated into increasing levels of complexity so that readers can focus on simple concepts, or if they choose, include all the realistic components such as no-zero targets, disturbance estimates and parameter uncertainty. The choice of whether to use OMPC/SOMPC is embedded in the choice of weights in the main script files containing the examples: $Q = Q_2$, $R = R_2$ will give OMPC.

8.7.1 Constructing invariant constraint sets/MCAS using MATLAB

The MAS/MCAS are computed using the file **findmas.m**. The use of this is embedded inside the file **ompc_simulate_constraints.m**, but readers will note that in effect the code produces a model/constraint pair equivalent to (8.27, 8.32) for a suitable augmented model and from these finds the corresponding set. Examples of the sets can be found for all the examples in the following subsection by interrupting the code at the relevant point. For example, a typical code snippet takes the form:

```
G=[-Kz;Kz;[Kxmax,zeros(size(Kxmax,1),nc*nu)]];
f=[umax;-umin;xmax];
[F,t]=findmas(Psi,G,f);
```

For cases where a non-zero target is desired, there are some subtleties not covered in this book to ensure the admissible set is bounded and exists; these mean that in practice the range of targets must be bounded before calling the admissible set algorithm. These additional constraints are catered for in an ad hoc manner in the alternative code in **findmas_tracking.m**.

STUDENT PROBLEMS

Revisit the illustrations in Section 8.5.6.4 which demonstrate the MCAS can change significantly with n_c. Generate your own examples to illustrate this and therefore to affirm that the choice of n_c is critically important to ensure a large enough feasible region.

Secondly, compare the MAS/MCAS that arise from using SOMPC with those from OMPC.

8.7.2 Closed-loop simulations of constrained OMPC/SOMPC using ompc_simulate_constraints.m: the regulation case

The file **ompc_simulate_constraints.m** is coded to be the simplest possible. Therefore it assumes perfect state measurement and no parameter uncertainty and thus deploys a cost function, predictions and constraint inequalities which do not utilise the offset measurement d_k. This allows students to focus solely on the core concepts without the distraction of other factors. The code focusses on the regulation case (target of the origin) and thus users can change the input and state constraints and the initial condition and indeed any other variables such as n_c, the model, run time and so on.

8.7.2.1   ILLUSTRATION: ompc_constraints_example1.m

Demonstrates that c_k is chosen to ensure the input limit is satisfied and reverts to zero once the state is inside the MAS so that the unconstrained optimal is feasible. The cost J_k is Lyapunov as expected and the state converges to the origin.

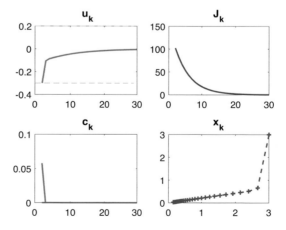

8.7.2.2   ILLUSTRATION: ompc_constraints_example2.m

This example demonstrates that a non-zero c_k is chosen over several samples to ensure both the input limit and state limits are satisfied. $c_k = 0$ once the state is inside the MAS so that the unconstrained optimal is feasible. The cost J_k is Lyapunov as expected and the state converges to the origin.

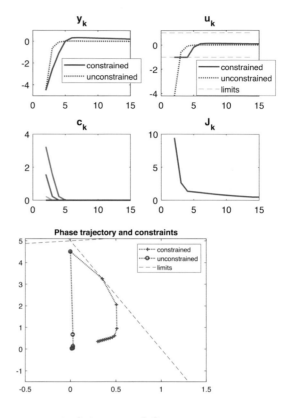

Also the phase plane plot overlays the unconstrained state trajectory, which clearly gives an input violation, and shows that this state trajectory is very different from the optimal constrained one.

8.7.2.3   ILLUSTRATION: ompc_constraints_example3.m

This example is similar to the previous and also demonstrates that c_k is chosen to ensure both the input and state limits are satisfied, the cost J_k is Lyapunov as expected and the state converges to the origin. Again, the overlays of the unconstrained state/input trajectories clearly show both an input violation and also the unconstrained state trajectory violates constraints (and exits the MCAS).

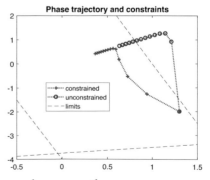

The input, output and perturbation plots are shown over the page.

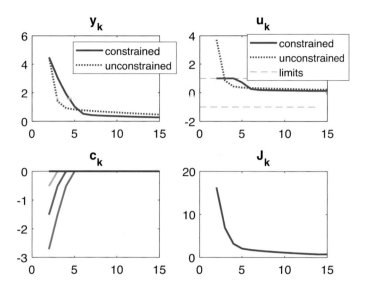

8.7.2.4 ILLUSTRATION: For OMPC, the optimal choice of input perturbation c_k does not change

Consider the previous illustration and inspect the optimum values of $\underset{\rightarrow k}{c}$ at each sample:

$$
\underset{\rightarrow j}{c} = \begin{bmatrix} -2.717 \\ -1.512 \\ 0.514 \\ 0 \end{bmatrix} ; \begin{bmatrix} -1.512 \\ 0.514 \\ 0 \\ 0 \end{bmatrix} ; \begin{bmatrix} 0.514 \\ 0 \\ 0 \\ 0 \end{bmatrix} ; \begin{bmatrix} 0 \\ 0 \\ 0 \\ 0 \end{bmatrix}
\tag{8.50}
$$
$$
\underbrace{}_{j=k} \quad \underbrace{}_{j=k+1} \quad \underbrace{}_{j=k+2} \quad \underbrace{}_{j=k+3}
$$

It is clear that in general, where the target/disturbance are fixed, $\mathbf{c}_{k+i|k+j} = \mathbf{c}_{k+i|k}$ and this reinforces the message that OMPC gives rise to consistent decision making from one sample to the next. NOTE: this observation relies on the optimal $\mathbf{c}_{k+n_c|k} = 0$, so that the optimal at sample k does not need more than n_c moves to find the constrained optimal solution.

STUDENT PROBLEMS

Using the script files **ompc_constraints_example1-3.m** as a base or otherwise, generate a number of examples of your own to demonstrate.

1. Scenarios where the optimal choice $\underset{\rightarrow k}{c}$ does not change from one sample to the next. How does this differ between OMPC and SOMPC?

2. Find examples which illustrate how the feasibility of an initial condition may vary with the choice of n_c, that is a specified $\mathbf{x}_o$ may be infeasible with low n_c but feasible with higher n_c.

3. For a number of examples, demonstrate that even in the presence of constraints, the cost function is Lyapunov.

8.7.3 Closed-loop simulations of constrained OMPC/SOMPC using ompc_simulate_constraintsb.m: the tracking case

The file **ompc_simulate_constraintsb.m** is coded to be the simplest possible with a non-zero target and therefore assumes the target is always unity. Also it assumes perfect state measurement and no parameter uncertainty and thus deploys a cost function, predictions and constraint inequalities which do not utilise the offset measurement $\mathbf{d}_k$; however, estimates of the steady-state input and state are required, as in effect, this is the OMPC algorithm based on deviation variables (thus keeping the state dimension as small as possible).

This allows students to focus solely on the core principles underlying OMPC/SOMPC when a target change is deployed. Users can change the input and state constraints and the initial condition and indeed any other variables such as n_c, the model, run time and so on.

8.7.3.1 ILLUSTRATION: ompc_constraints_example4.m

This example demonstrates that $\mathbf{c}_k$ is chosen to ensure the input limit is satisfied, the cost J_k is Lyapunov as expected and the state converges to the point corresponding to $\mathbf{y} = 1$ (not the origin). A display of the state trajectory also shows the MAS (terminal region/ shaded), MCAS (white) and original state constraints (dashed lines). As $\mathbf{x}_o \in MCAS$ the algorithm is always feasible, although clearly the MCAS is smaller than 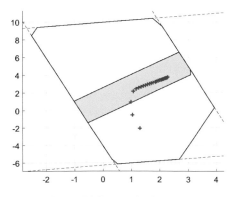 the region defined solely by the state limits. The MCAS may be bigger with larger n_c. Note that as the state enters the MAS, the optimal value of $\mathbf{c}_k$ goes to zero so, in effect, one is now using the unconstrained feedback. This is clear (overleaf) from around sample 5 onwards as $\mathbf{c}_k = 0$ and $\mathbf{u}_k$ is well inside its limits.

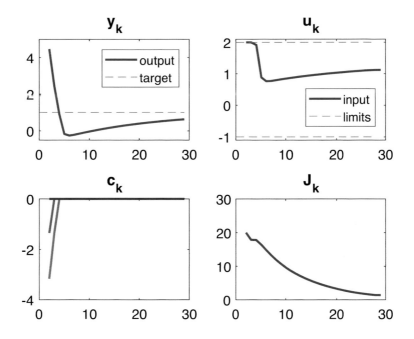

8.7.3.2 ILLUSTRATION: ompc_constraints_example5.m

This is a MIMO example based on SOMPC rather than OMPC. In this case it is clear that the values c_k do not converge to zero (as clear from control law (6.53)). However, constraints are handled effectively, there is no offset and the cost function is Lyapunov.

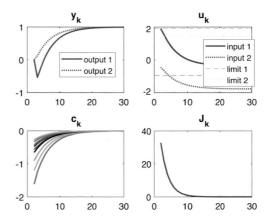

8.7.4 More efficient code for the tracking case and disturbance uncertainty

The difficulty with the approaches in the previous 2 subsections is simply that the Algorithm 8.2 for constructing the constraint inequalities can be very inefficient to use, both offline and on-line. This is because it does not remove redundant constraints.

While this issue is not mainstream, it does have a huge impact on dual-mode implementations and so in practice, the reader must engage with this to some extent. This section gives a very concise introduction to the potential issues and some possible proposals for developing more efficient algorithms. However, it also comes with a warning.

MAS realisations can be difficult to produce in the case where the target is not pre-specified

The autonomous model formulation of (8.22) acknowledges that due to the terminal control law and offset-free tracking, the target has to be included as an augmented state. However, within the predictions the target is constant and thus the associated mode does not change. Therefore, any admissible set algorithm may only converge if the target is artificially restricted because otherwise the feasible terminal set and so forth would need to be defined indirectly via all the other constraints and this may not happen in a reasonable number of iterations. A more pressing point is that a MCAS formulation which allows the asymptotic steady-state to change may require significantly more inequalities than one with a fixed target.

An algorithm that removes redundant constraints in a systematic fashion during the admissible set algorithm was proposed in [112] and has been deployed in the code **constructmas_tracking_nodisplay.m**; this does need reasonable limits on the target variation to be applied in advance. A simple comparison on one example reveals the following indicative data.

Algorithm	Number of inequalities	Time to compute
findmas_tracking.m	1644	100sec
construct_mas_tracking.m	35	10sec
Approach of Section 8.4	horizon $n_p \times 10$	fast

STUDENT PROBLEM

Compare the approaches of **findmas_tracking.m**, **construct_mas_tracking.m** and that of Section 8.4 for capturing the inqualities required to produce an OMPC/SOMPC algorithm for a number of scenarios of your choice. How do they compare by way of:

1. Ease and transparency of coding?

2. Efficiency of coding and computation?

3. Reliability of convergence?

Hence, the final piece of code offered in this section is **ompc_simulate_efficient.m** which allows the user to specify the target and an unknown disturbance signal freely, but under the assumption they do so sensibly. Unrealistic values will result in infeasibility and for this case the OMPC algorithm is not defined (the code gives **warnings** if this occurs). The MAS code uses the approach of [112] to determine the minimum number of inequalities required to capture the MCAS with the assumption that the target can change to some extent.

Summary: Using a fixed finite horizon for the constraint horizon such as in **sompc_constraints.m** is easy to code and fast off-line, but may result in a large number of redundant constraints and also, may not capture the entire MCAS. Using admissible sets allows us to capture the entire MCAS, but a simplistic algorithm such as 8.2 may be very inefficient and thus readers would need to engage with the literature on efficient realisations of these. An alternative approach which removes redundant constraints and builds the MCAS one row at a time is likely to be more efficient overall but may be more difficult to code and is beyond the remit of this book.

ILLUSTRATION: ompc_constraints_example7.m

A MIMO example utilising **ompc_simulate_efficient.m**.

It is clear that the inequalities can deal with both the time varying target and also the disturbance (around the 30th sample). Also the cost function is Lyapunov whenever the target/disturbance are constant.

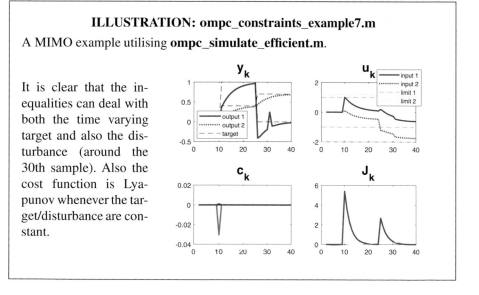

8.8 Discussion on the impact of cost function and algorithm selection on feasibility

This section gives a brief discussion of how the terminal constraint associated to a typical OMPC algorithm can imply difficulties maintaining feasibility whereas an alternative GPC type of approach suffers no such difficulties and thus could be a much safer choice, despite the lack of a mathematical convergence guarantee.

It was noted in Chapter 4 that there are two common choices of performance index:

1. A popular choice with MFD models takes the format:

$$J = \sum_{i=1}^{n_y} \mathbf{e}_{k+i}^T W_y \mathbf{e}_{k+i} + + \sum_{i=0}^{n_u-1} \Delta\mathbf{u}_{k+i}^T W_{\Delta u} \Delta\mathbf{u}_{k+i} \qquad (8.51)$$

2. A popular choice with state space models (and indeed dual-mode algorithms) takes the format:

$$J = \sum_{i=1}^{n_y} [\mathbf{x}_{k+i} - \mathbf{x}_{ss}]^T Q[\mathbf{x}_{k+i} - \mathbf{x}_{ss}] + \sum_{i=0}^{n_u-1} [\mathbf{u}_{k+i}^T - \mathbf{u}_{ss}] R[\mathbf{u}_{k+i} - \mathbf{u}_{ss}] \quad (8.52)$$

One might consider this choice arbitrary or made for convenience; however, the impact can be more significant than one might think. The latter choice *forces* the inputs towards their steady-state and does not penalise rates, whereas the other worries less about the absolute values of the input as long as they do not change too aggressively. In fact the popular choice for OMPC of (8.52) is more for algebraic convenience rather than any engineering reason and thus readers may wonder whether there would be a benefit in redefining the OMPC algorithm using cost function format of (8.51), but with infinite horizons. Some advantages could accrue from this change.

- Dual-mode MPC requires the terminal constraint to be feasible. However, where the initial state is a good distance from the target steady-state and the associated terminal region (MAS), it may require many samples to move from one to the other. If the required number of samples exceeds n_c, then there will not exist a suitable set of values $\underset{\rightarrow k}{\mathbf{c}}$ satisfying the corresponding constraints, e.g., (8.47). In such a case, OMPC becomes undefined, or in essence fails.

  Feasibility can be retained [43, 165] by moving the target closer to the initial condition, but this requires another layer of control design to be produced and coded (such as reference governing) and is a headache a typical user may prefer not to have.

- GPC uses a much less ambitious terminal mode of $\Delta\mathbf{u}_{k+n_u+i} = 0, i \geq 0$. This means that the asymptotic output in the prediction is not *forced* to a specified steady-state and in principle can select any steady-state it wishes to ensure feasibility.

Summary: The MCAS associated to GPC for a given n_u will often be larger than those associated to OMPC/SOMPC for $n_c = n_u$. Moreover, GPC caters for target/disturbance changes much more easily as it avoids the need for admissible sets which can be challenging to define and artifically restrictive.

STUDENT PROBLEMS

1. Investigate how the feasibility of a dual-mode approach compares with a finite horizon approach such as GPC for the same target and constraints. To what extent is GPC better able to retain feasibility automatically, and track successfully, in scenarios where OMPC becomes infeasible (with small n_c)?

2. In scenarios where the desired target/steady-state is infeasible [119] OMPC becomes undefined because the MCAS is empty by definition. Demonstrate by example with MATLAB that GPC often deals well with such a scenario and simply allows a steady-state offset.

3. For a number of examples of your choice, compare the unconstrained performance that arises from using either $\|\mathbf{u} - \mathbf{u}_{ss}\|_2^2$ or $\|\Delta\mathbf{u}\|_2^2$ in the performance index.

8.9 Chapter summary

This chapter has shown that while constraint handling with dual-mode predictions might, at first sight, need inequalities running up to an infinite horizon, in practice as long as the asymptotic steady-state does not lie on a constraint, one can use admissible set theory to represent this with a finite number of inequalities. In effect, it can be shown that predictions beyond a horizon of N_c, for a suitable N_c, cannot violate constraints.

This observation gives the very powerful result that dual-mode MPC can offer guarantees of stability (nominal case) even during constraint handling. However, there are a significant number of caviats.

1. The feasible region within which the OMPC/SOMPC algorithm is well defined may not be large and indeed is often notably smaller than the equivalent region for a GPC approach. The reason is that GPC does not have such a demanding terminal constraint and therefore is inherently more flexible.

2. Ensuring a large feasible region may require a large n_c and/or detuning of the terminal mode, that is a use of SOMPC as opposed to OMPC and/or careful integration with some form of supervisory loop/reference governing.

3. The algebraic requirements for forming the MCAS capturing the constraint information is non-trivial and is quite clumsy when handling changing targets and disturbance estimates. In practice, it may be easier and is often sufficient, just to select an arbitrary N_c and check constraints over that finite horizon.

8.10   Summary of MATLAB code supporting constrained OMPC simulation

This code is provided so that students can learn immediately by trial and error, entering their own parameters directly. Readers should note that this code is not *professional quality* and is deliberately simple, short and free of comprehensive *error catching* so that it is easy to understand and edit by users for their own scenarios. Please feel free to edit as you please.

Readers may wonder why such a diversity of files is provided when many are subsumed in others. The rationale is to keep any given file as simple as possible so that readers who wish to read and edit a file are not overwhelmed by unnecessary complexity and conditional expressions.

The default code is SOMPC but reduces to OMPC when $Q = Q_2, R = R_2$.

8.10.1 Code for supporting SOMPC/OMPC with constraint handling over a pre-specified finite horizon

sompc simulate.m	Performs SOMPC simulations with input constraints and state constraints using a specified prediction horizon n_p to determine the constraint inequalities.
sompc_data.m	Supports SOMPC simulations by building terminal control law, cost function parameters and constraint inequalities.
sompc_observor.m	Forms the terminal mode control law (and possible observer parameters (not currently used).
sompc_predclp.m	Forms the closed-loop predictions (8.11) with a finite horizon.
sompc_constraints.m	Forms constraint inequalities.
sompc_cost2.m	Forms the cost function parameters needed by SOMPC for use in the quadratic programming optimisation.
sompc_constraints_example1-3.m	Example files students can use as a base for developing scenarios of their own. Based on **sompc_simulate.m**

8.10.2 Code for supporting SOMPC/OMPC with constraint handling using maximal admissible sets

ompc_simulate_constraints.m	Simplest constraint handling possible with OMPC/SOMPC, zero target, no disturbances and no parameter uncertainty.
ompc_simulate_constraintsb.m	Simplest constraint handling possible with OMPC/SOMPC, but with a non-zero target of one, no disturbances and no parameter uncertainty.
ompc_simulate_efficient.m	Constraint handling with OMPC/SOMPC, but with a time varying target and an efficient computation of the MCAS. Also allows non-zero disturbances.

Example files students can use as a base for developing scenarios of their own.

ompc_constraints_example1-3.m	Based on **ompc_simulate_constraints.m**
ompc_constraints_example4-5.m	Based on **ompc_simulate_constraintsb.m** so with a non-zero target of one.
ompc_constraints_example6-8.m	Based on **ompc_simulate_efficient.m** so with a time varying target/disturbance.
ompc_suboptcost.m	Subroutine used to build cost function parameters and autonomous model for **ompc_simulate_constraints.m**.
finitehorizonenough1-2.m	Demonstrates with pictures the principles behind the definition of the MAS.

findmas.m **findmas_tracking.m**	Forms admissible set for systems of structure (8.32) assuming convergent dynamics with constant target and with a time varying target.
construct_mas_tracking.m	Efficient algorithm for forming the admissible set for systems of structure (8.32) assuming convergent dynamics and with a time varying target. Also removes redundant inequalities.
constructmas_tracking_nodisplay.m	Forms the MAS for OMPC and removes redundant inequalities from the MAS but without displaying algorithm progress in the command window.
feasibletargetcheck.m	Elementary checks/creation of limits on the target variation to ensure the admissible set algorithms are feasible and the targets are reachable.
ompc_suboptcost.m	Forms admissible set for systems of structure (8.32) assuming convergent dynamics.

9

Conclusions

CONTENTS

9.1 Introduction

This book has given a relatively detailed background in basic MPC algorithms, approaches and conceptual thinking. Consequently, it may be useful to summarise, briefly, what has been learnt and to consider how to apply this understanding. This chapter will attempt to capture the core points that a reader may want to consider before choosing which MPC algorithm to apply, but the guidance comes with a fundamental warning: there is no single correct solution and ultimately it is likely to be a question of trade-offs and different users may select the desired trade-off differently.

Summary: In the author's view it is critical that readers perform a number of simulation studies of their own to properly understand what is going on and the impact of different choices. With this insight, users are more likely to make wise decisions and/or understand the causes when behaviour is not as expected.

9.2   Design choices

This section gives a very brief summary of some of the design choices available and criteria that could be used to help in these choices. It will be assumed for simplicity that the design can be separated into a initial three-way decision.

1. Predictive functional control (PFC); the cheapest option.

2. Finite horizon MPC or equivalently GPC.

3. Infinite horizon or dual-mode control, that is OMPC/SOMPC.

We will take it for granted that constraints are a significant factor, as without these one could revert to a simple classical approach, standard robust control or optimal control. It is also assumed that tuning is carried out in such a way that, notwithstanding the lack of rigour, in practice GPC and PFC will give stable closed loops.

9.2.1   Scenario and funding

A first question is linked to the funding available to implement the control strategy and whether this is competitive with PID (relatively cheap and simple) or a much larger expense is required, perhaps running into hundreds of thousands of pounds. This decision will be partially linked to the dynamics of the process being studied.

• Simple first/second-order dynamics and a SISO system can often be handled fairly well with classical approaches and in this case it is likely that PFC will do a good job and thus is selected as it is relatively cheap.

• SISO systems with more challenging dynamics are not handled as well or as systematically by PFC and thus one will likely have a preference for either GPC or OMPC.

• MIMO systems, that is those with significant interaction, are unlikely to be amenable to PFC and thus one would choose GPC or OMPC.

• Open-loop unstable/highly under-damped systems should generally be tackled with a dual-mode approach to ensure the associated predictions are convergent and close enough to something reasonable.

9.2.2   Effective tuning

Depending on the choice of algorithm, effective tuning requires slightly different approaches.

PFC   PFC has two main tuning parameters, the coincidence horizon n_y and target closed-loop pole λ. The former should generally be chosen to match a point where the system step response has achieved around 30-70% of its steady-state. Implicitly, this puts a limit on an achievable closed-loop pole because one cannot ask the system to converge faster than something larger than the minimum reasonable value for the coincidence horizon n_y. In fact, with the exception of systems having step response behaviour close to first-order dynamics, with PFC it is likely that a closed-loop dynamic close to the open-loop dynamic is likely to be a reasonable choice.

GPC As discussed in Chapter 5, tuning GPC could be viewed as a somewhat iterative process. Begin by selecting very large horizons and then varying the performance index weights to achieve something close to what is a desirable trade-off between input activity and output convergence times; readers will note the implicit assumption here is that the performance index is somewhat arbitrary, a means to an end, and therefore should not be considered as overly precious. Having selected the weights and found the *optimal* unconstrained behaviour, reduce the horizons to smaller values close to the expected settling times for use on-line. Often users may again iterate with n_u to see how small it can be made while retaining good performance and prediction consistency, as smaller n_u implies smaller on-line computational load.

OMPC At one level tuning of OMPC to get good performance is far easier than with GPC/PFC and comes with the benefit of guaranteed properties (stability/recursive feasibility) for almost any choice of performance index. Thus, for those keen on rigour this seems an obvious algorithm to use. However, the tuning of OMPC/SOMPC is much more strongly interlinked with feasibility than GPC due to the terminal constraint implicit in the dual-mode structure. As with GPC, one can choose weights to gain ideal unconstrained behaviour, but following that, there is a need to find a suitably large n_c to ensure the feasible region is large enough. Unfortunately, with large state dimensions the computation and display of feasible volumes is difficult, so this latter step may not be straightforward in practice. Where the desired *weights* require an excessively large n_c to give reasonable feasible volumes, some trade-off will be required to reduce performance. However, there do not currently exist convenient tools for performing this trade-off, because the computation and display of feasible regions is non-trivial with large state dimensions.

9.2.3 Constraint handling and feasibility

The capability of the algorithms to do constraint handling differs.

PFC Practical PFC algorithms tend to use simple saturation type policies although there is some use of predictive constraint handling at times. In consequence this will likely outperform *ad hoc* approaches in that unforeseen violations are avoided but PFC still could be severely suboptimal compared to more systematic MPC approaches.

GPC The constraint handling is easy to set up, especially for input constraints where the required number of inequalities is relatively small, and moreover is likely to have fairly large feasible regions due to the lack of a terminal constraint; this means input rate constraints can be managed more easily due to the lack of a necessity to move quickly to the final steady-state. A weakness is that recursive feasibility proofs cannot be given in general.

OMPC The structure of the dual-mode predictions is far more complex than for GPC predictions and thus the associated constraint inequalities are also both more numerous (likely going up to the prediction settling time and perhaps beyond) and more difficult to define. Moreover, the addition of the terminal constraint can be a significant impediment to feasibility, especially during target/disturbance changes, even when there are no output constraints! However, on the positive side, OMPC allows a simple proof of recursive feasibility and thus provides rigour not available with finite horizon approaches.

9.2.4   Coding complexity

Industry needs to decide whether to code a controller in house which enables some flexibility and may ease maintenance, or buy in, at large expense, code which is already written and validated.

PFC PFC is known for simplicity of coding. Indeed, for simple second and third-order systems the algorithm control law can be coded in about 10 lines (as apparent from the supplied m-files such as **pfcsimulateim.m**) while requiring no matrix algebra if prediction is done as in Section 2.7.3. This simplicity facilitates cheap, transparent and flexible adoption in industry and with costs comparable to a PID implementation and benefits of some systematic constraint handling and effective management of delays.

GPC The formulation of predictions for GPC (see Chapter 2) can be done with relatively simple coding using Toeplitz/Hankel matrices for a MFD model or a state space model. Some minor complications ensue to include a T-filter or step response model or closed-loop prediction. In summary, while it cannot be described as trivial algebra, in fact formulation of predictions is straightforward enough to code easily on most hardware. Nevertheless, this subtlety means industry is more likely to *buy in* the code rather than produce in house. Finally, the requirement for both matrix handling and constraint handling in the formulation of the final control law (for example a quadratic programming optimiser is a common requirement) is something that some software/hardware will not automatically support without the creation of additional software libraries. These additional requirements will be a disincentive to in-house developments for many practitioners but are not so onerous as to be an obstacle where the industry is large enough.

OMPC Dual-mode predictions are algebraically more demanding than open-loop predictions with a number of subtleties to ensure offset-free tracking (unbiased prediction) and thus the coding of OMPC is a significant step up in demand compared to GPC. The formulation of the control law requires matrix handling and the solution of a Lyapunov equation and thus is significantly more demanding again than GPC. Finally, even with just input constraints, the number of inequalities for constraint handling is usually far more than GPC, as well as the definition of the required inequalities being somewhat more complex. A suggestion would be that software supporting this type of algorithm is likely to be maintained only by dedicated industries and thus will be expensive.

9.2.5 Sensitivity to uncertainty

Readers may find it surprising that sensitivity is little studied in the literature, even though uncertainty within MPC is a popular topic. This is likely because of the focus in the literature on guarantees of behaviour for a specified uncertainty class, as opposed to quantification of impacts in general. In consequence there is little obvious in the existing literature which would enable one to compare and contrast the sensitivity of PFC/GPC/OMPC to differing types of uncertainty. Indeed one may begin from the premise that the nominal algorithms are likely to be similar in this respect, as they are all underpinned by equivalent prediction models which are thus equally affected by uncertainty.

STUDENT PROBLEMS

1. Using examples of your choice and/or taken from the final chapter of problems, use the decision making tools above to decide which algorithm to apply.

2. Contrast the solutions that arise from making different choices.

(i) How easy was the problem to code?

(ii) How good was performance?

(iii) How easy was it to obtain feasibility for a good range of disturbances and targets?

(iv) How sensitive is the solution to different types of uncertainty?

(v) Do your observations change for different model types, different prediction choices and so forth?

9.3   Summary

One cannot say that algorithm X is better than algorithm Y in any generic sense. In practice, the user will have to quantify the attributes of their problem and then decide which approach best meets the numerous criteria that may apply. Nevertheless, a common rule of thumb is to use the simplest approach which gives good enough behaviour and to recognise that the lack of theoretical guarantees/rigour is not normally an obstacle as long as some *common sense* has been deployed in the choice of the horizons. Good luck with your endeavours!

A

Tutorial and exam questions and case studies

CONTENTS

In the author's view, assessment of predictive control must include aspects which utilise modern computing because it is a numerically intensive topic which means only very trivial numerical problems can be tackled using pen and paper. Therefore I suggest assessment which is separated into a few components.

1. One aspect where students design, tune, implement, evaluate, explore and so forth with a range of case studies having different attributes. This way students learn by doing and need to justify their design choices. It also means they are more likely to gain an appreciation of nuances such as the impact of horizon choices, the impact of constraints and so forth.

2. A second aspect, which could be assessed by written exam to minimise opportunities for collusion, could focus on conceptual aspects such as algebraic derivations and proofs. As far as possible, number crunching should be avoided during this aspect.

3. Numerical questions where students are asked to compute prediction equations, constraint equations, performance indices and/or control laws for low-order examples with small horizons. The MATLAB software discussed in this book can be used to check numerical answers.

This chapter is thus divided into questions of different types. Clearly it is straightforward for a lecturer to modify these to their own requirements.

A.1 Typical exam and tutorial questions with minimal computation

It is implicit, even where not stated, that matrix multiplication and indeed any other implied computations should be done algebraically and not numerically.

Some outline answers are available to lecturing staff for these questions by contacting the publisher.

1. Compare the efficacy of a classical (say PID or lead/lag) design approach with PFC using the following examples. You can vary the criteria such as achieving a bandwidth similar to the open-loop or achieving an increase in bandwidth.

    (a) DC servo: takes the form $[Js^2 + Bs]\theta = kv$ where θ is angular displacement and v is voltage and J, B, k are model parameters.

    (b) Cruise control of a car whose dynamics can be approximated by $M\frac{dv}{dt} + Bv = kf$, f the engine force/throttle position and v the velocity.

    (c) Model from rotor angle to forward speed for a helicopter (taken from [94], p.24).

    $$G = \frac{9.8(s^2 - 0.5s + 6.3)}{(s + 0.6565)(s^2 - 0.2366s + 0.1493)}$$

    (d) The systems in Table 3.1.

    (e) Add some dead-time to the examples you have used and explore the impact. [The code supplied in **pfcsimulateim.m** has an input which allows the addition of delay.]

    (f) Add some input constraints and input rate constraints to the examples and explore the impact of these; how does the impact depend upon the size of the constraints? [The code supplied in **pfcsimulateim.m** has inputs which allow the addition of input constraints.]

    (g) Code up and implement PFC for ramp targets.

2. You are given a system that has the following ARMA model:

    $$y_{k+1} + 1.2y_k - 0.32y_{k-1} = 0.2u_k$$

    (a) Using an output horizon of 4, an input horizon of 2, use Toeplitz and Hankel matrices to show how one could compute the system predictions for this model. Then use these matrices to formulate a GPC control law.

    (b) Using a control weighting of λ in the GPC performance index with the same horizons, show how one could define a GPC control law using these predictions.

    (c) Explain why using these predictions to form a GPC control law may give poor closed-loop performance and how one would go about forming a predictive control law which is expected to give good performance.

    (d) Illustrate with key equations the impact on parts (a) and (b) of including a T-filter. What impact would you expect the T-filter to have on closed-loop behaviour?

3. You are given a state space system of the form (C.2) and constraints;

$$\underline{u} \leq u_k \leq \bar{u}; \quad \underline{y} \leq y_k \leq \bar{y}$$

    (a) Build the constraint equation of the form $M u_{\rightarrow k} \leq d$ for input/output horizons of 3, assuming that the predictions are expressed in terms of u and not Δu. Hence show how the GPC control law is defined in the presence of constraints. [You can assume that the performance index takes the form $J = u_{\rightarrow k}^T S u_{\rightarrow k} + 2 u_{\rightarrow k}^T b\mathbf{x}$.]

    (b) Stability of predictive control can be assured in the nominal case by using infinite output horizons. Show, for a prediction and a cost of your choice: (i) how the infinite horizon cost can be computed; (ii) how a feasible terminal region can be computed and (iii) hence how the quadratic programme to be solved is defined.

4. This question uses as a base a CARIMA model of the form

$$y_{k+1} = \frac{0.2z^{-1} + 0.4z^{-2}}{1 - 0.85z^{-1}} u_k + \frac{1}{1 - z^{-1}} \zeta_k$$

    (a) Prediction equations can be represented in compact form as in (2.49). Show how all the terms in this prediction equation are defined and illustrate with an input horizon of 3 and an output horizon of 6.

    (b) Hence (without explicit numeric calculation), define a GPC control law with a control weighting of $\lambda = 2$ and show how the implied closed-loop poles would be computed.

    (c) A GPC control law of the form defined in part (b) will often have poor noise rejection properties. Show how a T-filter is introduced to counter this; include details of the impact on the predictions and hence the control law.

    (d) Discuss briefly why the prediction structure of (a) might be a poor choice when the control horizon is one.

5. (a) Give a conceptual overview of a dual-mode predictive control algorithm explaining the key components and why this structure has benefits. Discuss the impact of different choices for the asymptotic mode (mode 2).

    (b) Demonstrate why the use of infinite output horizons and the availability of the tail are sufficient to guarantee closed-loop stability, even in the presence of constraints.

(c) A key advantage of predictive control is the ability to incorporate constraints into the control law systematically. Show how input and output constraints can be combined into a single set of inequalities and hence show how the overall quadratic programming problem defining the control law is set up.

6. You are given a system of the form

$$\mathbf{x}_{k+1} = \begin{bmatrix} 1.2 & 0.5 & 0 \\ 0.2 & 0.1 & 2 \\ -0.3 & 0 & 0 \end{bmatrix} \mathbf{x}_k + \begin{bmatrix} 0 \\ 1 \\ 2 \end{bmatrix} u_k; \quad y_k = \begin{bmatrix} 1 & 0 & 1 \end{bmatrix} \mathbf{x}_k$$

and a performance index

$$J = \mathbf{x}_{k+10}^T W \mathbf{x}_{k+10} + \sum_{i=1}^{10} \mathbf{x}_{k+i}^T Q \mathbf{x}_{k+i} + u_{k+i-1}^T R u_{k+i-1}.$$

(a) Given that the degrees of freedom are $u_{k+i-1}, i = 1, 2, , 10$, show how this performance index can be expressed as a quadratic form in terms of a vector of these degrees of freedom (you do not need to do explicit numeric calculation).

(b) Hence, show how the optimal predictive control law would be derived in the constraint free case.

(c) Discuss, with justification, how a suitable value for W might be derived.

(d) What are the key strengths of predictive control, making it a popular choice in industry?

7. You are given a system model in CARIMA form.

(a) Demonstrate that a general form for the predictions (for any horizons n,r) can be given in the form of:

$$\underset{\rightarrow k+1}{y} = H\underset{\rightarrow k}{\Delta u} + P\underset{\leftarrow k-1}{\Delta u} + Q\underset{\leftarrow k}{y}$$

for suitable values of $H, P, Q, \underset{\rightarrow k+1}{y}, \underset{\rightarrow k}{\Delta u}, \underset{\leftarrow k-1}{u}, \underset{\leftarrow k}{y}$.

(b) Hence deduce and illustrate the unconstrained feedback structure of a GPC control law.

(c) What is the impact on these predictions and hence this structure of adding a T-filter?

(d) Discuss the advantages and disadvantages of using a T-filter.

(e) Show how simple input and input rate constraints are embedded into a GPC control law.

8. You are given a system of the form:

$$y_{k+1} - 0.8y_k = u_k + 0.4u_{k-1}$$

(a) Put this into incremental form and hence find the prediction equation for y_{k+1} in terms of future control increments Δu_k.

(b) Given that you are using a GPC algorithm, show how you could use Toeplitz and Hankel matrices to define output predictions for an output horizon of 10 and an input horizon of 2.

(c) Hence (without explicit numeric calculation), define a GPC control law with a control weighting of $\lambda = 0.1$. Show how the implied closed-loop poles would be computed.

(d) It is noted that this system gives poor noise rejection. Discuss, with relevant mathematical details, one method you could use to modify the algorithm to take account of this.

(e) The system is to be subject to input rate constraints. Show how the control law of part (c) should be modified to take account of these.

9. (a) It is known that the GPC algorithm does not, in general, have a guarantee of stability. Give some illustrations of good and bad choices of tuning parameters (output horizon and input horizon) giving explanations for why the choices might be good or bad.

(b) It is common practise to use infinite output horizons. Explain, with relevant proofs, why this is so. Also show how the performance index $J = \sum_{k=0}^{\infty} x_k^T x_k$ can be computed over an infinite horizon (you are given $x_{k+1} = Ax_k$).

(c) In this context, what do feasibility and infeasibility mean?

(d) What are the key strengths of predictive control making it a popular choice in industry?

10. (a) First, show how the robustness of a nominal predictive control law might be assessed. Second, introduce and explain methods which can be used to improve the loop sensitivity.

(b) Why can GPC with some tuning parameters give poor performance? Hence show how MPC can be tuned to give an a priori guarantee of the stability of the resulting closed-loop, even during constraint handling. Include some comment on potential weaknesses of these results.

11. You are given that a system has 3 inputs. These are subject to upper and lower limits of:
$$\begin{bmatrix} -2 \\ -3 \\ -1 \end{bmatrix} \leq u_k \leq \begin{bmatrix} 3 \\ 2 \\ 2 \end{bmatrix}$$

There are no rate constraints.

(a) For a control horizon of $n_u = 3$, define the inequalities (in matrix form) which ensure constraint satisfaction of the input predictions.

(b) State constraints are also to be included. Define: (i) the prediction equations for the state for a prediction horizon $n_y = 3$ and (ii) the matrix inequalities ensuring constraint satisfaction for $n_y = 3$. You are given the following state space model and limits:

$$\mathbf{x}_{k+1} = A\mathbf{x}_k + B\mathbf{u}_k; \quad \mathbf{x}_k \leq \begin{bmatrix} 4 \\ 5 \\ 3 \end{bmatrix}$$

(c) Hence define the quadratic programming problem whose solution gives the GPC control law for performance index

$$J = \sum_{i=1}^{3} \mathbf{x}_{k+i}^T Q \mathbf{x}_{k+i} + \mathbf{u}_{k+i-1}^T R \mathbf{u}_{k+i-1}.$$

(d) Discuss why GPC is ideally suited as a control design technique for the multivariable case and the constrained case.

12. You are given a system represented by the following ARMA model:

$$y_{k+1} - 0.8y_k = u_k + 0.5u_{k-1}$$

(a) Write this model in incremental form. Hence write down the prediction equations for output horizons of 2 and 3 and find y_3 given $y_0 = 0.2$, $y_{-1} = 0$, $\Delta u_{-1} = 1$, $\Delta u_0 = 1$, $\Delta u_1 = 1$, $\Delta u_2 = 1$.

(b) Without including the precise numbers, what is the general form of the prediction model used by GPC with a prediction horizon of 10 and a control horizon of 3.

(c) You are given that the system has input rate and output constraints of $|\Delta u_k| \leq 0.1$, $-1.3 \leq y_k \leq 2.4$, respectively. For an output horizon of 10 and control horizon of 3, define the inequalities (in matrix form) which ensure constraint satisfaction of the predictions.

(d) Use the results of (b) and (c) to define the quadratic programming optimisation which defines the GPC algorithm for control horizon of 3 and output horizons of 10, assuming unity weights and setpoint r.

13. (a) Show how the use of infinite horizons allows an a priori stability guarantee.

(b) Given this observation, why is it more common to use finite input horizons and in many cases finite output horizons?

(c) Why does the inclusion of a T-filter affect loop sensitivity? Illustrate your answer by defining sensitivity functions and also showing the impact the T-filter has on prediction equations and hence on the compensator.

(d) Discuss why GPC is ideally suited as a control design technique for the multivariable case and the constrained case.

14. What are the strengths and weaknesses of classical control approaches as compared to predictive control? How would you decide which control approach to apply?

15. What are the difficulties in controlling systems which include delays. How would you design a control law that copes with delays and how effective are the techniques you would consider?

16. Compare and contrast a PFC approach to a GPC approach.

17. How is a delay accounted for in a PFC design approach? For what classes of systems is PFC effective and for what classes is it more difficult to tune? Justify your answers.

18. Why does the explicit incorporation of constraints into a predictive control law make a difference in practice? Why is a 2-norm performance index a popular choice and what requirements are there on this for the associated control law to be sensible?

19. (a) Derive an unconstrained GPC control law from 1st principles giving all the key steps.
    (b) Show how the inclusion of a T-filter effects the closed-loop behaviour as compared to the GPC control law of part (a).

20. (a) For what scenarios is predictive control a good solution and why? How do you ensure appropriate tuning to give good behaviour?
    (b) Give some arguments for the use of infinite horizon predictive control as opposed to finite horizon variants.
    (c) Show how performance indices and optimisations are defined with infinite horizon algorithms?
    (d) Why is constraint handling with an infinite horizon algorithm usually more challenging than with a finite horizon algorithm?

21. (a) Show how unbiased predictions can be determined.
    (b) Explain the importance of an unbiased performance index. Use one of these in combination with the predictions from part (a) to determine an unconstrained predictive control law using finite horizons.
    (c) Demonstrate the importance of systematic constraint handling within a predictive control law. Show, using your working from parts (a,b), how typical constraints are included into the computation of the optimum predicted control trajectory.
    (d) Explain with as much detail as possible, the possible benefits and challenges of including terminal constraints and a terminal cost (equivalently using infinite horizons) into predictive control. Show how the predictive control law of parts (a,b,c) is modified to include an appropriate terminal weight and terminal constraints.

22. For what scenarios is predictive control a good solution and why? How do you ensure appropriate tuning to give good behaviour? When might a solution like PFC be a good option?

23. Explain the terminology of maximal admissible set and maximal controlled admissible set and n-step set.

    (a) Why are these definitions important for dual-mode control?

    (b) Why is the use of a MAS both an advantage and a disadvantage to dual-mode MPC algorithms.

    (c) How does the feasible region for a dual-mode algorithm compare with that for an algorithm using open-loop predictions?

24. Discuss the relative pros and cons of open-loop prediction MPC methods such as GPC as opposed to dual-mode (closed-loop) prediction methods such as OMPC.

25. What is meant by an autonomous model for prediction and give an example? For what reasons might this be preferable to alternative methods for formulating dual-mode predictions?

26. The code in this book ignored the role of an observer in estimating the values of the state. What impact will an observer have on the behaviour of a dual-mode algorithm such as OMPC?

A.2   Generic questions

Usually intelligent readers want to set their own questions and develop their own scenarios in order to test understanding. Ttherefore, this book contains only a few illustrative problems. Tutors could focus mainly on both the summary boxes and the numerous student problems in the various sections of this book and derive their own problems to test understanding.

1. **Summary boxes/student problems:** Choose any summary box or student problem and base a question on it. For example:

    •*Discuss with illustrations why the use of prediction is beneficial in creating good control strategies.*

    •*What guidelines should be used when selecting a prediction horizon?*

2. **Prediction equations:** Define a model, transfer function, state-space or FIR, then specify whether the d.o.f. are the inputs or the input increments and finally pose questions like:

    •*For the given model, find the prediction equations with $n_y = 3$ and $n_u = 2$.*

    •*What is closed-loop prediction and when is its usage advisable?*

3. **Control law and stability:** Define a prediction equation, for instance give the values of H, P as in Eqn.(2.2) or any other prediction equation from Chapter 2. Then ask:

- *For the performance index and predictions given, find the corresponding predictive control law and show how you would test stability in the nominal case.*

4. **Tuning:** This is a difficult topic to set a numerical exam question on, as it would usually require significant computation. However, it could be ideal for coursework assignments.

   - *Using the software provided, illustrate and explain the effects of the tuning parameters n_y, n_u, λ on several examples.*

5. **Sensitivity:** One could focus questions on what tools the designer has available to affect sensitivity. For instance:

   - *Derive the loop sensitivity with a GPC (or other) control law and discuss how the T-filter can be used to improve sensitivity.*

6. **Stability and performance guarantees:** Ask questions which test the students' ability of form a performance index with an infinite horizon and also to prove that infinite horizons allow stability guarantees. One could also ask more searching questions on tuning.

   - *Demonstrate why infinite output horizons facilitate stability and performance guarantees. How would the corresponding performance index be calculated?*

   - *Describe dual-mode control in the context of predictive control? Include detailed working on how a dual-mode controller could be computed.*

7. **Constraint handling:** Ask questions requiring students to construct the inequalities which ensure predicted constraint satisfaction. Ask for discussion on how this might impact on closed-loop stability or how the resulting optimisation might be solved. For instance:

   - *Given upper and lower limits on the input and input increments, define the constraint equation in the form $C\Delta\underset{\rightarrow k}{\mathbf{u}} - \mathbf{d} \leq 0$ for $n_u = 3$.*

A.3   Case study based questions for use with assignments

The author's view is that, for typical undergraduate or even master's courses, it is better to provide students with code for implementing MPC and allow students to explore. Hence, it may be convenient to give them some case studies or scenarios to explore. Several of these were discussed in Chapter 1, and that material could provide a good source for some assignment briefings.

Hereafter, this section gives a number of case studies which can be used as part of an assignment. For the more complicated cases, the models for these systems are provided in the partner m-files and in some cases, this file also includes an illustration of using MPC (not necessarily a good one though!). This means that readers can focus on the design, tuning and evaluation of MPC.

The author is not a believer in over emphasising problems with *fixed solutions* because in general there are many different effective control designs for the same scenario and an examiner is more interested in students demonstrating their thought processes and justifying their decisions.

NOTE: Remember, you will need to discretise any continuous time models before using in the provided MPC m-files.

Typical assignment briefings

Use case study XXX to:

1. Compare and contrast the efficacy of classical control design techniques with PFC or GPC or OMPC.

2. Demonstrate how a GPC/OMPC control should be tuned and the various trade-offs that this could entail.

3. Demonstrate the difficulties classical approaches have in handling constraints and, by contrast, how MPC approaches often give straightforward and simple solutions.

4. Demonstrate the impact of significant delays on a classical control design and contrast this with the efficacy of a predictive control approach.

5. Design classical controllers for MIMO systems with low and substantial interaction. Use case studies to contrast the ease and efficacy of design compared to predictive control.

A.3.1 SISO example case studies

A simple assignment brief would be something like: choose a SISO model to demonstrate the benefits of constraint handling/feedforward and impact of delays. Some typical scenarios are given next.

1. Cruise control: compare MPC with PID when there is limit on engine power. Is integral desaturation enough? How does the feedforward term change the comparison? A simple model for the speed of a car would be:

$$M\frac{dv}{dt} + Bv = kf$$

where for convenience the throttle f could be limited as $0 \leq f \leq 1$ and M, B, k are chosen based on a top speed or max acceleration. For an average car $M \approx 1000 kg$ and a reasonable maximum acceleration could be about $10ms^{-2}$ and maximum speed about $50ms^{-1}$.

2. House temperature: compare MPC with PID when there is limit on heating power. Is integral desaturation enough? How does the feed forward term change the comparison? Practical heating via a boiler and a radiator has a significant lag; suggest a model for this lag and include this in your study. What impact does this have the comparison between PID and predictive control? A simple model for the temperature in a house would take the form:

$$C\frac{d\theta}{dt} + k(\theta - \theta_e) = W$$

where C is the heat capacity of the house, θ is the internal temperature (assumed constant throughout), θ_e is the external temperature, k is a constant linked to the rate of heat loss and W is the heating (in Watts). Typical numbers could be of the order $C = 4 \times 10^6, k = 10^3, \theta_e = 0^o$ and W is limited as $0 \leq W \leq 3 \times 10^4$ although these can be varied as required. [**Remark:** How will you deal with a disturbance signal (external temperature) that is closer to a sinusoid than a step?]

3. A swimming pool temperature model and code is provided in [94] and takes the same form as the house temperature example above, but with different parameters.

4. Tank level control where the tank of fixed cross sectional area A has a variable inflow f (the input) and an outflow kh proportional to the depth h, hence:

$$A\frac{dh}{dt} + kh = f$$

Compare MPC with PID when there is limit on fluid flow into the tank (you can choose the limits on f according to A, k and the desired fill time). Is integral desaturation enough? How does the feed forward term change the comparison? What would be impact of a small delay and lag in changing the input fluid flow due to inertia from the main reservoir?

5. DC servos are very common and come with two alternative control problems: (i) position control or (ii) speed control. You can postulate sensible constraints and requirements before doing a PID design and comparing to MPC. A typical model takes the form:

$$v = \frac{Jk}{K}\frac{d^2\omega}{dt^2} + (\frac{Bk}{K} + \frac{JR}{k})\frac{dw}{dt} + (k + \frac{BR}{k})w$$

where v is the supply voltage (input), J is the load inertia, B is the load damping (units Nsm^{-1}), R is the electrical circuit resistance, k links the back emf to angular velocity w and K is a spring constant between the motor and the load.

6. Helicopter model taken from [94] (p24). The model links rotor angle to forward speed and is challenging for conventional strategies as both unstable and non-minimum phase.

$$G = \frac{9.8(s^2 - 0.5s + 6.3)}{(s + 0.6565)(s^2 - 0.2366s + 0.1493)}$$

7. Demonstrate using a simple instability such as given in $G(z)$ below, the potential dangers of using MPC strategies such as PFC/GPC which deploy open-loop prediction.

$$G = 0.3 \frac{0.2z^{-1} - .35z^{-2}}{(1 - 1.2z^{-1})(1 - 0.85z^{-1})}; \quad -2 \le u_k \le 2; \quad |\Delta u_k| \le 0.5$$

How are these dangers avoided and what is the price paid?

8. Choose a system with non-minimum phase dynamics. In this case you will find it difficult to determine an effective PID which gives reasonable bandwidth. Contrast this to what can be achieved with MPC. Next, add in some constraints and investigate how these impact on the PID performance. The situation is severely accentuated if the system also has a RHP pole. One possible example is the helicopter model above.

9. A two tank system (arranged side by side) where the input f is the flow into one tank and output is the depth in the second tank. The flow between the tanks is proportional to their relative depths $h_1 - h_2$ and flow out from the second tank is proportional to its depth h_2. A continuous time model takes the form:

$$\frac{d}{dt}\begin{bmatrix} h_1 \\ h_2 \end{bmatrix} = \begin{bmatrix} -\frac{c_1}{A_1} & \frac{c_1}{A_1} \\ \frac{c_1}{A_2} & -\frac{c_1+c_2}{A_2} \end{bmatrix} \begin{bmatrix} h_1 \\ h_2 \end{bmatrix} + \begin{bmatrix} 1 \\ 0 \end{bmatrix} f; \quad y = [0 \ 1] \begin{bmatrix} h_1 \\ h_2 \end{bmatrix}$$

where A_1, A_2 are the cross-sectional areas and c_1, c_2 constants which govern the dependence of flow on depth. Typical constraints would be on the depth in each tank (no overflows) and the input flow would be limited.

A.3.2   MIMO example case studies

Choose a MIMO system to demonstrate the difficulties of finding a suitable diagonal PID design to give good performance, low interaction and manage constraints. Then demonstrate that with MPC, finding a suitable control law is comparatively straightforward. The m-files for these are available on the website http://controleducation.group.shef.ac.uk/htmlformpc/introtoMPCbook.html and also via the book publisher.

1. Compressor model. This is taken from [16] but basic code is supplied for you in **compressor.m** (see MATLAB code in Chapter 4). The inputs

are a guide vane angle and valve position (limits $-2.75 < u_i < 2.75$), the outputs are air pressure and air flow rate (sensible targets are around 1).

$$G = \begin{bmatrix} \frac{0.1133e^{-0.715s}}{1.783s^2+4.48s+1} & \frac{0.3378e^{-0.299s}}{0.361s^2+1.09s+1} \\ \frac{0.9222}{2.071s1} & \frac{-0.321e^{-0.94s}}{0.104s^2+2.463s+1} \end{bmatrix}$$

2. Spey engine. These are common in aerospace applications but also more widely where gas turbines are used. Only the state space model (**speyengine.m**) is provided but it may be interesting just to compare what you can achieve with classical methods vs MPC even without constraints.

3. Fractionator model. Core data required to run via my GPC files is provided in the m-file **fractionator.m** with a sample rate of 4 min. Inputs are top draw, side draw and bottom reflux. Outputs are top composition, side composition and bottom reflux temperature. Meaningful targets are given as $\mathbf{r} = [0.5\ 0.3\ 0.1]^T$. Each input has the same limits so

$$-0.5 \le \mathbf{u} \le 0.5; \quad |\Delta\mathbf{u}| \le 0.2$$

Compare with what you can achieve with classical control.

4. For those who want an aerospace example, look at the Cessna aircraft [94] which has significant interaction. The m-file **makecita.m** defines the state space model and includes comments on constraints.

5. The model of a paper machine is also taken from [94] and is a simple MIMO example given in state space form. Constraints and variable descriptions are supplied in the m-file **papermachine.m**.

6. A stirred tank model is a simple MIMO model taken from [16] but comes without constraint information (**stirredtank.m**). The inputs are the feed flow rate and flow of coolant and the outputs are the concentration and temperature. This model is useful for simple comparisons of classical approaches with MPC.

7. A distillation column [16] is given in **distillationcolumn.m**. Output 1 is top product composition, output 2 is bottom product composition, input 1 is reflux and input 2 is boil up reflux. The model and constraints are given as:

$$G = \frac{\begin{bmatrix} 87.8(15s+1) & -86.4(12.1s+1) \\ 108.2(15s+1) & -109.6(17.3s+1) \end{bmatrix}}{(194s+1)(15s+1)}; \quad |\Delta\mathbf{u}_k| \le 0.2$$

In this case, a possible diagonal PID strategy is also provided and coded within the m-file.

$$K = \begin{bmatrix} \frac{s+0.005}{90s} & 0 \\ 0 & \frac{s+0.005}{100s} \end{bmatrix}$$

8. An oil powered powerstation model has output 1 as MW generated, output 2 is steam pressure, input 1 is fuel flow and input 2 is governor valve position. This system has significant interaction and thus is interesting to show MIMO issues. Also the system is *stiff* in that it displays both fast and slow time scales. The continuous time state space model is given in **powerstationmodel.m**.

9. Linearised model of a AV-8A harrier in hovering flight (implicitly uses deviation variables about a steady-state). This is useful for comparing classical approaches with unconstrained predictive control as no constraint information available.

$$\dot{\mathbf{x}} = \begin{bmatrix} 0 & 1 & 0 & 0 \\ 0 & -0.13 & 0.014 & 0.0056 \\ 0 & 0 & -0.105 & -0.011 \\ -9.8 & 0 & -0.02 & -0.035 \end{bmatrix} \mathbf{x} + \begin{bmatrix} 0 & 0 & 0 \\ .079 & 0 & .0016 \\ .063 & 0.28 & -.309 \\ 0 & -9.8 & .002 \end{bmatrix} \mathbf{u}$$

with states of pitch angle, pitch rate, vertical velocity and forward speed and inputs of longitudinal stick perturbation, nozzle angle perturbation and engine speed perturbation.

B

Further reading

CONTENTS

B.1   Introduction

This chapter provides a brief pointer towards some topics not covered in this book but which would be of interest to a reader pursuing research in MPC. The underlying motivation is that while a basic MPC algorithm suffices for many cases, there are always times where *tailoring* or *modification* of a basic algorithm may give some important advantages. This chapter cannot cover all such solutions, or even a majority, but hopefully will pique the reader's interest into investigating the research literature.

    Obvious examples to be discussed are:

1. The choice of the performance index and the parameterisation of the d.o.f. in the predictions/optimisation.

2. Parametric solutions of the quadratic programming problem.

3. The use of feedforward in MPC.

4. So-called *tube* or *back-off* methods for handling uncertainty.

5. Invariant sets and the use of LMIs.

B.2   Guidance for the lecturer/reader

This chapter is far beyond expectations of a typical taught module and is here for interest only. As such, algebraic detail will be minimal, there are no supporting numerical examples or tutorial questions, and readers are invited to use the literature for a full discussion.

B.3   Simple variations on the basic algorithm

B.3.1   Alternatives to the 2-norm in the performance index

In the late 1980s and early 1990s there was some investigation into the use of 1-norms and ∞-norms (e.g., [2], [131], [199]) in the performance index as opposed to the 2-norm. This amounts to minimising the worst case error or the sum of the error moduli. Such a change facilitated better a priori robustness results for certain classes of model uncertainty and moreover a reduction in computational load was possible as the optimisation implied was a linear program (LP) rather than quadratic program (QP). However, the typical control was not smooth and therefore such approaches were never really accepted as attractive in practice.

However, there will be cases where an alternative choice or norm to the standard 2-norm is more appropriate and readers therefore should consider what impact such choices may have on MPC behaviour and performance/convergence assurances. One obvious example is economic objectives which are often best represented with 1-norms. Another example is a contraction constraint used to ensure stability, where the contraction constraint could be presented with different norms.

Summary: MPC is flexible in terms of the models and performance indices that it can utilise. Readers should not be afraid to consider a performance index which most closely matches the precise need, but should recognise that some analysis may be required to ensure this leads to tractable optimisations and expectations of good behaviour.

B.3.2 Alternative parameterisations of the degrees of freedom

Classical MPC algorithms use the future control increments/values as the d.o.f. (e.g., Section 4.4.3) in the optimisation. This is also, in effect, the case with the dual-mode approaches (e.g., Eqn.(6.18)). However, it was noted very clearly in Chapter 5 that the use of small n_u often leads to poorly posed optimisations which are inconsistent with the actual desired behaviour. In order to get strong assurances of good behaviour, in general a high n_u is needed (or a dual-mode approach which in effect also uses high n_u) but in practice high values of n_u are not desirable because of the associated computational load, especially where a quadratic programming optimisation is required.

Historically the input moves were used as the d.o.f. within an MPC optimisation both for convenience and transparency, but not because in any sense these provided the best d.o.f. from which to form an MPC control law.

Chapter 6 on dual-mode approaches provided one solution to this conundrum where it was noted that the d.o.f. could be re-parameterised in terms of a loop perturbation c_k (e.g., Eqn.(6.45)). This enables strong performance and convergence assurances while retaining a low number of d.o.f., but as seen in Chapters 7 and 8, could introduce conflicts with constraints and thus limit feasibility.

There has been surprisingly little work in the literature asking whether alternative parameterisations of the d.o.f. in the predictions could be useful and offer properties allowing easier or improved trade-offs between computational loads and performance. Indeed, most research work [113, 198] seems focussed on improving the computational efficiency of QP solvers for a standard MPC formulation. Nevertheless, readers may be interested in the few works that have explored this avenue to some extent.

- Some early works investigated the potential of drawing on the properties of the unconstrained closed-loop with alternative choices of control law [65, 74, 82, 155, 164, 166, 179]. Typically these approaches had feasible regions linked to the predefined linear control laws and the computational complexity is linked to the number of alternative control laws allowed. However, a downside of these approaches is that it is awkward to incorporate *the tail* (Section 6.3.1), and thus although computational efficiency and feasibility may be good, stability assurances are often weak or clumsy.

- Another popular theme is to express the future control moves as linear combinations of predefined functions. Some authors focussed on arbitrary blocking

[48, 134] but these approaches suffered from the lack of *the tail* and thus had weak performance/stability properties. An alternative blocking type approach using Laguerre or other orthonormal functions was popularised in [190, 192] and later pursued in the context of dual-mode control [72, 166, 184]. It can be argued that this latter approach often gives significant benefits compared to a standard approach in that larger feasible regions can be achieved with similar numbers of d.o.f.; the downside is the optimisation changes its structure and the community has yet to consider the extent to which a tailored QP could exploit the resulting prediction structure.

Summary: Consideration of the parameterisation of the d.o.f. within an MPC optimisation is relatively under-studied. The author believes many opportunities exist here for improving feasible volumes for a given computational load/complexity. However, to some extent further research will be driven by need, in that, the community focus on efficient solvers for a standard approach alongside computing advances mean that for most cases, the online computational load is no longer a significant barrier to implementation.

B.4 Parametric approaches to solving quadratic programming

This book does not have space to look at parametric programming in any detail and thus readers are referred elsewhere, e.g., [8, 9, 12, 111]. Instead, this brief section will give an outline of the underlying principles and an explanation of why such an approach is worth pursing.

A key challenge with a conventional MPC law is the quadratic programme (QP) required, e.g., (7.32). In an online scenario, solving such a problem needs an efficient and reliable QP solver and these may be neither cheap nor easy to code in general; indeed this could be a particular challenge for controllers which are to be implemented via a cheap microprocessor (although some authors have worked on this [27]). A secondary difficultly is linked to auditing and sample times: it is difficult to prejudge what outcome will result from the use of a QP solver for a range of different scenarios in such a way that the behaviour can be strongly validated, say for safety critical systems. Fast sample times will increase the potential for incomplete convergence as many QP solvers such as active set methods have potentially high iteration counts for guaranteed convergence (even if rarely in reality).

B.4.1 Strengths and weaknesses of parametric solutions in brief

Parametric solutions provide a possible solution to many of the above issues:

- The solution is transparent to the user and thus can be validated formally.

- The implementation reduces to a lookup table and thus can be coded very simply and is amenable to very fast sample times, especially where the lookup table search can be parallelised.

It is not the place of this book to discuss the corresponding weaknesses (or possible mitigations), but it should be noted that in practice multiparametric QP (MPQP) solutions have not yet been widely adopted except on low order, fast sample time processes. This is partly because:

- The size of the lookup table can (not must) quickly run into millions and more for systems with realistic state dimensions. With large lookup tables to search through and the corresponding storage requirements, it is often quicker and easier to perform the original QP.

- It is usually only very low order systems which have lookup tables with a sensible number of variants (say less than a 100).

- The off-line optimisation required to build the lookup table can become very ill-conditioned and even fail, especially as the state dimension increases.

B.4.2 Outline of a typical parametric solution

In the following we give a quick outline of how a parametric solution can be derived. Consider the following QP optimisation where d.o.f. w is of dimension n_w and, for now, n_p dimensional parameter p is assumed fixed; the solution is called parametric as ultimately it varies as parameter p varies.

$$\min_{w} \quad w^T S w + w^T p \quad s.t. \quad Fw + Gp \le g \tag{B.1}$$

It is known that at the optimum, at most n_w of the inequalities are required to determine this solution as these inequalities could be active (we do not discuss here special cases). Next, determine the Lagrange multipliers[1] $\lambda_i, i = 1, \cdots, n_v$, $n_v \le n_w$ needed to identify the optimum constrained solution to (B.1).

$$\min_{w} \quad w^T S w + w^T p + \sum_{i=1}^{n_v} \lambda_i (F_i w + G_i p - g_i) \tag{B.2}$$

where F_i, G_i are the ith rows of F, G, respectively. The optimum is determined by setting the gradients wrt to both w, λ_i equal to zero and ensuring that $\lambda_i \ge 0, \forall i$. Hence, first setting the gradients equal to zero gives the conditions:

$$2Sw + p + \sum_{i=1}^{n_v} F_i^T \lambda_i = 0; \quad \{F_i w + G_i p - g_i = 0, \ i = 1, \cdots, n_v\} \tag{B.3}$$

---

[1] For convenience of notation using $\lambda_i, i = 1, \cdots, n_{v_i}$. In general we can take any n_v values of i from the range defined by the number of rows in $Fw + Gp \le g$.

Combining these for convenience:

$$
\begin{bmatrix}
2S & F_1^T & \cdots & F_{n_v}^T \\
F_1 & 0 & \cdots & 0 \\
\vdots & \vdots & \vdots & \vdots \\
F_{n_v} & 0 & \cdots & 0
\end{bmatrix}
\begin{bmatrix}
w \\
\lambda_1 \\
\vdots \\
\lambda_{n_v}
\end{bmatrix}
+
\begin{bmatrix}
I \\
G_1 \\
\vdots \\
G_{n_v}
\end{bmatrix}
p =
\begin{bmatrix}
0 \\
g_1 \\
\vdots \\
g_{n_v}
\end{bmatrix}
\tag{B.4}
$$

This implies that the optimum for w is linearly dependent on varying parameter p and has an affine dependence on the constraint information in g (which is fixed), for example:

$$
w = X_w p + q_w; \quad \lambda = X_\lambda p + q_\lambda \tag{B.5}
$$

for suitable $X_w, q_w, X_\lambda, q_\lambda$. In order for the optimum to be valid, we need the Lagrange multipliers to be positive and also for feasibility we need to satisfy $Fw + Gp \leq g$, and hence the region in which this solution is indeed the constrained optimum of (B.1) is given from the set R_i defined by:

$$
\{F(X_w p + q_w) + Gp \leq g \ \& \ X_\lambda p + q_\lambda > 0\} \quad \Rightarrow \quad p \in R_i \tag{B.6}
$$

Hence, the optimum is defined as $p \in R_i \quad \Rightarrow \quad w = X_w p + q_w$; with this particular set of Lagrange multipliers we cannot say anything about the optimal solution outside of R_i.

Off-line computation of full parametric solution: A full parametric solution is determined by finding all possible combinations of active constraints for all possible values of parameter p, determining the corresponding Lagrange multipliers (B.5) and then the corresponding region R_i where this combination is active. Efficient search routines to find all the possible regions R_i and associated optimums (B.5) are available in the literature (e.g., [106]), but are non-trivial in general.

On-line implementation of parametric solution: On-line the algorithm requires two steps.

1. Identify the region R_i in which the current state p lies.

2. Implement the corresponding optimum (B.5) from the look up table.

Remark B.1 *In the linear case it can be shown that valid regions R_i do not overlap, but the union of R_i also has no internal holes. The corresponding solutions (B.5) for w at the boundary between two regions are the same on the boundary so the control law can be shown to be piecewise linear.*

Summary: Where the required number of regions R_i is small, the online implementation of an MPQP solution to MPC can be very efficient and simple to code and thus has clear benefits over coding a full QP solver online.

An obvious weakness is that the solution is fixed (being computed offline) and cannot be updated online due to modelling changes, constraint changes, set point changes and so forth.

ILLUSTRATION: Typical picture of a parametric solution for a 2-dimensional example

This figure shows what a parametric solution might look like. Each region R_i is distinct with no overlaps and no holes. In each region R_i, a different set of values for X_w, q_w from (B.5) will define the constrained optimum.

Hence, identify in which region p lies, then implement the corresponding optimum.

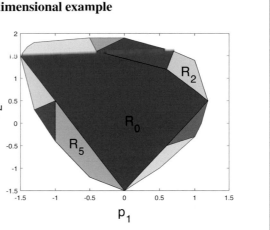

B.5 Prediction mismatch and the link to feedforward design in MPC

One of the major weaknesses of the original predictive control algorithms such as DMC, GPC, PFC, is the prediction structure used in the optimisation of performance. In simple terms the class of predictions over which the optimisation is performed is not necessarily closely matched to the resulting closed-loop responses. As such the optimisation could be ill-posed; that is, finding a minimum for the given objective need not imply good performance.

This issue was well discussed in Chapter 5, but assuming that $\mathbf{r}_{k+i} = \mathbf{r}_{k+1}, \forall i > 1$. A clear summary was that with finite horizons, n_y and n_u, the optimal open-loop predictions may not be well matched either to the consequent closed-loop behaviour arising from a receding horizon implementation or to the global optimum. In this case one may ask is minimising J meaningful? A partial solution was proposed by ensuring that both n_u and n_y are sufficiently large, as then the mismatch between open-loop predictions and closed-loop behaviour will be small and hence the minimisation will be well-posed. However computational limitations may imply that n_u (or the number of d.o.f. in the optimisation) must be small so some prediction mismatch is inevitable.

Nevertheless, a key point here is that the earlier chapters did not consider the impact, if any, of the prefilter P_r on this issue. It so happens that the impact is far greater and potentially more damaging than intuitively expected, and thus it is useful for the reader to have some awareness of the issue and straightforward mitigations.

B.5.1 Feedforward definition in MPC

One of the supposed advantages of MPC is that it can make systematic use of advance knowledge of set points and disturbances. However, as will be shown here, one must treat this claim with extreme caution [134]. Following the same lines as in Section 5.4, it will be shown here that the default choice of feedforward compensator (e.g., P_r of Eqn.(4.27)) is often poor. This is also due to the mismatch between the assumptions made on the open-loop predictions and the actual closed-loop behaviour.

The default prefilter P_r is anti-causal; that is, it contains powers of z rather than z^{-1}. Hence the current control action depends upon future set points. Assume that the default prefilter is given as

$$P_r = p_1 z + p_2 z^2 + \cdots + p_{n_y} z^{n_y} \tag{B.7}$$

If one wanted to reduce the advance information used by the control law to say n_r steps, this is equivalent to assuming that $r_{k+n_r+i} = r_{k+n_r}$, $i > 0$. One can achieve this most simply by rewriting the prefilter as

$$P_r = p_1 z + p_2 z^2 + \cdots + p_{n_r-1} z^{n_r-1} + [p_{n_r} + p_{n_r+1} + \cdots + p_{n_y}] z^{n_r} \tag{B.8}$$

If no advance knowledge is available, then $P_r = N_k(1) = \sum p_i$, i.e., a simple gain; this is in fact the most commonly adopted assumption in the literature.

B.5.2 Mismatch between predictions and actual behaviour

In the performance index minimisation, only n_u control moves are allowed even though in the closed-loop it is known that a control move is ultimately to be allowed at every sample. Because the optimisation is working with just a few moves, it will optimise tracking over the whole output horizon, assuming just those few input changes. If there is a set point change towards the latter end of the horizon, the optimisation will request some movement from the inputs now in order to prepare for that set point change. That is, it starts moving too soon because the current minimisation does not have the freedom to use later control moves. One will then get a slow creeping of the output towards the expected set point with a faster movement once the set point has actually occurred.

Summary: It is clear that for the simple illustration, one improves behaviour by using less advance knowledge about the future set point. This apparently counter intuitive result is because of prediction mismatch. With a small n_u, the prediction class is not sufficiently flexible to incorporate both transient tracking and information about target changes far into the future and thus will give *supposedly optimum* solutions which are not close to what is actually achievable.

It is not an objective of this book to pursue this issue in any more detail suffice to give the following advice: IN GENERAL, IT IS UNWISE TO PICK $n_r \gg n_u$ (for dual-mode $n_r \gg n_c$).

ILLUSTRATION: Poor use of advance knowledge by GPC

One can illustrate this with the following example:

$$y = \frac{0.2z^{-1} + 0.06z^{-2}}{1 - 1.2z^{-1} + 0.32z^{-2}} u; \quad n_y = 15, \ n_u = 1 \text{ or } 3, \ \lambda = 1 \qquad \text{(B.9)}$$

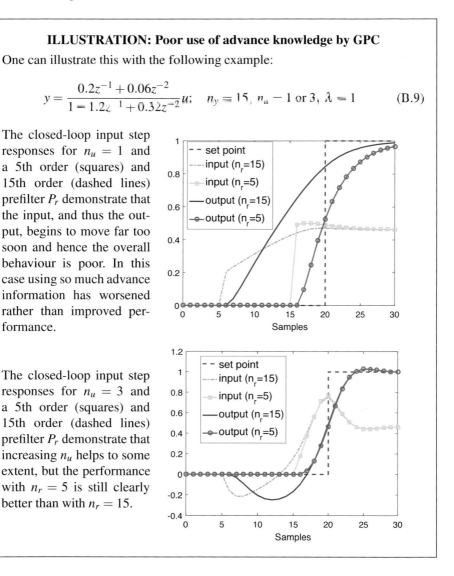

The closed-loop input step responses for $n_u = 1$ and a 5th order (squares) and 15th order (dashed lines) prefilter P_r demonstrate that the input, and thus the output, begins to move far too soon and hence the overall behaviour is poor. In this case using so much advance information has worsened rather than improved performance.

The closed-loop input step responses for $n_u = 3$ and a 5th order (squares) and 15th order (dashed lines) prefilter P_r demonstrate that increasing n_u helps to some extent, but the performance with $n_r = 5$ is still clearly better than with $n_r = 15$.

STUDENT PROBLEMS

With the GPC algorithm and using examples of your own choice, demonstrate how using n_r that is too large can result in performance which is far worse than choosing $n_r \approx n_u$.

More examples and background are also available at:

http://controleducation.group.shef.ac.uk/htmlformpc/feedforward.html

For interested readers, some further studies are available (e.g., [35, 185]), although in most cases the summary advice reduces to one of two options:

1. Choose $n_r \approx n_u$ (or n_c).

2. Consider an off-line choice of P_r and embed this as a fixed feedforward into the predictions (requires dual-mode predictions), thus ensuring that any feedforward information is in some sense already optimised in its use.

3. A number of examples and more insight are also available at: http://controleducation.group.shef.ac.uk/htmlformpc/introtoMPCbook.html

Of particular interest is the tight linking to n_u, n_c which means that contrary to the impressions given in many MPC papers, only a limited amount of future target information can be taken account of systematically.

Remark B.2 *Where future information (in essence feed forward) about disturbances is available, this will have a similar impact to future information about the target and thus similar insights are likely to follow.*

> **WARNING:** If you get poor performance from a MPC implementation on the nominal model, then there is likely to be a large mismatch between the *optimised* predictions used in the cost function and the desired or actual closed-loop responses. This mismatch could be due to a poor choice of parameters such as n_u, n_c, n_r and thus readers should be aware of the need to check carefully for the source of the problem.

B.6   Robust MPC: ensuring feasibility in the presence of uncertainty

The chapter on constraint handling dealt with the nominal case, so that one can assume the predictions are always correct and predictions satisfying constraints implies that the real system behaviour will satisfy constraints. In reality, the predictions and true behaviour will be different due to parameter uncertainty and disturbances amongst other things. This observation leads to a difficult conflict.

If the actual predictions violate constraints when the nominal predictions do not, what impact will this have on the actual system behaviour?

ILLUSTRATION: Driving on a road with black ice

The driver expects the braking and steering of an automobile to behave in a fairly well defined manner learnt over many hours of driving a specific vehicle. Using this knowledge, the driver approaches a corner and applies the braking and steering expected to negotiate the corner in a safe manner. Even with some uncertainty due to variations in the car weight and road surface, for drivers who are *not speeding*, this almost invariably works well, thus small differences between predictions and actual outcomes do not affect the result and constraints such as braking limits and avoiding kerbs are met satisfactorily.

What happens however if there is some black ice just before the corner? The braking and steering both fail to have the expected effect and the car goes off the road or spins or suffers some other failure mode.

Summary: While MPC will generally cope with small levels of uncertainty in the predictions, there is usually an uncertainty level at which the expected behaviour is affected catastrophically by this uncertainty. This could be due to inconsistency between *expected satisfaction of constraints* and actual behaviour which does not satisfy constraints.

The worrying problem for the reader is that the amount of uncertainty to turn a reliable control law into an unreliable one can be arbitrarily small; for example: (i) what if the critical constraint that is violated is a safety valve? Or (ii) what if a racing driver was deliberately pushing a car to the speed and track limits?

Summary: A control law being robust in the unconstrained case need not imply any robustness at all in the constrained case! There is a need to consider how MPC might be modified to improve robustness during constraint handling scenarios.

Remark B.3 *The reader should note a general point about robustness and stability/convergence assurances. The only way you can guarantee not to crash your car is to leave it in the garage! Clearly this is a useless solution. You could drive very slowly and cautiously, but this is equally an unhelpful solution as the performance is so poor as to be uncompetitive (and dangerous to other road users). The summary insight from this is that the lower a risk you want, the lower the performance you are likely to obtain.*

Hence although results in the literature offer gaurantees *of stability, these are premised on assumptions about risk/uncertainty and thus are not absolute guarantees. Moreover, almost invariably the price of the guarantee is a substantial loss of performance compared to an approach with a weaker guarantee and thus in practice a trade-off is required.*

B.6.1   Constraint softening: hard and soft constraints

It is important to distinguish hereafter between hard and soft constraints.

1. Hard constraints are those that must be satisfied. These could include input limits such as fully open and fully closed, pressure limits (safety valves), depth limits in tanks to avoid overflow and so forth.

2. Soft constraints are limits we would like to satisfy such as quality constraints, input rate limits to avoid excessive fatigue and so forth. However, where necessary, these could be relaxed for short periods. Define these as:

$$C_S \underset{\rightarrow k}{\mathbf{u}} + H_S \mathbf{x}_k \leq \mathbf{d}_S \qquad (B.10)$$

Constraint softening is a practically relevant strategy [115] to deal with uncertainty and in principle sounds straightforward; it is focussed on exploiting soft constraints. Nevertheless, a formal design needs careful tailoring to the specific scenario.

Relaxing constraints in an iterative manner

If the constraint set (e.g., (8.9) within the MPC QP optimisation (e.g., (7.31)) is inconsistent, that is not all the constraints can be satisfied, then some of the soft constraints must be either relaxed or removed. A process dependent strategy could be developed but the associated logic needs to be designed and tested carefully. For example, one could iterate through a procedure such as:

1. Relax (or remove) the least important soft constraints from (B.10) and test for feasibility.
2. Relax (or remove) next most important constraints and test for feasibility.
3. etc.

The hope is that once enough soft constraints have been relaxed, the whole constraint set (8.9) will become feasible and one can then implement the MPC algorithm. Assuming constraint inconsistencies are caused by some transient effect, one should be able to enforce the soft constraints again after a few samples.

Relaxing constraints in a norm based manner

The above strategy is a little crude in that the ordering of relaxation of a constraint is predetermined and hence may be more conservative than required. Also the use of an iteration could cause worrying delays. An alternative is to define a variable $\mathbf{s} \geq 0$ and replace the soft constraints of (B.10) by:

$$C_S \underset{\rightarrow k}{\mathbf{u}} + H_S \mathbf{x} - \mathbf{d}_S \leq L\mathbf{s} \qquad (B.11)$$

The vector $\mathbf{s}$ defines the magnitude of the constraint violations (if positive). A possible strategy is then to minimise the maximum (weighted) value of $\mathbf{s}$ s.t. hard and terminal constraints. This will minimise the predicted soft constraint violations and is a relatively simple optimisation to set up (requiring only a linear program).

The weakness of these approaches is that all the emphasis is placed on the soft constraints and none on the performance. An alternative strategy may wish to find some compromise between predicted performance and the violations of the soft constraints; that is, one may be able to reduce or avoid the expected constraint violations by reducing the closed-loop bandwidth. In effect this means changing the weights in the performance index and indeed such an approach is almost implicit in algorithms which deploy different possible control laws, e.g., [74, 155, 179].

Summary: Feasibility can often be ensured by a systematic relaxation of soft constraints or switching between alternative control laws. This would be determined at a supervisory level and is commonly used in practice. However, it may be difficult to apply generic guidelines to a specific process and so in practice some careful tailoring may be needed.

B.6.2 Back off and borders

The topic of *back off* (e.g., [22, 50]) is not a mainstream topic within the MPC literature although it is used in the process industry. The basic idea is not to drive the system predictions to the input limits during intermediate transients and the steady-state, but to leave some extra freedom for emergencies.

ILLUSTRATION: Back off when driving

When a driver plans the desired speed and trajectory for the road over the next

20-100m, they usually plan such that the car is not operating on the very limits of its braking and grip. This means that if the behaviour is slightly different from expected, the driver can call upon the built-in slack (extra braking and extra grip) to deal with the scenario. However, the immediate control actions can be deployed right up to the constraints, that is the slack is made available at the time of need.

This picture gives a simple representation of the sort of data a human is likely to use. The back off is larger in the future where there is greater uncertainty in the predictions, but zero at the current time where the maximum control action may be needed.

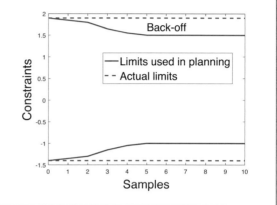

How much to back off the constraints and how exactly to incorporate this is process dependent. Clearly one should not do this unless necessary, as there will be a compromise with performance.

Mathematically, slack can be represented very simply and thus is easy to incorporate into a MPC law. For example, let the actual upper input limit be:

$$\mathbf{u}_{k+i} \le \bar{\mathbf{u}}, \quad \forall i \tag{B.12}$$

Now back off these constraints by a time dependent value $\mathbf{b}_i$, that is, replace (B.12) by:

$$\mathbf{u}_{k+i|k} \le \bar{\mathbf{u}} - \mathbf{b}_i, \quad \forall i, \quad \mathbf{b}_i \ge \mathbf{b}_{i-1} \tag{B.13}$$

It is assumed that $\mathbf{b}_i \ge \mathbf{b}_{i-1}$.

Theorem B.1 *Using constraints (B.13) on the predictions ensures that extra control capacity is introduced at each sample to deal with uncertainty.*

Proof: At sampling instant k the constraints are given in (B.13). At sampling instant $k+1$ they will be given by

$$\mathbf{u}_{k+i|k+1} \le \bar{\mathbf{u}} - \mathbf{b}_{i-1}, \quad \forall i \tag{B.14}$$

Comparing (B.13, B.14) one sees that the back off on the constraint on prediction $\mathbf{u}_{k+i|k}, \mathbf{u}_{k+i|k+1}$ has changed from $\mathbf{b}_i$ to $\mathbf{b}_{i-1}$; hence the predicted control $\mathbf{u}_{k+i|k+1}$ has an extra potential movement of $\mathbf{b}_i - \mathbf{b}_{i-1}$. $\square$

Summary: Using artificially tight constraints on future predictions automatically builds in some slack which can be used to retain feasibility in the presence of moderate uncertainty. The slack should be monotonically increasing with the horizon. However, the design of suitable back off parameters is not simple in general [22] and systematic methods such as tubes [114] are very involved/conservative.

B.7 Invariant sets and predictive control

It was noted in the previous chapter that a set is invariant [6] if, once a state enters that set, it can no longer leave. So, for instance, a set $\mathscr{S}$ is invariant iff:

$$\{\mathbf{x}_k \in \mathscr{S} \Rightarrow \mathbf{x}_{k+1} \in \mathscr{S}\} \Rightarrow \mathbf{x}_{k+i} \in \mathscr{S}, \ \forall i > 0 \tag{B.15}$$

Assuming that a system is subject to feedback, the shape of the invariant set depends upon several factors:

1. The system dynamics (or model).

2. The embedded feedback law.

3. The number and shaping of the d.o.f. within the predictions.

This will be clear from Section 8.5.6 on admissible sets (MAS and MCAS).

B.7.1 Link between invariance and stability

The reason why invariance is so popular as a tool is that an invariance condition is equivalent to a bounded stability test. Once the state is inside $\mathscr{S}$ it is guaranteed to remain inside and thus, in the worst case, the system is bounded output stable. In fact it is commonplace to determine stronger results which also enable convergence to be established.

A second key factor covered in the previous chapter is that invariance is invaluable in the case of constraint handling as, assuming sets are defined to include constraint information (that is $\mathbf{x}_k \in \mathscr{S}$ implies constraints (e.g., (8.1) are satisfied), then these sets also enable a bounded output guarantee in the presence of constraints.

This section will briefly indicate how these observations allow extension of convergence and stability assurances to the uncertain case, even in the presence of constraints.

Summary: The existence of an invariant set is equivalent to the existence of a Lyapunov function and hence is equivalent to a stability test. Such a test can be extended to include constraint handling scenarios in the presence of uncertainty.

B.7.2 Ellipsoidal invariant sets

Although the earlier chapter focussed on polyhedral invariant sets, in fact a good deal of literature considers ellipsoidal invariant sets because these allow a number of numerical and algebraic advantages when it comes to analysis, especially for the uncertain case.

Define an ellipsoidal set

$$\mathscr{S} = \{\mathbf{x} : \mathbf{x}^T W \mathbf{x} \leq 1\}; \quad W > 0 \tag{B.16}$$

Theorem B.2 *The set $\mathscr{S}$ of (B.16) is invariant for model $\mathbf{x}_{k+1} = \Phi \mathbf{x}_k$ if*

$$\Phi^T W \Phi - W \leq 0 \tag{B.17}$$

Proof: Substitute $\mathbf{x}_{k+1} = \Phi \mathbf{x}_k$ into invariance condition (B.15), i.e.,

$$\begin{aligned}
\mathbf{x}_k^T W \mathbf{x}_k = 1 &\Rightarrow \mathbf{x}_{k+1}^T W \mathbf{x}_{k+1} \leq 1 \Rightarrow \mathbf{x}_k^T \Phi^T W \Phi \mathbf{x}_k \leq 1 \\
&\Rightarrow \mathbf{x}_k^T [\Phi^T W \Phi - W] \mathbf{x}_k \leq 0, \quad \forall \mathbf{x}_k \in \mathscr{S}
\end{aligned} \tag{B.18}$$

from which the proof is obvious. □

One can use a set membership test to ensure predicted constraint satisfaction. This section assumes the sets exist and does not tackle the question (e.g., [74]) of how the invariant sets (B.16) could be computed.

Lemma B.1 *If there exists an ellipsoidal invariant set (B.15, B.16) for autonomous model/feedback $\mathbf{x}_{k+1} = \Phi \mathbf{x}_k$ $\mathbf{u}_k = -K\mathbf{x}_k$, then this set can always be scaled to be small enough so that non-zero input constraints are always satisfied.*

Proof: The input is given by the feedback $\mathbf{u}_k = -K\mathbf{x}_k$ and $\mathbf{x}_k$ is restricted to satisfy $\mathbf{x}_k^T W \mathbf{x}_k \leq 1$. Hence [74], we can always take W large enough, so that the allowable values for $\mathbf{x}_k$ ensure $\mathbf{u}_k = -K\mathbf{x}_k$ is within non-zero input limits ($\underline{\mathbf{u}} < 0, \bar{\mathbf{u}} > 0$). □

Summary: An ellipsoidal set $\mathbf{x}_k^T W \mathbf{x}_k \leq 1$ is invariant for model $\mathbf{x}_{k+1} = \Phi \mathbf{x}_k$ if

$$\Phi^T W \Phi - W \leq 0 \tag{B.19}$$

(Note: For simplicity assuming the origin is the asymptotic value of $\mathbf{x}_k$.)

If there exists an ellipsoidal invariant set, then one can also define an equivalent invariant set within which the closed-loop trajectories always satisfy non-zero constraints.

B.7.3 Maximal volume ellipsoidal sets for constraint satisfaction

Ellipsoidal invariant sets have the advantage of having a simple definition (B.16) which reduces complexity, but they have the disadvantage of being suboptimal in volume. Nevertheless, with some modification of objectives, ellipsoidal sets can also be used to computational advantage [80] and to handle robust problems [74] far more efficiently than with polyhedral sets. Hence this section outlines some methods by which ellipsoidal invariant sets can be computed.

Simple choices of ellipsoidal invariant set

A simple choice of invariant set arises from the level set of the performance index (for the infinite horizon case). It was shown in Section 6.5 that the unconstrained optimum could be written as $J = \mathbf{x}^T P \mathbf{x}$ and moreover that this was a Lyapunov function. Hence P is a suitable candidate for the matrix W of (B.16).

In an equally simple fashion, one could make use of the eigenvalue/vector decomposition of Φ and the observation that the eigenvalues are all modulus less than one to form a suitable ellipsoid.

Both of these choices may be quite conservative in volume by comparison with other ellipsoids and hence are not favoured in general.

The condition (B.17) for invariance can be represented as a linear matrix inequality (LMI) [13, 74].

$$\begin{bmatrix} W^{-1} & \Phi W^{-1} \\ W^{-1}\Phi^T & W^{-1} \end{bmatrix} \geq 0 \tag{B.20}$$

Input constraints ($\underline{\mathbf{u}} = -\overline{\mathbf{u}}$ for simplicity) can be handled using the following observation:

$$
\begin{aligned}
|K_i^T \mathbf{x}|^2 \leq |K_i^T W^{1/2} W^{-1/2} \mathbf{x}|^2 &\leq \|K_i^T W^{1/2}\|_2^2 |W^{-1/2}\mathbf{x}|^2 \\
&\leq (K_i^T W K_i)(\mathbf{x}^T W^{-1}\mathbf{x}) \leq K_i^T W K_i \leq \overline{\mathbf{u}}_i^2
\end{aligned} \tag{B.21}
$$

Hence the constraints $-K_i^T \mathbf{x} \leq \overline{\mathbf{u}}_i$ could be achieved via satisfaction of the LMIs:

$$\begin{bmatrix} W^{-1} & W^{-1}K_i \\ K_i^T W^{-1} & \overline{\mathbf{u}}_i^2 \end{bmatrix} \geq 0; \quad i = 1, 2, \dots \tag{B.22}$$

Finally, the set given in (B.16) is invariant and moreover constraints are satisfied if both LMIs (B.20, B.22) are satisfied and $W > 0$ (positive definite).

Theorem B.3 *The maximum volume invariant ellipsoid such that constraints are guaranteed to be satisfied can be computed from the following optimisation:*

$$max \log \det(W^{-1}) \quad s.t. \quad (B.20, B.22) \tag{B.23}$$

Proof: LMI (B.20) ensures invariance. The set of LMIs in (B.22) ensure constraint satisfaction inside the set. The volume of an ellipsoid is inversely proportional to the product of the eigenvalues, that is, the determinant. □

Remark B.4 *Additional constraints give rise to LMIs similar to (B.22). Hence each separate constraint will give rise to an additional LMI to be satisfied.* **Nevertheless** *it is emphasised that using LMIs (or ellipses) to represent linear constraints can lead to significant conservatism in general so membership of the resulting invariant sets is sufficient for invariance/feasibility but not necessary.*

Remark B.5 *LMI methods have become popular in the literature but still require quite significant computation compared, for instance, to a conventional quadratic programming (QP) optimisation. The real potential is in the application to model uncertainty and nonlinearity. There is not space in this book to discuss this properly and so the reader is referred elsewhere for a detailed study of LMI techniques.*

> **Summary:**
> 1. An ellipsoidal invariant set within which constraints are satisfied is given by $\mathbf{x}_k^T W \mathbf{x}_k \leq 1$ where conditions (B.20, B.22) and $W > 0$ all apply.
> 2. For the certain case the ellipsoidal set is suboptimal in volume and hence its use may unnecessarily restrict the regions within which an MPC algorithm is defined.

B.7.4   Invariance in the presence of uncertainty

One important motivation for introducing ellipsoidal invariant sets was to handle uncertainty; this section demonstrates how that can be achieved.

In practice all systems exhibit some uncertainty, by way of disturbances or parameter uncertainty. Clearly the invariance conditions such as (B.15) may no longer be valid in the presence of uncertainty and the conditions need reformulating.

Uncertainty affects the prediction assumption whereby the system update model could be replaced by:

1. Disturbance uncertainty:

$$\mathbf{x}_{k+1} = \Phi \mathbf{x}_k + \beta_k \tag{B.24}$$

    where β_k is unknown but possibly bounded.

2. Parameter uncertainty:

$$\mathbf{x}_{k+1} = [\Phi + \Delta_\Phi] \mathbf{x}_k \tag{B.25}$$

    where Δ_Φ is bounded.

For convenience hereafter, we will quantify the parameter uncertainty using linear differential inclusions; for instance, let the closed-loop state space matrix be described as:

$$\Phi = \sum_i \mu_i \Phi_i, \quad \sum_i \mu_i = 1, \quad \mu_i \geq 0 \qquad (B.26)$$

Summary: In the presence of uncertainty, the invariance conditions need reformulating so that they apply to the whole uncertainty class. To do this an uncertainty class must be defined such as in (B.24, B.25, B.26).

B.7.4.1 Disturbance uncertainty and tube MPC

The most important observation [22] here is that, perhaps counter to one's intuition, in the presence of disturbances small invariant sets are not possible. This in turn means that for either a well-tuned loop or one with tight constraints it may not be possible to define an invariant set.

The explanation is simple. The invariance condition (B.17) can be written as

$$\mathbf{x}_{k+1}^T W \mathbf{x}_{k+1} \leq \mathbf{x}_k^T W \mathbf{x}_k \qquad (B.27)$$

Now substitute in update model (B.24):

$$
\begin{aligned}
[\mathbf{x}_k^T \Phi^T + \beta^T] W [\Phi \mathbf{x}_k + \beta] &\leq \mathbf{x}_k^T W \mathbf{x}_k \\
\mathbf{x}_k^T \Phi^T W \Phi \mathbf{x}_k + 2\beta^T W \Phi \mathbf{x}_k + \beta^T \beta &\leq \mathbf{x}_k^T W \mathbf{x}_k \\
\mathbf{x}_k^T \Phi^T W \Phi \mathbf{x}_{k+1} &\leq \mathbf{x}_k^T W \mathbf{x}_k - 2\beta^T W \Phi \mathbf{x} - \beta^T \beta
\end{aligned}
$$

Finally consider the case where $\mathbf{x}_k$ is small (or even zero), then the invariance condition reduces to

$$0 \leq 0 - \beta^T \beta \qquad (B.28)$$

and clearly this is inconsistent.

Well-tuned controllers in conjunction with input constraints often result in small invariant sets so one can quickly set up a contradiction such that for certain input limits in the face of uncertainty, an invariant set will not exist for a given fixed state feedback. In the presence of disturbance uncertainty, the invariant set must be big enough so that for a state $\mathbf{x}$ on the boundary, then $\mathbf{x}_k^T [W - \Phi^T W \Phi] \mathbf{x}_k > \beta^T \beta$. Hence the larger the possible disturbance signal, the larger the invariant set needs to be.

It is quite possible that the size of invariant set required comes into conflict with the LMI requirements of Theorem B.3, in particular constraints (B.22); and then a simple invariant set cannot be defined.

It so happens that the most popular approach (so-called tube MPC) has strong analogies to the back off of the previous section [22, 50, 114]. A *tube* is a mathematical representation of the maximum variation, about a nominal, in the predictions due to unknown disturbances. By ensuring the *tube* of predictions satisfies constraints, the true unknown prediction will do so. The downside of the method is that the tubes can be quite large for representative disturbances and this is equivalent to deploying a large back off which then greatly restricts the control freedom for performance.

Summary: In the presence of disturbances and constraints, it may be difficult to define a useful invariant set. This is because invariance requires the natural *change* in the state to be larger than the effect of the disturbance which may only be possible for *large* sets.

B.7.4.2 Parameter uncertainty and ellipsoidal methods

The case of parameter uncertainty can be handled more easily than disturbances because the impact of parameter uncertainty is proportional to the magnitude of the state, whereas a disturbance signal is not. Hence one can obtain realistic invariant sets.

Consider the condition (B.17) for invariance for a certain process. This must be satisfied for each member of the class of uncertainty (B.26), that is:

$$\Phi_i^T W \Phi_i - W < 0; \quad \forall i \tag{B.29}$$

This gives rise to a number of LMI conditions analogous to (B.20). There exist [13] many efficient tools for solving the resulting LMI problems.

Remark B.6 *Although one can easily state the LMI conditions for an invariant ellipsoid to exist in the case of parameter uncertainty. However, this does not need to imply the conditions can be satisfied or that the implied computation is simple.*

ILLUSTRATION: Algorithm of Kothare et al [74]

In [74] the authors took the premise of a linear time varying (LTV) process where the state matrices $A(k), B(k)$ lay somewhere inside a polytope. However, its time variation was assumed unknown. Let the vertices of such a polytope be given by A_i, B_i, $i = 1, \ldots$ In order to guarantee convergence, the objective was to find a state feedback (full state knowledge was assumed) such that

$$|\lambda(A_i - B_i K)| < 1, \quad \forall i \tag{B.30}$$

Given that $\Phi_i = A_i - B_i K$, choosing K is non-trivial so the problem was replaced by an alternative test. Define an ellipse as

$$S = \{\mathbf{x} : \mathbf{x}^T P \mathbf{x} \le 1\} \tag{B.31}$$

where P is yet to be selected. Now S is invariant if $\mathbf{x}_k \in S \Rightarrow \mathbf{x}_{k+1} \in S$ for each model in the set. Therefore the test for stability of the uncertain system is equivalent to the test

$$\Phi_i^T P \Phi_i - P < 0, \quad \forall i; \quad P > 0 \tag{B.32}$$

The objective then is to parameterise K in such a way that one can optimise the predicted performance subject to inequalities (B.32) and subject to $\mathbf{u} = -K\mathbf{x}$ satisfying input constraints (see Eqn.B.22). The d.o.f. in the optimisation are both K and P, as the only requirement on P is that it is positive definite.

One then finds that the optimal K is time varying, so as the state moves nearer the origin K becomes gradually more highly tuned. The downsides of this algorithm are that:

1. The computation of K arises from a very involved LMI computation, at each sampling instant.

2. At each step, it is assumed that in the predictions the control law is linear. However, it is known that, in general, the optimal trajectory (during constraint handling) is nonlinear during the transients.

There are two main branches to the use of LMIs and robust ellipsoidal invariance:

1. The first branch [74] allowed the feedback K to be the d.o.f. and searched for a K such that the current state was inside an invariant set for the given constraints and model uncertainty. This could give rise to cautious control and requires a significant on-line computation.

2. The second approach predetermined the K and found the maximum volume invariant set (for the given number of d.o.f.) [81]. This could give better performance and required only a small on-line computation [80], but feasibility was restricted by the off-line assumption on the underlying K.

3. Most results are still restricted to ellipsoidal sets and this is a severe restriction given that realistic MAS are polyhedral.

Summary: One can formulate an MPC algorithm that handles constraints with a guarantee of recursive feasibility and convergence and also allows for parameter uncertainty. However, the on-line computational load may be large and the associated theory is demanding. Typical algorithms are based on ellipsoidal regions and thus are suboptimal in volume.

To obtain a guarantee of robust stability in the presence of constraints, it is likely that the associated algorithm will give conservative performance and be valid only within a quite restricted region.

B.8   Conclusion

This chapter has given a brief insight into some, but by no means all, of the topics that occupy the minds of MPC researchers. It will be clear from this that while the approaches and algorithms of Chapters 1-9 are largely sufficient for many industrial scenarios, they are still relatively limited in scope. It is still possible for a naive use of MPC to result in poor behaviour because of an inadequate understanding of the links between constraints and uncertainty, ineffective use of advance knowledge, poor choices of input parameterisations and/or objective functions and so on. A key point is to understand the limitations of the assumptions made and the repercussions of these assumptions failing.

Having said that, in the author's view there is rarely a need to adopt the more complex solutions in the literature which offer rigorous assurances of convergence and feasibility, but at the cost of much increased computation and potentially severe suboptimality. Often practical and commonsense modifications are sufficient as is evident from the widespread adoption of techniques such as DMC.

C

Notation, models and useful background

C.1   Guidance for the lecturer/reader

This appendix is for reference only and thus defines core notation and modelling assumptions adopted throughout the book. There is no associated assessment or learning outcomes and it is assumed that readers who have taken any control course will find the material standard.

C.2   Notation for linear models

This book focusses on predictive control with linear models and thus a brief summary of typical linear models is given for completeness. Simple manipulation and algebra requires linear models as then superposition can be used.

1. MPC deploys linear models whenever these are good enough.

2. These resources do not discuss nonlinear models and the associated MPC [55] as there is huge variety in that topic and it is far beyond normal taught course level.

Typical linear models are transfer function or state space or step response models. While processes operate in continuous time and indeed so do classical control laws, decision making tends to be more of a discrete process. Decision making requires processing time and thus cannot be instantaneous, especially where one is considering interacting inputs/outputs/constraints and performance. Therefore, common predictive control laws are implemented in discrete time. While continuous time variants exist in the literature, the author is not aware of much take up in industry.

C.2.1   State space models

This section gives the terminology adopted in this book for representing state space models and typical modelling assumptions used in MPC. The assumption will be made that the reader is familiar with state space models.

Using the notation of $(.)_k$ and/or $(.)(k)$ to imply a value at the kth sampling instant, the state space model is given as:

$$
\underbrace{\begin{bmatrix} x_1(k+1) \\ x_2(k+1) \\ \vdots \\ x_n(k+1) \end{bmatrix}}_{\mathbf{x}_{k+1}} = \underbrace{\begin{bmatrix} a_{1,1} & a_{1,2} & \cdots & a_{1,n} \\ a_{2,1} & a_{2,2} & \cdots & a_{2,n} \\ \vdots & \vdots & \vdots & \vdots \\ a_{n,1} & a_{n,2} & \cdots & a_{n,n} \end{bmatrix}}_{A} \underbrace{\begin{bmatrix} x_1(k) \\ x_2(k) \\ \vdots \\ x_n(k) \end{bmatrix}}_{\mathbf{x}_k} +
$$
$$
\underbrace{\begin{bmatrix} b_{1,1} & b_{1,2} & \cdots & b_{1,m} \\ b_{2,1} & b_{2,2} & \cdots & b_{2,m} \\ \vdots & \vdots & \vdots & \vdots \\ b_{n,1} & b_{n,2} & \cdots & b_{n,m} \end{bmatrix}}_{B} \underbrace{\begin{bmatrix} u_1(k) \\ u_2(k) \\ \vdots \\ u_m(k) \end{bmatrix}}_{\mathbf{u}_k} \tag{C.1}
$$

$$
\underbrace{\begin{bmatrix} y_1(k) \\ y_2(k) \\ \vdots \\ y_l(k) \end{bmatrix}}_{\mathbf{y}_k} = \underbrace{\begin{bmatrix} c_{1,1} & c_{1,2} & \cdots & c_{1,n} \\ c_{2,1} & c_{2,2} & \cdots & c_{2,n} \\ \vdots & \vdots & \vdots & \vdots \\ c_{l,1} & c_{l,2} & \cdots & c_{l,n} \end{bmatrix}}_{C} \underbrace{\begin{bmatrix} x_1(k) \\ x_2(k) \\ \vdots \\ x_n(k) \end{bmatrix}}_{\mathbf{x}_k}
$$

$$
+ \underbrace{\begin{bmatrix} d_{1,1} & d_{1,2} & \cdots & d_{1,m} \\ d_{2,1} & d_{2,2} & \cdots & d_{2,m} \\ \vdots & \vdots & \vdots & \vdots \\ d_{l,1} & d_{l,2} & \cdots & d_{l,m} \end{bmatrix}}_{D} \underbrace{\begin{bmatrix} u_1(k) \\ u_2(k) \\ \vdots \\ u_m(k) \end{bmatrix}}_{\mathbf{u}_k}
$$

In abbreviated form the model is

$$
\mathbf{x}_{k+1} = A\mathbf{x}_k + B\mathbf{u}_k; \quad \mathbf{y}_k = C\mathbf{x}_k + D\mathbf{u}_k \tag{C.2}
$$

or equivalently

$$
\mathbf{x}(k+1) = A\mathbf{x}(k) + B\mathbf{u}(k); \quad \mathbf{y}(k) = C\mathbf{x}(k) + D\mathbf{u}(k) \tag{C.3}
$$

where $\mathbf{x}$ denotes the state vector (dimension n), $\mathbf{y}$ (dimension l) denotes the process outputs (or measurements) to be controlled, $\mathbf{u}$ (dimension m) denotes the process inputs (or controller output) and A, B, C, D are the matrices defining the state space model.

Remark C.1 *Ordinarily for real processes $D = 0$. Also, in this book we will normally replace the output equation by*

$$
\mathbf{y}(k) = C\mathbf{x}(k) + \mathbf{d}(k) \tag{C.4}
$$

where $\mathbf{d}(k)$ represents an output disturbance. Typically this value is unknown and indeed the parameters are uncertain and thus $\mathbf{d}_k$ will be estimated to ensure offset-free predictions in the presence of some uncertainty (e.g., see Section 2.4.3).

C.2.2 Transfer function models single-input-single-output and multi-input-multi-output

A popular model [25] is the so-called controlled auto-regressive integrated moving average (CARIMA) model.

1. This subsumes in its structure many other popular forms.
2. It is used because the uncertainty is included in a way that is a good representation of slowly varying disturbances that could have a non-zero steady-state (ζ_k is assumed to be a zero mean random variable).

A CARIMA model is given as:

$$a(z)y(z) = b(z)u(z) + T(z)\underbrace{\frac{1}{\Delta(z)}\zeta(z)}_{d(z)} \tag{C.5}$$

where ζ_k is an unknown zero mean random variable which can represent disturbance effects and measurement noise simultaneously. Although there exist modelling techniques to compute *best fit* values for the parameters of $a(z), b(z)$ and $T(z)$, in MPC it is commonplace to treat $T(z)$ as a design parameter (e.g., [25, 196]) since it has direct effects on loop sensitivity and so one may get better closed-loop performance with a $T(z)$ which is notionally not the best fit. It is common to write transfer function models in the equivalent difference equation form. For instance with $T = 1$, (C.5) is given as

$$y_{k+1} + a_1 y_k + \cdots + a_n y_{k-n+1} = b_1 u_k + b_2 u_{k-1} + \cdots + b_n u_{k-n+1} + d_{k+1} \tag{C.6}$$

where $a(z) = 1 + a_1 z^{-1} + \cdots + a_n z^{-n}$, $b(z) = b_1 z^{-1} + \cdots + b_n z^{-n}$ and d_k is the unknown disturbance term derived from $d(z) = \dfrac{T(z)}{\Delta(z)}\zeta(z)$ where ζ_k is assumed to be a zero mean unknown variable.

The choice of $T(z) = 1$ gives an equivalent disturbance model:

$$d(z) = \frac{1}{\Delta(z)}\zeta(z) \quad \equiv \quad d_{k+1} = d_k + \zeta_k \tag{C.7}$$

which is in essence a random walk (integrated white noise).

C.2.3 Author's MATLAB notation for SISO transfer function and MFD models

The author uses a particular notation in his MATLAB® code which may not always match that used by others, but has the advantage of being convenient and simple to code/understand.

In the multivariable or MIMO case, this book uses a matrix fraction description (MFD) as follows:

$$D(z)\mathbf{y}(z) = N(z)\mathbf{u}(z) + T(z)\underbrace{\zeta(\mathbf{z})/\Delta(z)}_{\mathbf{d}(z)} \tag{C.8}$$

where $D(z)$ and $N(z)$ are matrix polynomials. It is possible for $T(z)$ to be a matrix polynomial, but this book does not pursue modelling to that level of detail and assumes $T(z)$ is scalar for convenience.

1. For a SISO polynomial $a(z)$, the matching of coefficients of z^{-i} and parameters in a MATLAB array is as follows:

$$a(z) = 1 + a_1 z^{-1} + \cdots + a_n z^{-n} \quad \equiv \quad \underbrace{a = [1, a_1, \cdots, a_n]}_{MATLAB} \tag{C.9}$$

Hence, in terms of MATLAB, the first coefficient $a(1)$ is always the coefficient a_0 of z^0, the second coefficient $a(2)$ is always the coefficient a_1 of z^{-1} and so forth. Model order is not needed!

2. The notation for the MFD case is almost identical to the SISO case with one minor exception, the array in MATLAB needs to be thought of as a series of matrices stacked together:

$$D(z) = D_0 + D_1 z^{-1} + \cdots + D_n z^{-n} \quad \equiv \quad \underbrace{D = [D_0, D_1, \cdots, D_n]}_{MATLAB} \quad \text{(C.10)}$$

Hence, in terms of MATLAB, for an $m \times m$ system, the first block of the array D, that is $D(:, 1:m)$, represents the coefficient D_0 of z^0, the second block of the array, that is $D(:, m+1 : 2*m)$, represents the coefficient D_1 of z^{-1} and so forth. So, for example:

$$D(z) = \begin{bmatrix} 1 & 0 \\ 0 & 1 \end{bmatrix} + \begin{bmatrix} -1.4 & 0 \\ 0.2 & 1.2 \end{bmatrix} z^{-1} + \begin{bmatrix} 0.52 & 0 \\ -0.1 & 0.6 \end{bmatrix} z^{-2}$$

$$Q(z) = \begin{bmatrix} 3 & 0 \\ 0 & 2 \end{bmatrix} + \begin{bmatrix} 1 & 0.3 \\ -2.1 & 0 \end{bmatrix} z^{-1} + \begin{bmatrix} -0.2 & 1.1 \\ 1 & 0.4 \end{bmatrix} z^{-2}$$

IN MATLAB,
$D = [1, 0, -1.4, 0, 0.52, 0; 0, 1, 0.2, 1.2, -0.1, 0.6];$
$Q = [3, 0, 1, 0.3, -0.2, 1.1; 0, 2, -2.1, 0, 1, 0.4];$

Remark C.2 *The author assumes strictly proper processes only, so where a polynomial is* **known** *to represent a process numerator, the first coefficient is always zero and hence this is excluded from the code to avoid carrying around spurious zeros:*

$$G = \frac{2z^{-1} + 3z^{-2}}{1 - 1.6z^{-1} + 0.64z^{-2}} \quad \begin{cases} a(z) = 1 - 1.6z^{-1} + 0.64 \\ b(z) = 2z^{-1} + 3z^{-2} \end{cases}$$

In MATLAB, the author would use:

$$a = [1, -1.6, 0.64]; \quad b = [2, 3];$$

Remark C.3 *For a system with a delay, add zero coefficients to the numerator polynomial, so for example:*

$$b(z) = b_3 z^{-3} + \cdots + b_n z^{-n} \quad \equiv \quad \underbrace{b = [0, 0, b_3, \cdots, b_n]}_{MATLAB} \quad \text{(C.11)}$$

C.2.4 Equivalence between difference equation format and vector format

Readers will have noted that the difference equation

$$y_{k+1} + a_1 y_k + \cdots + a_n y_{k-n+1} = b_1 u_k + b_2 u_{k-1} + \cdots + b_n u_{k-n+1} + d_{k+1} \quad (C.12)$$

can equivalently be written in terms of z-transforms (and for simplicity here ignoring the unknown d_{k+1} term) as:

$$[1 + a_1 z^{-1} + \cdots a_n z^{-n}] y(z) = [b_1 z^{-1} + \cdots + b_n z^{-n}] u(z)$$

This equivalence is exploited in the definition of the GPC control law of Chapter 4 where an analogy is used as follows:

$$[f, g, h, \cdots] \begin{bmatrix} x_k \\ x_{k-1} \\ x_{k-2} \\ \vdots \end{bmatrix} \equiv [f + g z^{-1} + h z^{-2} + \cdots] x(z) \quad (C.13)$$

C.2.5 Step response models

A step response model is a particular form of transfer function model. This is popular because the step response is a readily available (measureable) characteristic for many process systems. However, it has the weakness of requiring an infinite number of parameters as with non-zero steady-state gain, the parameters H_i tend to the steady-state gain.

$$\mathbf{y}_k = H(z)\Delta\mathbf{u}_k + \mathbf{d}_k; \quad H(z) = \sum_{i=0}^{\infty} H_i z^{-i} \quad (C.14)$$

C.2.6 Sample rates

The sample rate dictates the frequency at which control decisions are updated. This must be fast enough to detect and react to disturbances and/or changes in target. In simple terms, the time between samples is the time in which no detection and/or reaction can take place. Conversely, the sample rate must not be too fast as this can result in over specifying the computation requirements and also pushes all the discrete poles towards the unit circle, which means excessive prediction horizons are needed to capture the key dynamics.

ILLUSTRATION: Typical sample rate for MPC

A typical argument is that one wants around 10 sample points within a typical response settling time or rise time as seen in this figure. [Vertical lines represent sampling points.]

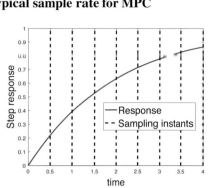

ILLUSTRATION: Faster than typical sample rate for MPC

Using too high a sample rate is pointless as the system output cannot respond to fast input changes; here it is noted that the output response is largely unresponsive to the rapidly changing input signal. Moreover, fast sampling increases the number of decision variables, which is generally undesirable.

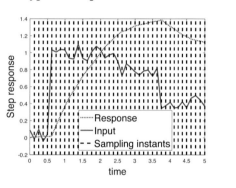

A common requirement is that the prediction horizon should be greater than the settling time (or key system dynamics) and thus typically horizons around 10-20 samples are chosen as this means the inter-sample delay is relatively small but there is sufficient gain in each input value to make a difference. If faster disturbance rejection/detection is required, one can sample faster but this likely to conflict with the other functionality of MPC and a better solution might be some form of cascade with an inner continuous time controller to give fast disturbance rejection.

C.2.7 Summary

The more modern literature tends to focus mainly on state space models as they handle multivariable systems more cleanly and indeed the algebra for constraint handling and dual-mode prediction tends to be much easier. Nevertheless, transfer function models may be more convenient in the SISO case or where ease of modelling suggests a preference for step responses, and thus are still a good start point for understanding basic principles.

Readers may also come across the term independent model. This represents a particular way of doing disturbance estimation and prediction, but the independent model itself can take whatever form is convenient.

C.3   Minimisation of functions of many variables

A typical predictive control law is defined using a gradient operator and thus it is useful for readers unfamiliar with this to see a quick derivation. For full explanations any advanced engineering mathematics textbook is recommended. Readers may also find it useful to remind themselves of the simple requirements for a minimum of a function of a single variable (i.e., 1st derivative is zero and 2nd derivative is positive).

C.3.1   Gradient operation and quadratic functions

Consider a function $f(.)$ of many variables:

$$f(x_1, x_2, \cdots, x_n); \quad \mathbf{x} = [x_1, x_2, \cdots, x_n]^T \tag{C.15}$$

The gradient of f is defined by constructing a vector of partial derivatives with respect to each independent variable, and hence:

$$\text{grad} f(x_1, x_2, \cdots, x_n) = \begin{bmatrix} \frac{\partial f}{\partial x_1} \\ \frac{\partial f}{\partial x_2} \\ \vdots \\ \frac{\partial f}{\partial x_n} \end{bmatrix} \tag{C.16}$$

At a minimum or maximum point, the gradient must be zero in every direction, thus a necessary (but not sufficient) condition for a minimum is that grad $f = 0$.

In general, this book is going to consider only multivariable quadratic functions of the type:

$$f = \mathbf{x}^T S \mathbf{x} + \mathbf{x}^T \mathbf{p} + q; \quad S > 0 \tag{C.17}$$

Remark C.4 *A function which is defined as the sum of squares of terms e_i where each term is linear in the variables x_i will give a function of the form of (C.17).*

$$f = \sum_i \underbrace{(\alpha_i - \mathbf{v}_i^T \mathbf{x})^2}_{e_i^2} = \mathbf{x}^T \underbrace{\left[\sum_i \mathbf{v}_i^T \mathbf{v}_i\right]}_{S} \mathbf{x} - 2\mathbf{x}^T \underbrace{\left[\sum_i \mathbf{v}_i \alpha_i\right]}_{0.5\mathbf{p}} + \underbrace{\sum_i \alpha_i^2}_{q} \tag{C.18}$$

If is clear that $f \geq 0$ and also $S > 0$.

Remark C.5 *The condition that $S > 0$ means that f represents an n-dimensional el- lipsoid (for example a function something like $f \equiv k_1 (x_1 - a_1)^2 + k_n (x_n - a_n)^2 + \cdots +$*

$k_n(x_n - a_n)^2 + K)$ *and therefore there is known to be a single stationary point which is also a minimum. This book will not prove this as it is elementary matrix/function algebra.*

Summary: The function f given in (C.17) has a unique minimum which is computed where $\operatorname{grad} f = 0$.

C.3.2 Finding the minimum of quadratic functions of many variables

This section will demonstrate that the computation of the gradient (C.16) of function (C.17) is straightforward. To do this, a few simple results are derived first.

- Gradient of a linear term of $\mathbf{x}$.

$$\operatorname{grad}(\mathbf{v}^T \mathbf{x}) = \operatorname{grad}(\mathbf{x}^T \mathbf{v}) = \operatorname{grad}(v_1 x_1 + v_2 x_2 + \cdots + v_n x_n) = \begin{bmatrix} v_1 \\ v_2 \\ \vdots \\ v_n \end{bmatrix} = \mathbf{v} \quad (C.19)$$

- Gradient of a quadratic term of $\mathbf{x}$ uses a standard product rule in conjunction with (C.19) (readers can also derive this by expanding out in full noting that by definition $S = S^T$).

$$\operatorname{grad}(\mathbf{x}^T S \mathbf{x}) = \operatorname{grad}(\mathbf{x}^T \underbrace{[S\mathbf{x}]}_{\mathbf{v}}) + \operatorname{grad}(\underbrace{[\mathbf{x}^T S]}_{\mathbf{v}^T} \mathbf{x})$$

$$= S\mathbf{x} + [\mathbf{x}^T S]^T = 2S\mathbf{x} \quad (C.20)$$

- Combining the results (C.19, C.20)

$$\operatorname{grad}[\mathbf{x}^T S \mathbf{x} + \mathbf{x}^T \mathbf{p} + q] = 2S\mathbf{x} + \mathbf{p} \quad (C.21)$$

Summary: The minimum of a quadratic function f of many variables is determined by solving for the gradient is zero. This reduces to the following algebra:

$$\operatorname{grad}[\mathbf{x}^T S \mathbf{x} + \mathbf{x}^T \mathbf{p} + q] = 0 \quad \Rightarrow \quad \mathbf{x} = -0.5 S^{-1} \mathbf{p}; \quad S > 0 \quad (C.22)$$

where q is independent of $\mathbf{x}$ and hence does not affect the optimisation.

Common Abbreviations	
Abbreviation	Definition
ASM	Active set method
CARIMA	Controlled auto-regressive integrated moving average
CLP/OLP	Closed-loop paradigm/Open-loop paradigm
d.o.f.	degrees of freedom within an optimisation
DMC	Dynamic matrix control
FIR	Finite impulse response
GPC/GPCT	Generalised predictive control without/with a T-filter
LP/QP	Linear program/Quadratic programming
LQR	Linear quadratic regulator
OMPC/SOMPC	Optimal/Suboptimal model predictive control
MAS	Maximal admissible set
MCAS	Maximal controller admissible set
MFD	Matrix fraction description
MIMO/SISO	Multi-input-multi-output/Single-input-single-output
MPC	Model predictive control
MPQP	Multiparametric quadratic programming
PFC	Predictive functional control
s.t./w.r.t.	subject to/ with respect to

TABLE C.1
Typical abbreviations adopted in the book.

C.4 Common notation

The most commonly used notation adopted in this book is given in Tables C.1 and C.2. Bold lowercase is used to denote vectors and non-bold lowercase implies a scalar quantity. Capitals are used for matrices. Arguments such as $(.)(z)$, $(.)_k$ and $(.)(k)$ are often omitted to improve readability where their presence is implicit.

Notation and Variable Names		
Description	Typical Notation	Alternates
Laplace transform operator	s	
Continuous time transfer function	$G(s)$	
Unit delay operator	z^{-1}	
Process outputs, inputs, states and set point	y, u, x, r	$\mathbf{y}, \mathbf{u}, \mathbf{x}, \mathbf{r}$
Process incremental input ($u_k - u_{k-1}$)	Δu	$\mathbf{\Delta u}$
Process disturbance/Measurement noise	d, v	$\mathbf{d}, \mathbf{v}$
State space matrices	A, B, C, D	
Open-loop pole polynominal (SISO)	$a(z)$	a
Closed-loop pole polynomial	$p_c(z)$	p_c
Open-loop zero polynomial (SISO)	$b(z)$	b
Open-loop pole polynominal (MIMO)	$D(z)$	a
Open-loop zero polynomial (MIMO)	$N(z)$	N
Vector of coefficients of polynomial $n(z)$	$\mathbf{n}$	
Controller numerator/denominator	$N_k(z), D_k(z)$	N_k, D_k
Difference operator	$\Delta(z) = 1 - z^{-1}$	Δ
State feedback gain	K	
Process model	$G(z)$	G
Toeplitz/Hankel matrices of $n(z)$	C_n/H_n	Γ_n/H_n
Value of signal $\mathbf{x}$ at sampling instant k	$\mathbf{x}_k$	$\mathbf{x}(k)$
Vector of future/past values of $\mathbf{x}_k$	$\underset{\rightarrow}{\mathbf{x}}_k / \underset{\leftarrow}{\mathbf{x}}_k$	
ith standard basis vector $\mathbf{e}_i$	$\mathbf{e}_1 = [1,0,0,\cdots]^T$	$\mathbf{e}_i$
Matrix E_1	$E_1 = [I,0,0,\cdots]^T$	
Matrix L	$L = [I,I,\cdots,I]^T$	
Output prediction horizon	n_y	
Input prediction horizon/number of d.o.f.	n_u	n_c
Input weighting	λ	
Upper and lower limits	$\overline{(.)}, \underline{(.)}$	
Expected value of signal x	$E[x]$	
Terminal region	S_T	

TABLE C.2
Typical notation adopted in the book. The dimensions of some of these matrices and vectors are modified automatically to the context.

References

[1] M. Abdullah, J.A. Rossiter, Utilising Laguerre Function in Predictive Functional Control to Ensure Prediction Consistency, UKACC Control 2016 (available in IEEE Xplore).

[2] J.C. Allwright, On mini-max model based predictive control, in workshop on Advances in Model Based Predictive Control, Oxford, 246-255, 1993.

[3] J. Alvarez-Ramirez and R. Suarez, Global stabilisation of discrete time linear systems with bounded inputs, Int. J. ACSP, 10, 409-416, 1996.

[4] T.A. Badgwell, A robust model predictive control algorithm for stable linear plant, American Control Conference, 1997.

[5] R.R. Bitmead, M. Gevers and V. Werrtz, Adaptive optimal control: The thinking man's GPC, Prentice Hall International, 1990.

[6] F. Blachini, Set invariance in control, Automatica, 35, 1747-1767, 1999.

[7] A. Bemporad, A. Casavola and E. Mosca, Nonlinear control of constrained linear systems via predictive reference management, IEEE Trans. on Automatic Control, 42, 3, 340-349, 1997.

[8] A. Bemporad, M. Morari, V. Dua and E.N. Pistokopoulos, The explicit linear quadratic regulator for constrained systems, Automatica, 38, 1, 3-20, 2002.

[9] A. Bemporad, F. Borrelli and M. Morari, The explicit solution of constrained LP-based receding horizon control, European Control Conference, Oporto, 2001.

[10] T.J.J. van den Boom and R.A. J de Vries, Cascade predictive controllers, IFAC World Congress, 1996.

[11] T.J.J. van den Boom and R.A. J de Vries, Constrained predictive control using a time varying youla parameter: a state space solution, European Control Conference, 1995.

[12] F. Borrelli, Constrained Optimal Control of Linear and Hybrid Systems, Springer, 2003.

[13] S. Boyd, L. El Ghaoui, E. Feron and V. Balakrishnan, Linear Matrix Inequalities, System and Control Theory, SIAM, Philadelphia, 1996.

[14] R. Boucher, Simulink model kindly supplied by Prismtech Ltd., Gateshead, UK, 1999.

[15] A.E. Bryson, Optimal Control:1950 to 1985, IEEE Control Systems, 16, 3, 26-33, 1996.

[16] E.F. Camacho and C. Bordons, Model Predictive Control, Springer, 1999.

[17] M. Cannon and B. Kouvaritakis, Fast suboptimal predictive control with guaranteed stability, Systems and Control Letters, 35, 19-29, 1998.

[18] M. Cannon, B. Kouvaritakis and J.A. Rossiter, Efficient active set optimisation in triple mode MPC, IEEE Transactions on Automatic Control, 46(8), 1307-1313, 2001.

[19] A. Casavola, M. Giannelli and E. Mosca, Global predictive regulation of null-controllable input saturated linear systems, IEEE Trans. on Automatic Control, 44, 11, 2226-2230, 1999.

[20] T. Chen and L. Qiu, $\mathcal{H}_\infty$ design of general multirate sampled-data control systems, Automatica, 30, 7, 1139-1152, 1994.

[21] L. Chisci and G. Zappa, Robust predictive control via controller invariant sets, IEEE Mediterranean Conference on Control and Automation, 1998.

[22] L. Chisci, J. A. Rossiter, G. Zappa Systems with persistent disturbances: Predictive control with restricted constraints, Automatica, 37, 7, 2001.

[23] L. Chisci, J.A. Rossiter and G. Zappa, Robust predictive control with restricted constraints to cope with estimation errors, ADCHEM 2000 (Pisa), 2000.

[24] D.W. Clarke and P. Gawthrop, Self tuning controller, *Proc. IEE*, 122, 922–934, 1975.

[25] D.W. Clarke, C. Mohtadi and P.S. Tuffs, Generalised predictive control, Parts 1 and 2, Automatica, 23, 137-160, 1987.

[26] D.W. Clarke and R. Scattolini, Constrained receding horizon predictive control, Proceedings IEE, Pt. D, 138, 4, 347-354, 1991.

[27] J. Currie, Practical applications of industrial optimization: from high-speed embedded controllers to large discrete utility systems, PhD thesis, University of Auckland, 2014.

[28] C.R. Cutler and B.L. Ramaker, Dynamic matrix control - a computer control algorithm, American Control Conference, 1980.

[29] J.A. De Dona, M.M. Seron, D.Q. Mayne and G.C. Goodwin, Enlarged terminal sets guaranteeing stability of receding horizon control, Systems and Control Letters, 47, 57-63, 2002.

[30] B. De Moor, P. Van Overschee and W. Favoreel, Numerical algorithms for subspace state space system identification – an overview, Birkhauser, 1998.

[31] Y. Ding and J. A. Rossiter, Compromises between feasibility and performance within linear MPC, IFAC World Congress, 2008.

[32] R.C. Dorf and R.H. Bishop, Modern control systems, Prentice Hall, 2001.

[33] G. De Nicolao, L. Magni and R. Scattolini, Stabilising receding horizon control of non-linear time varing systems, IEEE Transactions on Automatic Control, 43, 1030-1036, 1998.

[34] Y. Ding, J.A. Rossiter, Compromises between feasibility and performance within linear MPC, IFAC World Congress, 2008.

[35] S.S. Dughman and J.A. Rossiter, Systematic and effective embedding of feedforward information into MPC, International Journal of Control, 2018, http://dx.doi.org/10.1080/00207179.2017.1281439.

[36] W. Favoreel, B. De Moor and M. Gevers, SPC: subspace predictive control, Proceedings IFAC World Congress, Beijing, 235-240, 1999.

[37] A. Feuer and G.C. Goodwin, Sampling in Digital Processing and Control, Wiley, 1997.

[38] M. Fikar, S. Engell and P. Dostal. Design of infinite horizon predictive LQ controller, European Control Conference, Brussels, 1997.

[39] C.E. Garcia and M. Morari, Internal Model control 1. A unifying review and some new results, I&EC Process Design and Development, 21, 308-323, 1982.

[40] C.E. Garcia and A.M. Morshedi, Quadratic programming solution of dynamic matrix control (QDMC), Chem. Eng. Commun, 46, 73-87, 1986.

[41] C.E. Garcia, D.M. Prett and M. Morari, Model predictive control: theory and practice, a survey, Automatica, 25, 335-348, 1989.

[42] E.G. Gilbert and K.T. Tan, Linear systems with state and control constraints: the theory and application of maximal output admissable sets, IEEE Trans. AC, 36, 9, 1008-1020, 1991.

[43] E.G. Gilbert and I. Kolmanovsky, Discrete-time reference governors and the non-linear control of systems with state and control constraint, Int. J. Rob. and Non-linear Control, 5, 487-504, 1995.

[44] E.G. Gilbert and I. Kolmanovsky, Discrete time reference governors for systems with state and control constraints and disturbance inputs, Conference on Decision and Control, 1189-1194, 1995.

[45] E.G. Gilbert, I. Kolmanovsky and K.T. Tan, Nonlinear control of discrete time linear systems with state and control constraints: a reference governor with global convergence properties, Conference on Decision and Control, 144-149, 1994.

[46] E. G. Gilbert and I. Kolmanovsky, Discrete-time reference governors and the non-linear control of systems with state and control constraint, Int. J. Robust and Non-linear Control, 5, 487-504, 1995.

[47] E.G. Gilbert and I. Kolmanovsky, Fast reference governors for systems with state and control constraints and disturbance inputs, Int. Journal of Robust and Non-linear Control, 9, 1117-1141, 1999.

[48] R. Gondhalekar, J. Imura and K. Kashima, Controlled invariant feasibility? A general approach to enforcing strong feasibility in MPC applied to move-blocking, Automatica, 45, 12, 2869-2875, 2009.

[49] C. Greco, G. Menga, E. Mosca and G. Zappa, Performance improvements of self-tuning controllers by multistep horizons: the MUSMAR approach, Automatica, 20, 681-699, 1984.

[50] J.R. Gossner, B. Kouvaritakis and J.A. Rosssiter, Stable Generalised predictive control in the presence of constraints and bounded disturbances, Automatica, 33, 4, 551-568, 1997.

[51] J. R. Gossner, B. Kouvaritakis and J.A. Rossiter, Cautious stable predictive control: a guaranteed stable predictive control algorithm with low input activity and good robustness, 3rd IEEE Symposium on New Directions in Control and Automation, Cyprus, 243-250, Vol. II, 1995.

[52] J. R. Gossner, B. Kouvaritakis and J.A. Rossiter, Cautious stable predictive control: a guaranteed stable predictive control algorithm with low input activity and good robustness, International Journal of Control, 67, 5, 675-697, 1997.

[53] M.J. Grimble, Polynomial systems approach to optimal linear filtering and prediction, International Journal of Control, 41, 1545-1566, 1985.

[54] M.J. Grimble, H_∞ optimal multichannel linear deconvolution filters, predictors and smoothers, International Journal of Control, 63, 3, 519-533, 1986.

[55] L. Grune and J. Panneck, Nonlinear model predictive control, Springer, 2017.

[56] P.-O. Gutman and M. Cwikel, An algorithm to find maximal state constraint sets for discrete time linear dynamical systems with bounded controls and states, IEEE Trans. on Automatic Control, 32, 251-254, 1987.

[57] R. Haber, R. Bars, and U. Schmitz, Predictive control in process engineering: from the basics to the applications, Chapter 11: Predictive Functional Control, Wiley-VCH, Weinheim, Germany, 2011.

[58] H. Hjalmarsson, M. Gevers and F. De Bruyne, For model based control design closed loop identification gives better performance, Automatica, 32, 1659-1673, 1996.

[59] H. Hjalmarsson, M. Gevers and S. Gunnarsson, Iterative feedback tuning: theory and applications, IEEE Control Systems Magazine, 16, 26-41, 1998.

[60] H. Hjalmarsson, Iterative feedback tuning: an overview, International Journal of Adaptive Control and Signal Processing, 16, 373-395, 2002.

[61] Workshop on design and optimisation of restricted complexity controllers, Grenoble, January 2003.

[62] M. Hovd, J.H. Lee and M. Morari, Model requirements for model predictive control, European Control Conference, 2428-2433, 1991.

[63] K. Hrissagis and O.D. Crisalle, Mixed objective optimisation for robust predictive controller synthesis, Proceedings CPC-V, California, 1996.

[64] K. Hrissagis, O.D. Crisalle and M. Sznaier, 1995 Robust design of unconstrained predictive controllers, American Control Conference, 1995.

[65] Lars Imsland, J.A. Rossiter, Bert Pluymers, Johan Suykens, Robust triple mode MPC, International Journal of Control, 81, 4, 679-689, 2008.

[66] P.P. Kanjilal, Adaptive prediction and predictive control, IEE Control Engineering Series 52, 1995.

[67] M. Kamrunnahar, D.G. Fisher and B. Huang, Model predictive control using an extended ARMarkov model, Journal of Process Control, 12, 123-129, 2002.

[68] S. Keerthi and E.G. Gilbert, Computation of minimum time feedback control laws for systems with state control constraints, IEEE Transactions on Automatic Control, 36, 1008-1020, 1988.

[69] E.C. Kerrigan and J.M. Maciejowski, Invariant sets for constrained nonlinear discrete-time systems with application to feasibility in model predictive control, Conference on Decision and Control, 2000.

[70] E.C. Kerrigan, Robust constraint satisfaction: Invariant sets and predictive control, Ph.D. Thesis, Cambridge, UK, 2000.

[71] M.T. Khadir, and J.V. Ringwood, Extension of first-order predictive functional controllers to handle higher-order internal models, Int. J. Appl. Math. Comput. Sci., 18(2), 229-239, 2008.

[72] B. Khan and J.A. Rossiter, Alternative Parameterisation within predictive Control: a systematic selection, International Journal of Control, 86, 8, 1397-1409, 2013.

[73] P.P. Khargonekar, K. Poolla and A. Tannenbaum, Robust control of linear time-invariant plants using periodic compensation, IEEE Transactions on Automatic Control, 30, 1088-1096, 1985.

[74] M.V. Kothare, V. Balakrishnan and M. Morari, Robust constrained model predictive control using linear matrix inequalities, Automatica, 32, 1361-1379, 1996.

[75] B. Kouvaritakis, J.A. Rossiter and A.O.T. Chang, Stable generalized predictive control: an algorithm with guaranteed stability, Proceedings IEE, Pt. D, 139, 4, 349-262, 1992.

[76] B. Kouvaritakis, J.R. Gossner and J.A. Rossiter, Apriori stability condition for an arbitrary number of unstable poles, Automatica, 32, 10, 1441-1446, 1996.

[77] B. Kouvaritakis, J.A. Rossiter and J.R.Gossner, Improved algorithm for multivariable stable GPC, Proceedings IEE Pt. D, 144, 4, 309-312, 1997.

[78] B. Kouvaritakis and J.A. Rossiter, Multivariable stable generalized predictive control, Proc IEE Pt.D., 140, 5, 364-372, 1993.

[79] B. Kouvaritakis, J.A. Rossiter and M. Cannon, Linear quadratic feasible predictive control, Automatica, 34, 12, 1583-1592, 1998.

[80] B. Kouvaritakis, M. Cannon and J.A. Rossiter, Removing the need for QP in constrained predictive control, ADCHEM 2000 (Pisa), 2000.

[81] B. Kouvaritakis, J.A. Rossiter and J. Schuurmans, Efficient robust predictive control, IEEE Transactions on Automatic Control, 45, 8, 1545-1549, 2000.

[82] B. Kouvaritakis, M.C. Cannon and J.A. Rossiter, Efficient active set optimisation in triple mode MPC, IEEE Transactions on Automatic Control, 46, 8, 1307-1313, 2001.

[83] G.M. Kranc, Input output analysis of multi-rate feedback system, IRE Transactions on Automatic Control, 3, 21-28, 1957.

[84] W.H. Kwon and A.E. Pearson, A modified quadratic cost problem and feedback stabilisation of a linear system, IEEE Transactions on Automatic Control, 22, 5, 838-842, 1978.

[85] I.D. Landau and A. Karimi, Recursive algorithms for identification in the closed loop: a unified approach and evaluation, Automatica, 33, 8, 1499-1523, 1997.

[86] J.H. Lee, Recent advances in model predictive control and other related areas, CPC V (California), 1-21, 1996.

[87] J.H. Lee and M. Morari, Robust inferential control of multi-rate sampled data systems, Chem. Eng Sci., 47, 865-885, 1992.

[88] Y.I. Lee and B. Kouvaritakis, Linear matrix inequalities and polyhedral invariant sets in constrained robust predictive control, International Journal of Non-linear and Robust Control, 1998.

[89] D. Li, S.L. Shah and T. Chen, Identification of fast-rate models from multirate data, International Journal of Control, 74, 7, 680-689, 2001.

[90] W.H. Lio, J.A. Rossiter and B. Ll. Jones, Predictive control layer design on a known output-feedback compensator for wind turbine blade-pitch preview control, to appear, Wind Energy Journal, 2018.

[91] D. Liu, S.L. Shah and D.G. Fisher, Multiple prediction models for long range predictive control, IFAC World Congress (Beijing), 1999.

[92] L. Ljung, System identification: theory for the user, Prentice Hall, 1987.

[93] A.G.J. MacFarlane and B. Kouvaritakis, A design technique for linear multivariable feedback systems, International Journal of Control, 25, 837-874, 1977.

[94] J.M. Maciejowski, Predictive control with constraints, Prentice Hall, 2001.

[95] J.M. Maciejowski, Multivariable feedback design, Addison-Wesley, 1989.

[96] J.M. Martin Sanchez and Rodellar, J., Adaptive predictive control; from concepts to plant optimisation, Prentice Hall International, 1996.

[97] D.Q. Mayne, Control of constrained dynamic systems, European Journal of Control, 7, 87-99, 2001.

[98] D.G. Meyer, A parametrization of stabilizing controllers for multirate sampled-data systems, IEEE Transactions on Automatic Control, 35, 233-236, 1990.

[99] H. Michalska and D. Mayne, Robust receding horizon control of constrained nonlinear systems, IEEE Transactions on Automatic Control, 38, 1623-1633, 1993.

[100] D.Q. Mayne, J.B. Rawlings, C.V. Rao and P.O.M. Scokaert, Constrained model predictive control: Stability and optimality, Automatica, 36, 789-814, 2000.

[101] J.A. Mendez, B. Kouvaritakis and J.A. Rossiter, State space approach to interpolation in MPC, International Journal of Robust and Nonlinear Control, 10, 27-38, 2000.

[102] M. Morari and J.H. Lee, Model predictive control: Past present and future. Computers and Chemical Engineering, 23, 4, 667-682, 1999.

[103] E. Mosca and J. Zhang, Stable redesign of predictive control, Automatica, 28, 6, 1229-1233, 1992.

[104] E. Mosca, Optimal predictive and adaptive control, Prentice Hall, 1995.

[105] E. Mosca, G. Zappa and J.M. Lemos, Robustness of a multipredictor adaptive regulator: Musmar, Automatica, 25, 521-529, 1989.

[106] MPQP software toolbox http://control.ee.ethz.ch/mpt.

[107] K.R. Muske and J.R. Rawlings, Model predictive control with linear models, AIChE Journal, 39, 2, 262-287, 1993.

[108] N.S. Nise, Control systems engineering, Wiley, 2011.

[109] J.P. Norton, An Introduction to Identification, Academic Press, 1992.

[110] T. Perez, G.C. Goodwin and M.M. Seron, Cheap fundamental limitation of input constrained linear systems, IFAC World Congress, 2002.

[111] E. N. Pistikopoulos, M.C. Georgiadis and V. Dua, Multi-parametric programming: theory, algorithms, and applications, Wiley-VCH, 2007.

[112] B. Pluymers, J.A. Rossiter, J.A.K. Suykens, B. De Moor, The efficient computation of polyhedral invariant sets for linear systems with polytopic uncertainty, American Control Conference, 2005.

[113] Y. Pu, M.N. Zeilinger and C. Jones, Fast Alternating Minimization Algorithm for Model Predictive Control. IFAC World Congress, 8-24, 2014.

[114] S.V. Rakovic, A.R. Teel and D.Q. Mayne and A. Astolfi, Simple robust control invariant tubes for some classes of nonlinear discrete time systems, Conference on Decision and Control, 6397-6401, 2006.

[115] C.V. Rao, S.J. Wright and J.B. Rawlings, Application of interior point methods to model predictive control, Journal of Optimisation Theory and Applications, 99, 3, 723-757, 1998.

[116] J.B. Rawlings and K.R. Muske, The stability of constrained receding horizon control, Trans. IEEE AC, 38, 1512-1516, 1993.

[117] J.B. Rawlings, E.S. Meadows and K.R. Muske, Nonlinear model predictive control: A tutorial and survey, Proceedings ADCHEM, 1994.

[118] J.B. Rawlings, Tutorial overview of model predictive control, IEEE Control Systems Magazine, 20, 3, 38-52, 2000.

[119] J.B. Rawlings, D Bonne, J.B. Jorgensen, A.N. Venkat and S.B. Jorgensen, Unreachable Setpoints in Model Predictive Control, IEEE Transactions on Automatic Control, 53, 9, 2209-2215, 2008.

[120] J. Richalet, A. Rault, J.L. Testud and J. Papon, Model predictive heuristic control: applications to industrial processes, Automatica, 14, 5, 413-428, 1978.

[121] J. Richalet, Pratique de la commande predictive, Hermes, 1993 and Commande Predictive, in Techniques de l'Ingenieur, 2014 (in French).

[122] J. Richalet J. and D. O'Donovan, Predictive functional control? Principles and industrial applications, Springer-Verlag, 2009, London, England.

[123] J. Richalet, G. Lavielle and J. Mallet, Commande predictive, Eyrolles, 2004.

[124] J. Richalet, Plenary lecture at UKACC 2000, UK, 2000.

[125] O. J. Rojas and G. C. Goodwin, A simple antiwindup strategy for state constrained linear control, IFAC World Congress, 2002.

[126] J.A. Rossiter and B.Kouvaritakis, Constrained stable generalized predictive control, Proc IEE Pt.D, 140, 4, 243-254 , 1993.

[127] J.A. Rossiter, Notes on multi-step ahead prediction based on the principle of concatenation, Proceedings IMechE., 207, 261-263, 1993.

[128] J.A. Rossiter, B. Kouvaritakis and J.R. Gossner, Feasibility and stability results for constrained stable generalised predictive control, 3rd IEEE Conference on Control Applications (Glasgow), 1885-1890, 1994.

[129] J.A. Rossiter and B. Kouvaritakis, Numerical robustness and efficiency of generalised predictive control algorithms with guaranted stability, Proceedings IEE Pt. D., 141, 3, 154-162, 1994.

[130] J.A. Rossiter, B. Kouvaritakis and J.R. Gossner, Feasibility and stability results for constrained stable generalised predictive control, Automatica, 31, 6, 863-877, 1995.

[131] J.A. Rossiter, B. Kouvaritakis and J.R. Gossner, Mixed objective constrained stable predictive control, Proceedings IEE, 142, 4, 286-294, 1995.

[132] J.A. Rossiter, J.R. Gossner and B. Kouvaritakis, Infinite horizon stable predictive control, IEEE Transactions on Automatic Control, 41, 10, 1522-1527, 1996.

[133] J.A. Rossiter, J.R. Gossner and B. Kouvaritakis, Guaranteeing feasibility in constrained stable generalised predictive control, Proceedings IEE Pt.D, 143, 5, 463-469, 1996.

[134] J.A. Rossiter and B.G. Grinnell, Improving the tracking of GPC controllers, Proceedings IMechE., 210, 169-182, 1996.

[135] J.A. Rossiter, J.R. Gossner and B.Kouvaritakis, Constrained cautious stable predictive control, Proceedings IEE Pt.D, 144, 4, 313-323, 1997.

[136] J.A. Rossiter, Predictive controllers with guaranteed stability and mean-level controllers for unstable plant, European Journal of Control, 3, 292-303, 1997.

[137] J.A. Rossiter, L. Chisci and A. Lombardi, Stabilizing predictive control algorithms in the presence of common factors, European Control Conference, 1997.

[138] J.A. Rossiter, M.J. Rice and B. Kouvaritakis, A numerically robust state-space approach to stable predictive control strategies, Automatica, 34, 65-73, 1998.

[139] J.A. Rossiter and B. Kouvaritakis, Youla parameter and robust predictive control with constraint handling, Workshop on Non-linear Predictive Control (Ascona), 1998.

[140] J.A. Rossiter, L. Chisci and B. Kouvaritakis, Optimal disturbance rejection via Youla-parameterisation in constrained LQ control, IEEE Mediterranean Conference on Control and Automation, 1998.

[141] J.A. Rossiter and L. Chisci, Disturbance rejection in constrained predictive control, Proceedings UKACC, 612-617, 1998.

[142] J.A. Rossiter and B. Kouvaritakis, Reducing computational load for LQ optimal predictive controllers, UKACC Control Conference, 606-611, 1998

[143] J.A. Rossiter, M.J.Rice, J. Schuurmanns and B. Kouvaritakis, A computationally efficient constrained predictive control law, American Control Conference, 1998.

[144] J. A. Rossiter and B. Kouvaritakis, Reference Governors and predictive control, American Control Conference, 1998.

[145] J.A. Rossiter, L. Yao and B. Kouvaritakis, Identification of prediction models for a non-linear power generation model, UKACC Control Conference (Cambridge), 2000.

[146] J.A. Rossiter, L. Yao and B. Kouvaritakis, Application of MPC to a non-linear power generation model, Proceedings UKACC Control Conference, 2000.

[147] J.A. Rossiter, B. Kouvaritakis and M. Cannon, Triple mode control in MPC, American Control Conference, 2000.

[148] J.A. Rossiter, B. Kouvaritakis and M. Cannon, Triple Mode MPC for enlarged stabilizable sets and improved performance, Conference on Decision and Control (Sydney), 2000.

[149] J.A. Rossiter, B. Kouvaritakis and L. Huaccho Huatuco, The benefits of implicit modelling for predictive control, Conference on Decision and Control, 2000.

[150] J.A. Rossiter, Lectures and resources in modelling and control, http://controleducation.group.shef.ac.uk/indexwebbook.html

[151] J. Schuurmanns and J.A. Rossiter, Robust piecewise linear control for polytopic systems with input constraints, IEE Proceedings Control Theory and Applications, 147, 1, 13-18, 2000.

[152] J.A. Rossiter and B. Kouvaritakis, Modelling and implicit modelling for predictive control, International Journal of Control, 74, 11, 1085-1095, 2001.

[153] J.A. Rossiter, Re-aligned models for prediction in MPC: a good thing or not?, Advanced Process Control 6, York, 63-70, 2001.

[154] J.A. Rossiter, B. Kouvaritakis and M. Cannon, Computationally efficient predictive control, ISSC, Maynooth, Ireland, 2001.

[155] J.A. Rossiter, B. Kouvaritakis and M. Cannon, Computationally efficient algorithms for constraint handling with guaranteed stability and near optimality, International Journal of Control, 74, 17, 1678-1689, 2001.

[156] J.A. Rossiter and L. Chisci, An efficient quadratic programming algorithm for predictive control, European Control Conference, 2001.

[157] J.A. Rossiter, P.W. Neal and L. Yao, Applying predictive control to fossil fired power station, Proceedings Inst. MC, 24, 3, 177-194, 2002.

[158] J.A. Rossiter and J. Richalet, Handling constraints with predictive functional control of unstable processes, American Control Conference, 2002.

[159] J.A. Rossiter and J. Richalet, Predictive functional control: alternative ways of prestabilising, IFAC World Congress, 2002.

[160] J.A. Rossiter, Improving the efficiency of multi-parametric quadratic programming, Workshop on Non-linear Predictive Control (Oxford), 2002.

[161] J.A. Rossiter, Model-based predictive control: a practical approach, CRC press, 2003.

[162] J.A. Rossiter, T. Chen and S.L. Shah, Developments in multi-rate predictive control, ADCHEM, 2003.

[163] J.A. Rossiter, T. Chen and S.L. Shah, Improving the performance of dual rate control in the absence of a fast rate model, internal report, Dept. ACSE, University of Sheffield, 2003.

[164] J.A. Rossiter, B. Kouvaritakis, M. Bacic, Interpolation based computationally efficient predictive control, International Journal of Control, 77, 3, 290-301, 2004.

[165] J.A. Rossiter, A global approach to feasibility in linear MPC, International Control Conference, 2006.

[166] J. A. Rossiter, L. Wang, G. Valencia-Palomo, Efficient algorithms for trading off feasibility and performance in predictive control, International Journal of Control, 83, 4, 789, 2010.

[167] J.A. Rossiter, Input shaping for PFC: how and why?, Journal of Control and Decision, DOI: 10.1080/23307706.2015.1083408, 2015.

[168] J.A. Rossiter, R. Haber and K. Zabet, Pole-placement PFC (Predictive Functional Control) for systems with one oscillatory mode, European Control Conference, 2016.

[169] J. A. Rossiter, A priori stability results for PFC, International Journal of Control, 90, 2, 305-313, 2016.

[170] P.O.M. Scokaert and J.B. Rawlings, Infinite horizon-linear quadratic control with constraints, Proceedings IFAC World Congress, Vol. M, 109-114, 1996.

[171] P.O.M. Scokaert and J.B. Rawlings, Constrained linear quadratic regulation, IEEE Transactions on Automatic Control, 43, 8, 1163-1168, 1998.

[172] J. Sheng, Generalized predictive control of multi-rate systems, Ph.D. thesis, University of Alberta, 2002.

[173] J. Sheng, T. Chen and S.L. Shah, Generalized predictive control for non-uniformly sampled systems, Journal of Process Control, 12, 875-885, 2002.

[174] D.S. Shook, C. Mohtahdi and S.L. Shah, Identification for long range predictive control, Proceedings IEE Pt-D, 138, 1, 75-84, 1991.

[175] D.S. Shook, C. Mohtahdi and S.L. Shah, A control relevant identification strategy for GPC, IEEE Transactions on Automatic Control, 37, 7, 975-980, 1992.

[176] S. Skogestad, K. Havre and T. Larsson, Control limitations for unstable plants, IFAC World Congress, 2002.

[177] R. Soeterboek, Predictive control: a unified approach, Prentice Hall, 1992.

[178] M. Sznaier and M.J. Damborg, Heuristially enhanced feedback control of constrained discrete time linear systems, Automatica, 26, 3, 521-532, 1990.

[179] K.T. Tan and E.G. Gilbert, Multimode controllers for linear discrete time systems with general state and control constraints, in Optimisation Techniques and Applications (World Scientific, Singapore), 433-442, 1992.

[180] A.K. Tangirala, D. Li, R.S. Patwardhan, S.L. Shah and T. Chen, Ripple free conditions for lifted multi-rate control systems, Automatica, 37, 1637-1645, 2001.

[181] P. Trodden and A. Richards, Robust distributed model predictive control using tubes, American Control Conference, 2006.

[182] J. Tse, J. Bentsman and N. Miller, Minimax long range parameter estimation, Conference on Decision and Control, 277-282, 1994.

[183] T.T.C. Tsang and D.W. Clarke, Generalised predictive control with input constraints, IEE Proceedings Pt. D, 6, 451-460, 1988.

[184] G. Valencia-Palomo and J.A. Rossiter, An efficient suboptimal parametric solution for predictive control CENG Practice, 19, 7, 732-743, 2011.

[185] G. Valencia-Palomo, J.A. Rossiter and F.R. Lopez-Estrada, Improving the feed-forward compensator in predictive control for setpoint tracking, ISA Transactions, 53, 755-766, 2014.

[186] R.A. J de Vries and T.J.J. van den Boom, Robust stability constraints for predictive control, European Control Conference, 1997.

[187] R.A. J de Vries and T.J.J. van den Boom, Constrained robust predictive control, European Control Conference, 1995.

[188] B. Wams and T.J.J. van den Boom, Youla like parameterisations for predictive control of a class of linear time varying systems, American Control Conference, 1997.

[189] Z. Wan and M.V. Kothare, A computationally efficient formulation of robust predictive control using linear matrix inequalities, Proceedings of CPC-6, 2000.

[190] L. Wang, Use of orthonormal functions in continuous time MPC design, UKACC, 2000.

[191] L. Wang and J.A. Rossiter, Disturbance rejection and set-point tracking of sinusoidal signals using Generalized Predictive Control, Conference on Decision and Control, 2008

[192] L. Wang, Model predictive control design and implementation using MAT-LAB, Springer-Verlag, 2009.

[193] A.G. Wills and W.P. Heath, Using a modified predictor corrector algorithm for model predictive control, IFAC World Congress, 2002.

[194] G.F. Wredenhagen and P. R. Belanger, Piecewise linear LQ control for systems with input constraints, Automatica, 30, 403-416, 1994.

[195] D.C. Youla and J.J. Bongiorno. A feedback theory of two-degree-of-freedom optimal Wiener-Hopf design, IEEE Transactions on Automatic Control, 30, 652-664, 1985.

[196] T.-W. Yoon and D.W. Clarke, Observer design in receding horizon predictive control, International Journal of Control, 61, 171-191, 1995.

[197] K. Zabet, R. Haber, J.A. Rossiter, J. Richalet, Pole-placement Predictive Functional Control for over-damped systems with real poles, ISA Transactions, 229-239, 2016.

[198] M.N. Zeilinger, D. M. Raimondo, A. Domahidi, M. Morari and C. N. Jones, On real-time robust model predictive control, Automatica, 50,3, 683-694, 2014.

[199] Z.Q. Zheng and M. Morari, Robust stability of constrained model predictive control, American Control Conference, 379-383, 1993.

Index